FUNDAMENTALS *of* BIOLOGY *of* ANIMALS

AF817739

FUNDAMENTALS *of* BIOLOGY *of* ANIMALS

[Evolution, Genetics, Ecology, Adaptation, Embryology, and History of Zoology]

[Textbook for Degree Students]

Simananda Adhikari MSc, PhD

Arup Kumar Sinha MSc, PhD

New Central Book Agency (P) Ltd

LONDON

HYDERABAD ERNAKULAM BHUBANESWAR

NEW DELHI KOLKATA PUNE GUWAHATI

NCBA

REGD OFFICE

8/1 Chintamoni Das Lane, Kolkata 700 009, India
email: ncbapvtltd@eth.net

OVERSEAS

NCBA (UK) Ltd, 149 Park Avenue North, Northampton, NN3 2HY, UK
email: ncbauk@yahoo.com

EXPORT

NCBA Exports Pvt. Ltd, 212 Shahpur Jat, New Delhi 110 049
email: ncbaexports@ncbapvtltd.com

BRANCHES

2/27 Ansari Road, 1st Floor
Daryaganj, **New Delhi** 110 002
email: ncbadel@ncbapvtltd.com

Shop Nos. 3 & 4, Vinayak Towers
681/B Budhwar Peth, Appa Balwant Chowk
Pune 411 002
email: ncbapune@ncbapvtltd.com

Shop No. 15, Bharati Towers, Ground Floor
Forest Park, **Bhubaneswar** 751 001
email: ncbabub@ncbapvtltd.com

House No. 3-1-315, 1st Floor
Nimboliadda, Kachiguda, **Hyderabad** 500 027
email: ncbahydb@ncbapvtltd.com

GSS Shopping Complex, 1st Floor
Opposite College Post Office, Convent **Road**
Ernakulam 682 035
0484-4021054/53

Radhamayee Complex (Second Floor)
Opp. SBI, Panbazar, Jaswanta Road
Guwahati 781 001
email: ncbaguwahati@ncbapvtltd.com

FUNDAMENTALS OF BIOLOGY OF ANIMALS • Simananda Adhikari, Arup Kumar Sinha

© Copyright reserved by the Authors

Publication, Distribution, and Promotion Rights reserved by the Publisher

All rights reserved. No part of the text in general, and the figures, diagrams, page layout, and cover design in particular, may be reproduced or transmitted in any form or by any means—electronic, mechanical, photocopying, recording, or by any information storage and retrieval system—without the prior written permission of the Publisher

First Published: 1973, Second Edition: 1976
Third Edition: 1980, Fourth Edition: 1991
Reprinted: 1993, 1995, 2001, 2002, 2006, 2010, 2013, 2016

PUBLISHER AND TYPESETTER
New Central Book Agency (P) Ltd
8/1 Chintamoni Das Lane, Kolkata 700 009

PRINTER
New Central Book Agency (P) Ltd
Web-Offset Division, Dhulagarh,
Sankrail, Howrah

ISBN: 978 81 7381 117 3

Price: ₹ 235.00

TO

OUR TEACHERS

OF ZOOLOGY

Simananda
Arup

Foreword

Sivatosh Mookerjee

DEAN,
SCHOOL OF LIFE SCIENCES

JAWHARLAL NEHRU UNIVERSITY
NEW MEHRAULI ROAD, NEW DELHI 57

The writing of a viable text-book is not as easy as it appears to be. Without being endowed with a natural sense of communication, clarity of exposition and certain degree of originality in organizing the material, the attempt to write a text would be a costly misadventure. The prospective writer must consciously control the osmotic gradient of fresh knowledge flowing into the pool of the existing one. The process of science is revealed in its unceasing progress; this means that the horizon of knowledge is continuously receding. The greatest obligation of a text-book writer is to remain alive to this fact and to minimize the gap by constant vigilance. Texts, written once cannot be perpetuated indefinitely. Texts are not intended to enslave the mind but to prepare it for more knowledgeable things hidden beyond the present contemplated boundaries. Its purpose is to arouse a sense of aptness for the subject concerned and to embolden curiosity for its material contents, in order to lead the way to the further progress. The purpose of a book lies in the mental preparedness for creating a new sense of dedication for the unknown.

To attempt now to write a text in Zoology for the Degree Course would be a matter of great adventure because modern Zoology has undergone a radical ecdysis and has emerged out into a new conceptual entity, both in contents and implications. Zoology is no longer a mere plethora of disjointed morphological descriptions. It is now a field of coherently integrated and biochemically analysable entity. Biological diversities have become resolved into a common theme of endless activity of the wonder molecule of DNA. Throughout the new vista of Zoology, causal interpretations of functional dynamism, adaptiveness in time, exploration of new ecological niches, biometry of forms have firmly been established.

The writers of the present volume were my close collaborators in research. The image of modern biology is clearly inscribed in their minds. It is heartening to see that they have been able to assemble the main ingredients of the science of Zoology in a most coherent way and organised their material into a principle oriented discourse. Our syllabi, in most of the Universities, even today suffer from serious limitations. Often the contemporary outlook of Biology is not fully reflected in our curricula. However, within the available framework of the present-day curricula, the authors have done their best to convey the message of modern Zoology. I have no doubt that the present volume will generate impact on the modernization of the subject which we have cherished so much and for so long.

Preface to the First Edition

It is a reality that students at the undergraduate level of life sciences do not have a text book dealing, in a simple and concise way, with the unitary processes of Biology such as Genetics, Evolution, Adaptation, Ecology and Developmental Biology. The aim of the present volume '*Fundamentals of Biology of Animals*' is to cater to the students an up-to-date information on these important unitary topics. In compiling the book we have tried our level best to reflect the contemporary outlook on the subjects limiting ourselves within the framework of the prescribed syllabi of most Indian Universities. Despite our best efforts, we feel no hesitation to admit that the desired goal is far apart. We were severely handicapped in following the true contemporary outlook of the subjects by the rigidity of the syllabi of our Universities.

We are grateful to a large number of very competent persons who have helped us in the preparation of this book. We are indebted to them for their keen interest and friendly co-operation which have rendered this book presentable. We are specially thankful to Professor Dr. Sivatosh Mookerjee, Ph.D. (Edin), F.N.A. (formerly, Senior Professor and Head of the Department of Zoology, Presidency College, Calcutta and now Dean, School of Life Science, Jawharlal Nehru University, New Delhi) for the genuine interest that he has shown to us. We also acknowledge with deep respect the apt and excellent foreword that he has written.

We shall heartily appreciate receiving criticism and suggestions for correction from learned teachers and enthusiastic students. Suggestions and comments will improve the book in future editions. All such attitude of co-operation will be given due honour.

Last but not the least, we highly appreciate the courage of Shri Jogendranath Sen, B.Sc. of Messrs. New Central Book Agency, Calcutta for entrusting us to complete this book in their series of Life Sciences. We are happy in being able to fulfil the responsibility assigned to us.

SIMANANDA ADHIKARI
Assistant Professor of Zoology,
Darjeeling Government College,
Darjeeling, India.

ARUP KUMAR SINHA
Assistant Professor of Zoology,
Presidency College,
Calcutta, India.

Preface to the Second Edition

The liking of the book *Fundamentals of Biology of Animals* on the part of the teachers and students has prompted us to present a second edition of the book.

In compiling a text-book specially on Biology an author confronts with a peculiar situation. The prescribed syllabi, on one hand, lack plasticity while the knowledge, on the other hand, progresses at a faster rate. This keeps the author in two minds. He is always tempted to come out of the rigid framework of the syllabus but he simply can't do that because that may tentamount to violation of 'set-study'. We are indeed victims of that feeling. We earnestly hope that a recasting of the syllabus be made so that the students may get an opportunity to be in touch with the contemporary advances made by the subject in question.

The present edition contains the relevent and recent information on those unitary aspects of Biology such as Genetics, Evolution, Adaptation, Developmental Biology, Ecology and the History of Biology. Some of the parts have been re-oriented and in some the themes that are uppermost these days have been incorporated. We hope that these improvements would escalate the utility of the book to the teachers and students alike

We still look forward for receiving criticism and concrete suggestions from learned teachers and enthusiastic students so that the future edition may come out as an evolved edition.

SIMANANDA ADHIKARI
Assistant Professor of Zoology,
Maulana Azad College,
Calcutta, India.

ARUP KUMAR SINHA
Assistant Professor of Zoology,
Presidency College,
Calcutta, India.

Preface to the Third Edition

It is with much pleasure we are presenting the third edition of *Fundamentals of Biology of Animals*. Some recent informations in some of the chapters of the book have been incorporated to make it up-to-date. We presume that the book will prove useful to the students.

SIMANANDA ADHIKARI
Head, Deptt. of Biology
Krishnagar College,
Krishnagar, Nadia,
West Bengal.

ARUP KUMAR SINHA
Head, Deptt. of Zoology,
Jhargram Raj College,
Jhargram, Midnapore,
West Bengal.

Acknowledgements

The authors wish to acknowledge with gratitude the help rendered by the following persons:

Professor Dr. Sivatosh Mookerjee (Jawharlal Nehru University, New Delhi), Dr. Banku Behari Ganguli (Deptt. of Science Education, NCERT, New Delhi), Professor Kumud Shankar Das (Surendranath College, Calcutta), Professor Dr. Biswapati Dasgupta (Presidency College, Calcutta), Dr. Dhrubojyoti Lahiri (NCERT, New Delhi), Dr. Ajit Banerjee (Presidency College, Calcutta), Mrs. Nurjehan Adhikari (Bethune College, Calcutta), Prof. Debaprasad Chakraborty(Taki Govt. College), Dr. Himangshu Banerjee (Presidency College, Calcutta), Dr. Jaganmoy Mitra (NCERT, New Delhi), Dr. Ajit Aditya (Hooghly Mohsin College), Dr. Asok Bose (University of Kalyani, Kalyani), Dr. Biswanath Mitra (Visva Bharati, Santiniketan) Prof. Amarendra Nath De (R. G. Kar Medical College, Calcutta), Dr. Amitava Shankar Bhaduri (Bangabasi College, Calcutta), Professor Bijnananda Roy (Bangabasi College, Calcutta), Professor Subir Ganguly (Barisha Vivekananda College, Calcutta), Miss Snigdha Das (Victoria College, Calcutta), Dr. Somes Sanyal (Medical Faculty, Rotterdam University, Rotterdam. The Nedarland), Dr. Samarendranath Chatterjee (Bhaba Atomic Research Centre, Trombay), Dr. Sujit Dasgupta (Hooghly Mohsin College), Professor Keshab Mookerjee (Asutosh College, Calcutta), Dr. Sunil Datta (Bidhan Chandra Agricultural University, Kalyani), Dr. Satyabrata Bhattacharyya (Darjeeling Govt. College), Dr. Runu Bhattacharyya (Darjeeling Govt. College), Professor Subhas Ranjan Chakraborty (Maulana Azad College, Calcutta), Dr. Dipak Chatterjee (CIBA Research Centre, Bombay), Dr. Nitisranjan Das (Serampore College,. Serampore), Professor Samar Ghosh (Sri Chaitanya College), Professor Arunava Dutta (City College,—day, Calcutta), Prof. Bhowbani Sankar Joardar (Asutosh College, Calcutta) and many others.

The authors also appreciate the assistance of the staff of Messrs. New Central Book Agency and Books & Allied (P.) Ltd. (Printing Section) of Calcutta in the preparation of this book.

Our Outstanding Publications on

ZOOLOGY

and ALLIED SUBJECTS

BIOLOGY OF ANIMALS (Vol. I)	*Ganguly, Sinha & Adhikary*
BIOLOGY OF ANIMALS (Vol. II)	*Ganguly, Sinha, Adhikary & Bharati Goswami*
FUNDAMENTALS OF BIOLOGY OF ANIMALS (Vol. III)	*Sinha & Adhikary*
CELL BIOLOGY	*K. K. De & S. C. Roy*
INTRODUCTION TO BIOLOGY OF ANIMALS	*Ganguly, Sinha & Adhikary*
MICROBIOLOGY AND IMMUNOLOGY —an Encyclopedic Approach	*M. K. Majumdar*
INTRODUCTION TO GENERAL ZOOLOGY	*Chaki, Kundu & Sarkar*
ZOOLOGY FOR I.A.S. (Main)	*S. Prasad*
PRACTICAL ZOOLOGY	*Ghosh & Manna*
TEXT BOOK OF ZOOLOGY	*K. Ghosh*
FUNDAMENTALS OF ZOOLOGY	*Ghosh & Manna*
A SHORT TEXT BOOK OF ZOOLOGY	*Ghosh & Manna*
FUNDAMENTALS OF BIOCHEMISTRY	*A. C. Deb*
FUNDAMENTALS OF TOXICOLOGY	*Pandey, Shukla, Trivedi*
CONCISE MEDICAL PHYSIOLOGY	*S. K. Chowdhury*
APPLIED AGRICULTURAL ENTOMOLOGY	*L. K. Jha*
PROBLEM ON GENETICS, MOLECULAR GENETICS AND EVOLUTIONARY GENETICS	*Pranab Kr. Banerjee*
MICROBIOLOGY AND IMMUNOLOGY	*A. K. Banerjee & N. Banerjee*

Contents

PART ONE

ORGANIC EVOLUTION

Since time immemorial, the enigma of mankind was to know how life originated on earth and the ways in which the diverse forms of animals and plants came into existence. This part of the book, organic evolution, deals with the changes undergone by the animals in course of time and also embodies the different theories and evidences extant on the field to explain the mechanics of evolution. The theory of organic evolution relates that the present-day animals and plants are actually the modified descendants of somewhat different animals and plants which lived in times past. The ancestors are, in their turn, thought to be the descendants of predecessors which also differed from them. The evidences advocate that modern animals inherited their similarities from common ancestors.

CONCEPT OF EVOLUTION

Although many people regard that the concept of evolution started with the publication of the epoch-making book of Charles Darwin in 1859, many Greek philosophers afford evidences of organic evolution. Anaximander, in sixth century B. C., advanced that men were first formed as fishes which eventually cast off their fish-skin and adapted to life on dry land. It is a fact that the distant ancestor of man was indeed fishes. Xenophanes, in fifth century B. C., was the pioneer to recognise the fossils and regarded them as being the remains of animals that once lived. Existence of fossils of marine animals on dry land, in the opinion of Xenophanes, indicated that the dry land was once covered by ocean. In the fifth century B. C., many Philosophers, like Empedocles and others are regarded as "the father of the Evolution idea". The fourth century B.C. saw "the greatest investigator of antiquity", Aristotle whose scientific observation and contribution to evolutionary thinking demand admiration. Since the days of Aristotle many workers have added convincing evidences of evolutionary thought and it is not possible to furnish a complete survey of evolutionary thought. Thinking about evolution is as ancient as human thought.

ORIGIN OF LIFE

The origin of life on earth is one of the most elusive and oldest problems of biology. For about one half of the long span of earth's history, the earth was barren and lifeless. But at some indeterminate time the entity called life came into existence. What form it took and what circumstances brought it into being remain ever a problematic issue. The attention of the keenest minds of science and philosophy tried to answer this question. From time immemorial several explanations have been put forward to explain this phenomenon. But recent researches and experimentations have abandoned most of the explanations.

Theories on the origin of life

By the turn of this century several main approaches have been made on the origin of life. These approaches are the vitalism, special creation, panspermia, mechanistic theory and materialism.

Vitalism. This concept attributes the distinctive properties of the living things to a supernatural life force.

Special creation. This concept is nothing but the literal interpretation of the Biblical aspects of genesis.

Panspermia. By assuming the eternal and universal existence of life, this concept avoids the question of the origin of life on this earth.

These three ideas regarding the origin of life, *viz.*, vitalism, special creation and panspermia, need no detailed consideration, because they are based on assumption without having any scientific background. But the other two theories, the mechanistic theory and the materialistic doctrine need tremendous consideration.

Mechanistic theories. These theories are based on the assumption that (*i*) the origin and activities of living things are derived from natural laws and (*ii*) that the living things are derived from inorganic matters in accordance with those natural laws. But these ideas were confronted with serious difficulties. The real potent question is how can a living thing be formed from inorganic matters? These theories believed that the earth was totally devoid of organic compounds in prebiological days and that only living organisms can produce organic compounds. The organic compounds constitute the most vital part of the living organisms and as such their origin in the absence of organic compounds remains a formidable problem for ever.

The mechanistic theories resolved that the first living thing was a macromolecule or "living molecule" that was originated by the accidental coming together of the components that composed the living molecule in appropriate and proper arrangements and proportions. It is also believed that the living molecule was probably a molecule of protein—the most important part of the living things. The mechanistic concepts advocate the origin of first living things by the chance combination of these elements. This has come about by trail and error through an enormously long span of time.

Materialistic theories. These theories hold a different view by applying the natural laws to explain the origin of life and reject the contention of the mechanistic theories that only living things can synthesise organic compounds. On the contrary these theories propose that the organic compounds were formed *abiogenecally*, (i.e. without the intervention of living things) prior to the living organisms came into existence. Instead of an accidental get-together of the elements to produce living things all at once as advocated in mechanistic doctrine, materialistic theories advocate the origin of life as result of series of steps towards increasing complexities and eventually leading up to the living state. One of the most important steps is the abiogenic synthesis of the organic compounds. Every step leads into a higher grade of organisation of matter that did not exist at the lower level. With the gradual formation of molecules, new physical as well as chemical properties appeared in the scene which were non-existing in the realm of uncombined elements. By this way and with the advent of living organisms, biological laws came into operation which did not exist before. The materialistic theories hold that the origin of life was not an accident, but it was the end result of matters evolving into higher and higher forms step by step.

Evolution of living things from non-living matters was the speculation of the thinkers of the past. To name a few, they were F. Engels, H. Spencer, T. H. Huxley, C. Mitchell and E. A. Schafer. F. Engels in his "Dialectics of Nature" condemned the concept of the spontaneous generation and vitalism and maintained that the living things have originated from a continuous evolution of matters. A Russian Biochemist, A. I. Oparin in 1922, holds that the organic compounds had existed on the earth before the origin of living things and the living things have evolved from these pre-existing compounds. He views the evolutionary sequences (such as the formation of inorganic compounds, organic compounds, primitive and higher living things) not as separate events, but as merely the steps in the process of unfolding of the same biological process.

J. B. S. Haldane in 1928 was also of the opinion of the abiogenic formation of organic substances in the prebiological history of the earth. He postulated that the organic compounds must have accumulated before the origin of life.

He assumed that the primitive atmosphere contained only carbon dioxide, ammonia and water vapour. There was no oxygen. Haldane was of the opinion that such a combination, exposed to ultra-violet light, can evolve into a large variety of organic compounds.

Oparin's hypothesis. In his attempt to solve the problem of the origin of organic compounds on this earth, Oparin based his concept on the then accepted theory on the origin of earth proposed by Sir James Jeans. Although Sir Jean's theory of the fiery origin of the planets is no longer acceptable, it provided a basis for the formulation of Oparin's theory of the origin of organic compounds on the earth.

Oparin assumes that the first carbon compounds were the hydrocarbons and not the carbon dioxide. To put the question in other way, he assumes the first appearance of carbon in the reduced state and not in the oxidised state because the primordial atmosphere was devoid of oxygen. With the cooling down of the atmosphere to 1000°C. and lower, the highly reactive free radicals (*viz.* CH and CH_2) combined to produce a variety of simple saturated and unsaturated hydrocarbons. The crust of the earth had begun to form but the temperature was still too hot to allow condensation of steam to water. But the atmosphere above the crust contained carbon as well as nitrogen in reduced state. The reduced state of carbon was the hydrocarbons while that of nitrogen was ammonia. Such reducing atmosphere forms the basis of Oparin's theory of the origin of life.

The simplest substances, such as hydrogen, steam, methane, hydrogen sulfide, ammonia and nitrogen are continuously producing simple hydrocarbons and their derivatives—the alcohols, aldehydes, sugars, ketones and acids. Of the derivatives there are the amino-acids and the polymerisation of the amino-acids results in the formation of protein. The molecules of protein have the capability of coming together and form large complexes. By this way colloidal hydrophilic complexes are produced and these complexes play the active role in the transformation of inanimate things into living systems.

CONCEPT OF ORGANIC EVOLUTION

The most important question confronting biologists is how the diverse forms of living organisms came into existence. Many theories have been advanced to explain the emergence of the various living forms existing on earth today. The theories can possibly be reduced to two broad categories. They are: (1) Concept of Special Creation and (2) Concept of Organic Evolution.

Concept of Special Creation. The creationists believed that the existing forms of living organisms were brought in their present forms by Divine command. One of the greatest supporters of the special creation doctrine, Father Suarez (1548–1617) advocated that in six days the earth with animals and plants was brought into being. At the end of the period creation was closed and the living things remained almost as such as they were originally created. This concept is a static theory in which the living organisms stand still. Certain minor variations may be admitted, but such variations are not enough to produce new forms or species. Until the middle of the nineteenth century European people believed in the special creation and hold the orthodox belief with tenacity.

Concept of Organic Evolution. Opposing the static idea stands the evolutionary concept. Ever since man started to contemplate nature, exponents of evolutionary ideas came into being but only in the nineteenth century that serious observation and experimentation in this line were carried out. The evolutionary concept holds that nothing is static, but all animate objects are dynamic and are subjected to changes. The animals and plants that we see today, have come into existence through a slow and gradual process of transition from old pre-existing forms. The process of changing is still continuing and the evolutionary concept is dynamic. The concept of organic evolution is regarded by many to be comparatively a recent idea. It is believed that Charles Darwin is the originator or promulgator of the theory of organic evolution. The doctrine of static creation seemed wholly inadequate to many original thinkers and the idea of evolutionary concept can be traced back to the early Greeks, because the germ of the evolutionary concept had its inception with the Greeks.

Empedocles (495–435 B.C.) is usually regarded as the father of evolutionary concept. He emphasises on the attempts of nature to produce the fittest forms through chance and the elimination of the unfits. Next to Empedocles comes another Greek intellect, Democritus (460–357 B.C.) who advances the idea of adaptations of structures and organism as a whole. But the most notable Greek figure is Aristotle (384–322 B.C.) who assumes the existence of an intelligent design in nature which acts as the perfecting principle. Aristotle advances the notion that from nonliving minerals plants arose and the plants in turn gave origin to simple animals like sponges. From these lower forms of living organisms arose the higher animals and finally man at the peak through a long and continuous series of transformations. He emphasises the operation of natural law in the inanimate and animate forms. Gregory of Nyssa, a bishop and a theologian of the fourth century A.D. attempts to make a compromise between the ideas of the special creationists and the Greek philosophers. He believed that God imparted the fundamental properties to the matters. The matters were at first chaotic and without definite forms. Gradually from such disorganised materials evolved complete and organised forms of the universe under Divine command. Saint Augustine (353–430 A.D.), a church father rejected the concept of special creation and advanced developmental concept. He was the believer of change as involved in the formation of complete objects and creatures. Thomas Aquinas (1225–1274 A.D.) embraced Augustine's ideas with much confidence. From Aquinas on, a host of workers expressed their views regarding the genesis of living objects. Most of the views are discarded at present, because they add nothing substantial but some of them may be briefly placed here as historical importance.

Linnaeus (1707–1778) believed the origin of species through special creation but he advocated the production of post-creation forms by hybridisation. Buffon (1707–1788) regarded that variations in species are caused by the direct influence of environment and the modifications of the structures are conserved by heredity. He puts significance on geographical isolation, formation of varieties by domestication (artificial selection), influence of climate particularly on human races. He throws much light on the existence of struggle for existence to stop overcrowding and thus the balance of nature is maintained. Erasmus Darwin (1731–1802), grandfather of Charles Darwin, did not emphasise on the influence of environment in the production of changes, but believed that structural modifications arose as the resultant of reactions within the body of the organisms. He regarded that all animals undergo transformations and almost followed the Lamarckian principle of transmission. He first stated the factors operating in the process of inheritance of acquired modifications. Jean Baptise Lamarck (1744–1829) is one of first biologists to develop a theory of biological transformation of living organisms and interpretes life in term of evolution. He explains evolution on the principle of use and disuse supplemented by heredity. He believes that acquired characters pass from parents to offsprings. This theory is under great dispute and is unacceptable to modern biologists. Another Frenchman, E Geoffroy St. Hilaire (1772–1844), a contempory of Lamarck, believes on the direct influence of environment as the primary cause of evolution. He anticipates the idea of de Vries. Charles Darwin (1809–1882) is undoubtedly the foremost figure in the evolutionary history. His prominency is not only for the originality of his ideas but he paved the way for the general acceptance of the truth of evolution by Natural Selection as a force in evolution. The publication of his celebrated book entitled "On the Origin of Species by Means of Natural Selection" in 1859 is the crowning event in biology. His twenty years observations and researches are shaped into a reasonable theory of evolution. Subsequent investigations have pin-pointed certain shortcomings of Charles Darwin but the central idea of evolution remains unaltered.

Evolution is possible because living organisms have three primary attributes. The first is the capacity of *reproduction*, the second is the progressive force which produces *mutation* and the third is the conservative process which preserves and continues the mutations is the *heredity*. If there is no reproduction life would cease to exist, without heredity there would be no continuity from generation to generation and without mutation there would be no variety and living organisms would never have evolved beyond its original form.

ORGANIC EVOLUTION—EVIDENCES

With the establishment of the idea of evolution, attempts have been made from all possible corners to explain the ways how it happened. There are various evidences to prove the process of evolution and all of them justify the idea that the living organisms in course of time become adapted through an orderly sequence of gradual change and also indicate the fact that new species evolved by constant and progressive changes from old pre-existing forms.

EVIDENCES OF EVOLUTION

There is a large body of positive evidences to explain the phenomenon of Organic Evolution. One particular source of evidence may not always be conclusive, but if it is considered in conjunction with other sources, a complete evolutionary picture becomes obvious. The various evidences discussed in this part are as follows:

(1) Biogeographical evidences.
(2) Morphological evidences.
(3) Embryological evidences
(4) Geological evidences.
(5) Taxonomical evidences.
(6) Evidences from comparative Physiology and Biochemistry.
(7) Cytogenetical evidences.
(8) Evidences from domestication.
(9) Evidences from specific adaptations.

BIOGEOGRAPHICAL EVIDENCES

The study of biogeography or the geographical distribution of animals and plants on the surface of the earth throws some light on the possibilities of the origin of species by the process of evolution.

Geographical distribution

No animal species enjoys uniform distribution over the whole surface of the world, each has a restricted range or area of distribution. Every species of animals produces offspring in excess of the numbers that can survive. The rapid rate of multiplication of all animals causes population pressure by which individuals tend to expand their boundaries. Many other factors (competition, enemies, disease, adverse seasonal weather, shortage of food, decrease in available shelter) also act to reduce the population. The distribution of animals on the surface of the earth is dynamic and always subject to change. The factors which limit distribution are designated as the **barriers.** The barriers are:

(1) PHYSICAL BARRIERS: High and extensive mountain ranges, land mass, large bodies of water, etc.
(2) CLIMATIC BARRIERS: Temperature, moisture, sunlight and others.
(3) BIOLOGICAL BARRIERS: Absence of appropriate food or presence of competitors, enemies or diseases, etc.

Each and every species of living organism has a *limit of tolerance* to each factor in its environment. The tolerance may be maximum as well as minimum. Changes in a factor in an environment beyond the threshold of tolerance cause in migration or survival or death. The individuals which can survive in the altered conditions are more tolerant i.e., better suited. The limitation in distribution of a species depends on the sum total of external influences which are mostly interdependent and is subjected to *Liebig's Law of minimum* (Limited by the presence of essential factor in least amount).

Factors regulating distribution of animals

Environmental changes sometimes compel animal species to migrate or disperse into new territories because the "old home" is rendered unsuitable for their survival. But there are barriers of various sorts which stand on their ways and the dispersal of animals is governed by these barriers.

Barriers to dispersal

Physical barriers. High and extensive *mountain ranges* act as barriers and limit the distribution of many terrestrial animals. The mountain ranges become an effective barrier if they are situated parallel with the equator as seen in Europe and Asia. When the mountain ranges extend north and south their influence upon the distribution is not marked. The animal species occupying the northern part differ markedly from those inhabiting the southern part. The great Himalaya Range with its summits beyond the limits of perpetual snow is a notable physical barrier. The south of the barrier is the hot and moist plains of India with characteristic tropical animals resembling those of Africa in many respects. On the north of the barrier, the conditions of temperature and moisture are entirely different and as a consequent the animals inhabiting the region are totally different from the southern counterparts, resembling nearly to those of Europe. The mountain ranges in the western part of United States exert indirect influence on animal distribution. On the western side of the barrier, the winds from the Pacific ocean laden with moisture deflected higher and condensed in the form of rain. The winds become moistureless after crossing the barrier and aridity prevails east of the mountain including dry plains to desert. The differential climatic condition on the two sides of the mountain range influence upon the growth of vegetation. The nature of vegetation exerts direct influence upon herbivorous animals and indirectly upon the carnivores. Mountain ranges cause variation in rainfall and temperature which govern the distribution of animals to a large extent. Mexican Plateau in North America is an effective and important topographic barrier.

Large bodies of water, when not frozen, form barriers for terrestrial vertebrates except the forms having powers of sustained flight (birds, bats, flying lemur, etc.). Barriers of one group of animals become highways for others. Extensive bodies of water are inseperable barriers for terrestrial vertebrates but highways for the primary and secondary adapted aquatic animals. Large bodies of salt water are effective barrier for the exclusively freshwater fishes. Some salt-water fishes (Salmon, Sturgeon, Shad, etc.) migrate to fresh water annually (anadromous migration) for breeding. The reverse migration occurs in eels which pass from fresh water to salt water (catadromous migration) for the same purpose. Salt water constitutes the most effective barrier for the modern amphibians, especially for the larval forms. For many reptiles, especially crocodiles and sea turtles, seas do not afford an obstacle for movement. Many serpents excepting sea-snakes are usually incapable of passing large bodies of salt water. The flightless birds (Ostrich, Rhea, Cassowary, etc.) are debarred from trans-oceanic migrations. Impurity and lack of salinity of sea water afford barrier to the dispersal of marine invertebrates.

Land masses, just as bodies of water,

form effective barriers for the sea animals. Cape Cod which separates cold waters of Massachusetts Bay from the sea is a typical instance. Isthmus of Panama is another example of land barrier. Caspian, a relic sea, contains many marine vertebrates (seals, porpoises) which is not off from the sea by land barrier. The forefathers of the marine vertebrates entered into it when there was open communication with the sea. Caspian became geographically isolated from the sea with the severance of communication. Lake Nicaragua contains a species of Shark whose forefather entered into it when there was communication with the sea.

Climatic barriers. Degree of heat and moisture exert influence on the distribution of animals. Temperature acts as a limiting factor for the distribution of cold-blooded animals. The amphibians and reptiles are quite abundant in tropical and temperate regions, gradually diminishing towards the poles. Lack of moisture may lead to desert conditions which become barrier to animals excepting desert adapted forms. Excess of moisture renders a swampy condition which becomes impassable to animals which are not adapted to that condition of life.

Biological barriers. Food supplies limit distribution of animals. The animals depend directly or indirectly on the vegetation for food. The growth of vegetation depends largely upon the climatic factors, like degrees of temperature, moisture, sunlight, etc. Dense growth of forest renders large terrestrial animals incapable of penetrating it. Lack of trees limits the distribution of primates. Arboreal primates require trees for safety, food and progression. Besides food, presence of competitors, enemies and diseases are effective biological barriers limiting dispersal of animals.

Means of dispersal

Despite effective barriers, there are numerous migratory routes for the dispersal of animals. The means of dispersals are:

Land bridges. Some continents have been connected by land bridges at times when admixture of faunas of different continents takes place. Panamanian bridge was a land connection between North and South America at least up to late Eocene period. When this connection existed free intermigration of faunas took place between the continents. The land bridge is marked by the position where the Isthmus of Panama now lies.

Natural rafts and drift wood. Many terrestrial animals, either intentionally or accidentally, took shelter upon drifting material which help them to migrate over-water journeys of considerable extent. Natural rafts are potent factor in dispersal of animals. If the factors of repression and expansion remain equal, the area occupied by each species will remain unchanged.

Discontinuous distribution. Every region of the globe has its own quota of animals and plants. Diverse forms of organisms inhabit the different regions of the earth. But the concept of the doctrine of specific centres reveals that a particular species arose at a definite place and then migrated to different regions from the centre of origin. The scattering movements of the animals are due to overproduction and overcrowding in a particular area. Geographical distribution shows that sometimes a particular species or closely similar species becomes widely separated by its migratory power into different regions of the earth. Once migrated to different regions, they are separated by barriers and owing to geological or climatic variations prevailing in these regions they become different. Animals, once alike and sharing the same ancestry, become in course of years, decidedly unlike for living in two regions having quite different climatic or life conditions.

Representatives of a group may inhabit two widely different and separated places. A good example is the camel family. One branch of the group comprising of true camels is found in central Asia and Arabia; while the other branch of the family represented by the much smaller and graceful *Llama* and *Alpaca* inhabiting the South America. These two branches of the camel family are closely related forms. This fact is tested by anatomical as well as serological studies. How did they become so diverse? Palaentological study reveals that camels appeared first in North America and then migrated to Asia across a land bridge which connected America with Asia in the prehistoric days. Elephants are found in India and Africa. The Indian elephants are slightly different from their

African brethren. Great Britain and New Zealand have almost similar climatic conditions but the living forms inhabiting these two regions are quite unlike.

The distributional anomalies in space make it quite clear that the present-day animals have a common ancestry and the migratory power enables them to diverge from the original home. Once migrated to widely different regions, the animals are prevented to come back by physical or biological barriers. Thus isolated, the animals underwent structural and functional modifications to cope with the prevailing environmental conditions and became adapted accordingly. The animals which were successful to overcome the obstacles, became completely changed and transform into new species.

Island fauna. Island fauna also provide convincing evidence of such distributional effects. Animals inhabiting the islands show marked deviation from the nearest mainland and greater the distance between the places, higher is the degree of differences. The Galapagos islands may be cited as an example. This group of islands is situated at about a distance of 500 miles apart from the coast of America. The animals present in these islands are either migrated from the mainland or have been blown there through air or could have been drifted there. The older the islands are, the greater are the degrees of differences. Taking all these facts into consideration that the ancestral forms of the animals have migrated to the islands from the mainland and after arriving at these isolated regions they became gradually transformed into new species. The cumulative evidences of the geographical distribution of animals lead us to the plausible interpretation that a definite change in living organisms is observed which are nothing but the evolutionary changes.

Biogeographical realms

Besides the familiar political boundaries that separate nations from each other, a remarkable attempt has been put forward to divide the land masses of the globe into different life realms. Such life realms are characterised by the distribution and likeness or unlikeness of their contained fauna. Several plans have been furnished by many workers. These plans have many resemblances but differ in certain minute details. Variations are due to the selection of animal inhabitants as determinants. First serious attempt to divide the earth into different zoological realms was based upon the distribution of birds. This scheme was confronted with serious objections because of the utilisation of vagrant and barrier-defying creatures like birds as the determinants. The recent scheme of division of the surface of the earth into different zoological realms is based chiefly on the distribution of mammals. Use of mammals as the determinants has several advantages, because being warm-blooded, they are capable of occupying a wide range of habitats and lastly they are the most recently evolved group of animals and got comparatively less time to radiate from the center of origin.

According to most naturalists (Sclater, 1958; Huxley, 1868; Wallace, 1876), the land surface of the globe is divided into six primary realms (Fig. 1.1) which in turn have been subdivided into subrealms. The land surface of the globe is divided into **Nearctic, Neotropical, Palearctic, Ethiopian, Oriental and Australian realms.**

Nearctic realm. This region includes the whole of North America to the edge of Maxican Plateau, all the islands of the North together with Greenland. It has the following subrealms:

(1) Californian, (2) Rocky Mountain, (3) Alleghanian and (4) Canadian.

Nearctic region has characteristic fauna, such as the Opossum, Racoon, Blue Jay, Turkey buzzards, Rattle snakes, Axolotl, Necturus, *Amia calva*, etc.

Neotropical realm. This region consists of Central America, south of the Maxican Plateau, whole of South America and the Antilles or West Indian islands. It has the following subrealms:

(1) Brazilian, (2) Chilean, (3) Maxican and (4) Antillean.

Neotropical realms have the following characteristic fauna. Prehensile tailed monkeys (cebidae), Marmoset, Chinchillas, Llama, Rhea, Gigantic Tortoise, *Lepidosiren*, Vampire bats, etc.

Palearctic realm. This region embraces the whole of Eurasian continent excepting the portion lying south of the northern line of Afghanisthan and Persia, the Himalayan Mountain and the Nanling range in China. Africa, north of

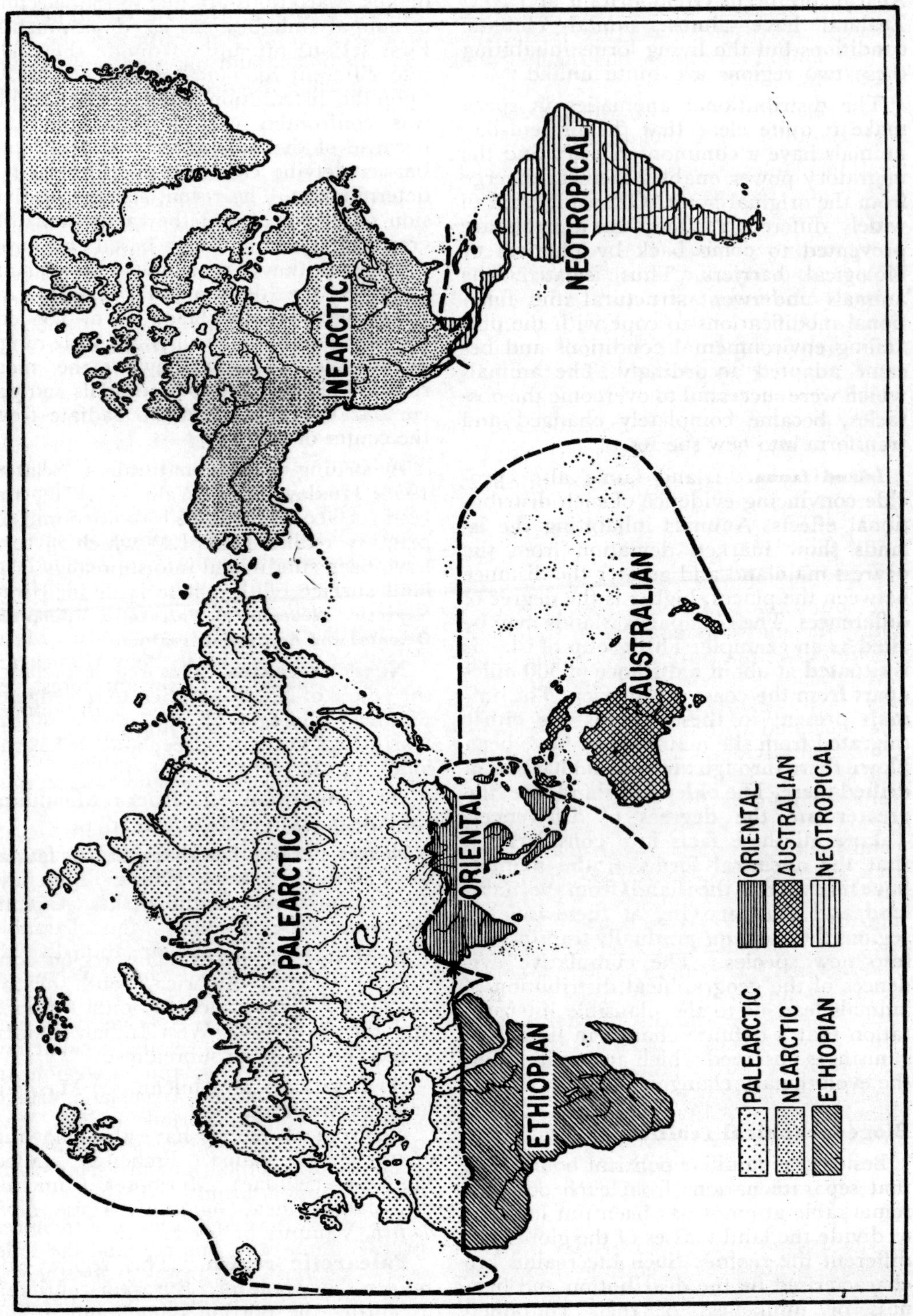
NEARCTIC
NEOTROPICAL
PALEARCTIC
ORIENTAL
ETHIOPIAN
AUSTRALIAN
PALEARCTIC
NEARCTIC
ETHIOPIAN
ORIENTAL
AUSTRALIAN
NEOTROPICAL

Sahara, Iceland, Spitzbergen, the Arctic islands, north of Siberia are included in this realm. This region has the following subrealms:

(1) European, (2) Mediterranean, (3) Siberian and (4) Manchurian.

It has Moles, Sheep and Goats, Dormice, Rheasant, Robin, Magpie as the characteristic fauna.

Ethiopian realm. This region includes Africa and Arabia, south of Tropic of Cancer, although some authorities extend it up to north to the Atlas Mountain. Madagascar and other small adjacent islands also come within this realm. It has the following subrealms:

(1) West African, (2) East African, (3) South African and (4) Malagasy.

The diagnostic fauna are Gorilla, Chimpanzee, Baboons, Zebra, Secretary bird, *Protopterus*, Giraffe, Lion and Hippopotomus.

Oriental realm. This region includes the south coast of Asia, east of Persian Gulf, the entire peninsula of India, the portion of China south of Nan-ling Range (Malaysia). The islands of Sumatra, Borneo, Java, Celebes and the Philippines are also included within this region. This region is subdivided into the following subregions. (1) Indian, (2) Ceylonese, (3) Indochinese and (4) Indomalayan. This region is temperate with moderate climatic conditions. Luxuriant growth of forest with rich biotic fauna characterises this region. The characteristic animals are: FISHES: silurids, notopterids, anabantids, cyprinoids, etc. Primitive fishes are absent. AMPHIBIANS: caecilians, rachophorids, tree frogs and toads, few urodeles. REPTILES: vipers, sea-snakes, pythons, *Crocodylus*, *Gavialis*, geckos, iguanas, etc. BIRDS: kingcrows, sunbirds, passers, woodpeckers, cuckoos, kingfishers, pigeons, doves, fowls, peacock, etc. MAMMALS: hedgehogs, shrews, "flying foxes", old world monkeys, cats, dogs, bears, elephants, rhinoceros, rodents, orang-utan, gibbons, tarsiers, etc.

The Indian subregion includes whole of Indian peninsula extending from the Himalyan slopes to Cape Comorin. The Ceylonese subregion includes the island of Srilanka, Indochinese subregion comprises of China, south of Palaearctic boundary, Burma, Thailand, the island of Andamans, Formosa, Haiaan, etc. The Indomalayan subregion embraces the Malayan Peninsula and the islands of Malayan Archipelago (Borneo, Java, Sumatra, Nicobar, etc.).

Australian realm. This region includes Australia, New Guinea, Tasmania, New Zealand and the oceanic islands of the Pacific. The Australian region is characterised by both tropical and temperate climates. The inner part of this region has arid climates having deserts. The animals and plants are peculiar because of their isolation from other continents of the world. Amongst the mammals, all the famlies of monotremes and marsupials are present. They are the predominant mammalian fauna. It is subdivided into the following subregions: (1) Austro-Malayan, (2) Australian, (3) Polynesian and (4) New Zealand. The following animals, Dingo, Parrots, *Neoceratodus*, etc., are the characteristic fauna. New Zealand contains *Sphenodon* and Kiwi. The *Austro-Malayan subregion* comprises of the islands of Malayan Archipelago (Aru, Mysol, Waigion and Mollucas, New Guinea and Solomon islands). Luxuriant forest, excessive moisture and stable high temperature are the characteristic features of this subregion. The peculiar animals of this subregion are: *Dendrolagus*, *Petaurus*, paradise birds, cassowaries, tree frogs, narrow mouthed toads, *Dasyureres* (native cat), *Gouaidae* (crowned pigeons), honey eaters, etc. The Australian subregion includes Australian mainland and Tasmania. Wombats, Duckbilld platypus, Marsupial moles are some of the peculiar animals. The Polynesian subregion includes Polynesian and Sandrich islands (Fiji, Caroline, Society, New Caledonia, New Hebrides, etc.) Tooth-billed pigeons are the peculiar faunal forms. New Zealand subregion includes New Zealand, Auckland Island, Campbell Island, Norfolk Island, Macquaric Islands, etc. *Sphenodon*, *Liopelma*, typical and free-tailed bats are some of the peculiar animal forms.

Owing to the resemblances of their respective fauna, the Nearctic and the Palearctic regions are grouped together as *Holarctic realm*. Lydekker (1896) has proposed the division of the entire terrestrial surface into three major divisions. According to him Nearctic, Palearctic, Ethiopian and Oriental regions are included under one region—the *Arctogaeic realm* or the *North land*. Neotropical is known as *Neogaeic* and the Australian region is called as *Notogaeic realm* as shown in Fig. 1.2.

Six biogeographical or zoogeographical regions are separated from one another by nearly impassable barriers. The Nearctic and Palearctic regions are separated by the Atlantic and Pacific oceans. The Ethiopian region is separated from the Palearctic region by the vast stretch of Sahara desert. The Oriental region is separated from the Palearctic by the Himalayas and Nan-Lings mountain chains. South America and North America are connected now by the Isthmus of Panama, but during geological eras this connection was submerged. As a consequence South America was completely isolated from the land masses. Sometimes the physical conditions prevailing within one region may be much alike to those of another. South America and Africa have similar physiography and climate. These two regions have few organisms in common, which represent the living relics of once world-wide groups. The present-day distribution of dipnoans reveals the same truth. The African genus (*Protopterus*) and South American genus (*Lepidosiren*) belong to the same family while the Australian genus (*Neoceratodus*) is placed in another family. Considering the distribution of the surviving genera of the dipnoans, it is clear that these three continents of the globe have special relation although these are separated by oceanic barriers. This is testified by the fact that the fossil dipnoans had once enjoyed world-wide distribution.

The geographical distribution of animals on the surface of the earth is easily understandable if one assumes that each group of animals has been originated in one of the central regions of the globe, then they spread to inhabit different zones separated by ecological barriers. The divergence from the central region results into adaptation of the animals to various conditions. This phenomenon signifies the main theme of evolution.

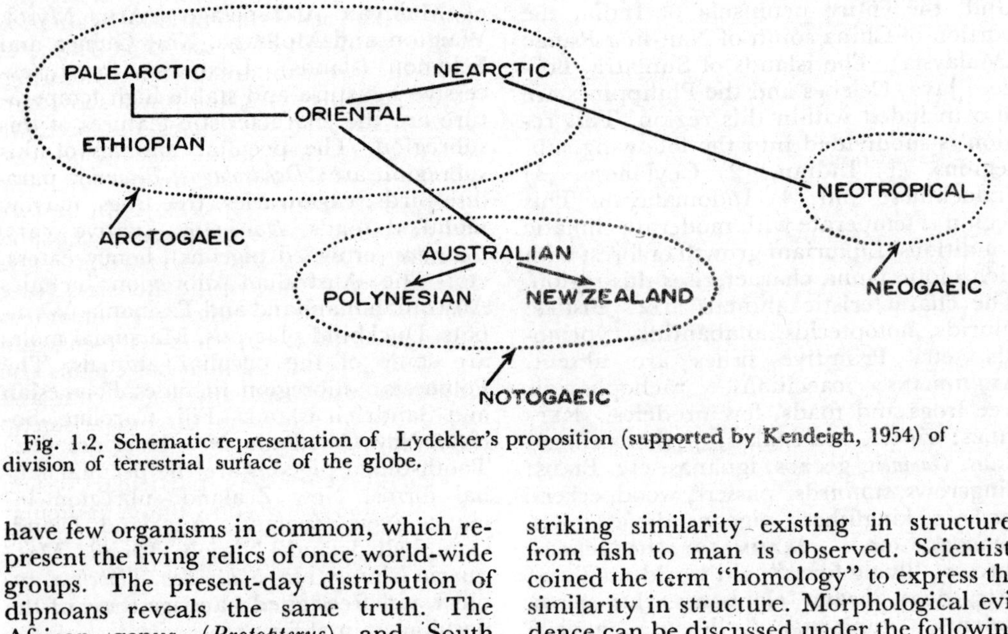

Fig. 1.2. Schematic representation of Lydekker's proposition (supported by Kendeigh, 1954) of division of terrestrial surface of the globe.

MORPHOLOGICAL EVIDENCES

Morphology is a special branch of biology which deals with the form and structure of living organisms. Morphological study of the vertebrate bodies reveals the truth that the different structures are built on a basic plan. Minor variations encountered in some forms are the effect of adaptation to diverse mode of living. If vertebrates are examined in detail, a striking similarity existing in structures from fish to man is observed. Scientists coined the term "homology" to express the similarity in structure. Morphological evidence can be discussed under the following heads:

Comparative study of different structures in vertebrates

A comparative study of the organ systems in different vertebrates shows that they are built on the same structural plan and at the same time shows gradual blossoming up of complexities. There are many examples from the comparative anatomy of vertebrates which provide source of evidences for organic evolution. The vertebral column in all vertebrates originates from four mesodermal masses in each somite. The vertebral column is composed of vertebrate, each having its centrum, neural arch, neural canal, neural spine, transverse processes and attaching processes. The heart in vertebrate series relates the same story (Fig. 1.3). The two-chambered heart of fishes become three-chambered in amphibia in response to the

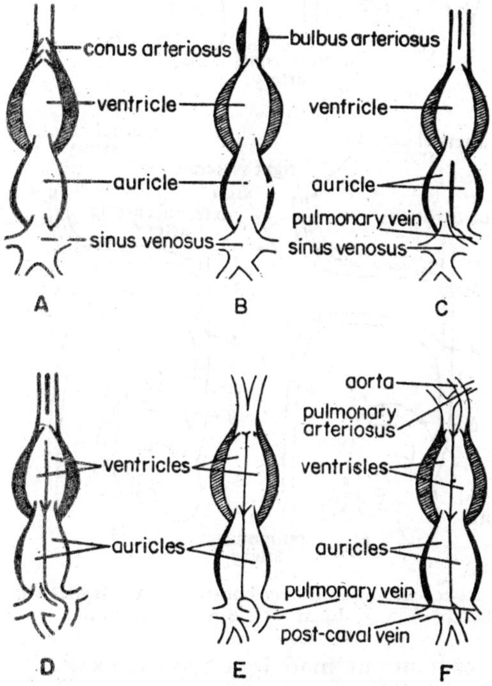

Fig. 1.3. Showing the diagrammatic longitudinal sectional views of heart in vertebrate series (after Walter and Sayles). A. Cartilaginous fish. B. Bony fish. C. Amphibian. D. Lizard. E. Crocodile. F. Birds and mammals.

change of habitat. The main purpose is to prevent the admixture of oxygenated and deoxygenated blood. In reptiles the heart is also three-chambered, but the ventricle is incompletely divided into two by an incomplete median vertical partition. In crocodiles the heart is almost four-chambered. But in birds and mammals the heart is completely four-chambered and there is a complete separation of oxygenated and deoxygenated blood. The aortic arches in different vertebrates also relate the same story. In all the vertebrates, the adult arches develop from six embryonic arches. The variation and modification in different vertebrates (Fig. 1.4) are due to specialisation of respiratory system and heart. The brain in different vertebrates also furnishes the same truth. Starting with fish to mammal, the brain has five subdivisions and shows progressive evolutionary changes, especially in the development of the cerebral hemispheres and cerebellum (Fig. 1.5). Other homologous structures also relate the same story. Let us take for example of the forelimbs in vertebrates. The fundamental part of the forelimbs is its bony structure In all the vertebrates excepting fishes the upper portion of the forelimb is called the humerus. The humerus is attached to the shoulder girdle. Below the humerus lie the radius and ulna. The radius and ulna are followed by carpal bones, metacarpals and phalanges. If comparison is made with the corresponding structure of vertebrates a striking similarity exists throughout the series. Main bony elements are comparable section for section. Modifications are due to adaptation to diverse mode of living (Fig. 1.6). In birds, the forelimbs are modified into wings for flying. In horses, the limbs show cursorial adaptation. In whales, the forelimbs are modified into paddle-like flippers for swimming in water. In human being, the forelimbs are pentadactylous in nature and are adapted for grasping objects and for manipulation. But examination of the skeletal architecture of the aforesaid types of limbs indicate that they are fundamentally composed of same bony elements and are also built on the similar basic plan. The hindlimbs also reveal the same truth and exhibit slight deviation in comparison to the forelimbs.

The survey of any particular system discussed above will show that a particular system is based upon a prototype which becomes modified in class to class.

Vestigial organs. Vestigial organs are the useless rudimentary remains of the organs that are believed to be fully formed and functional. These structures are useless to their possessors but resemble very closely and correspond to the useful organs of other forms. A large number of examples of the vestigial organs is found in animal kingdom. In Greenland whale, the hindlimbs had no use and have become abortive.

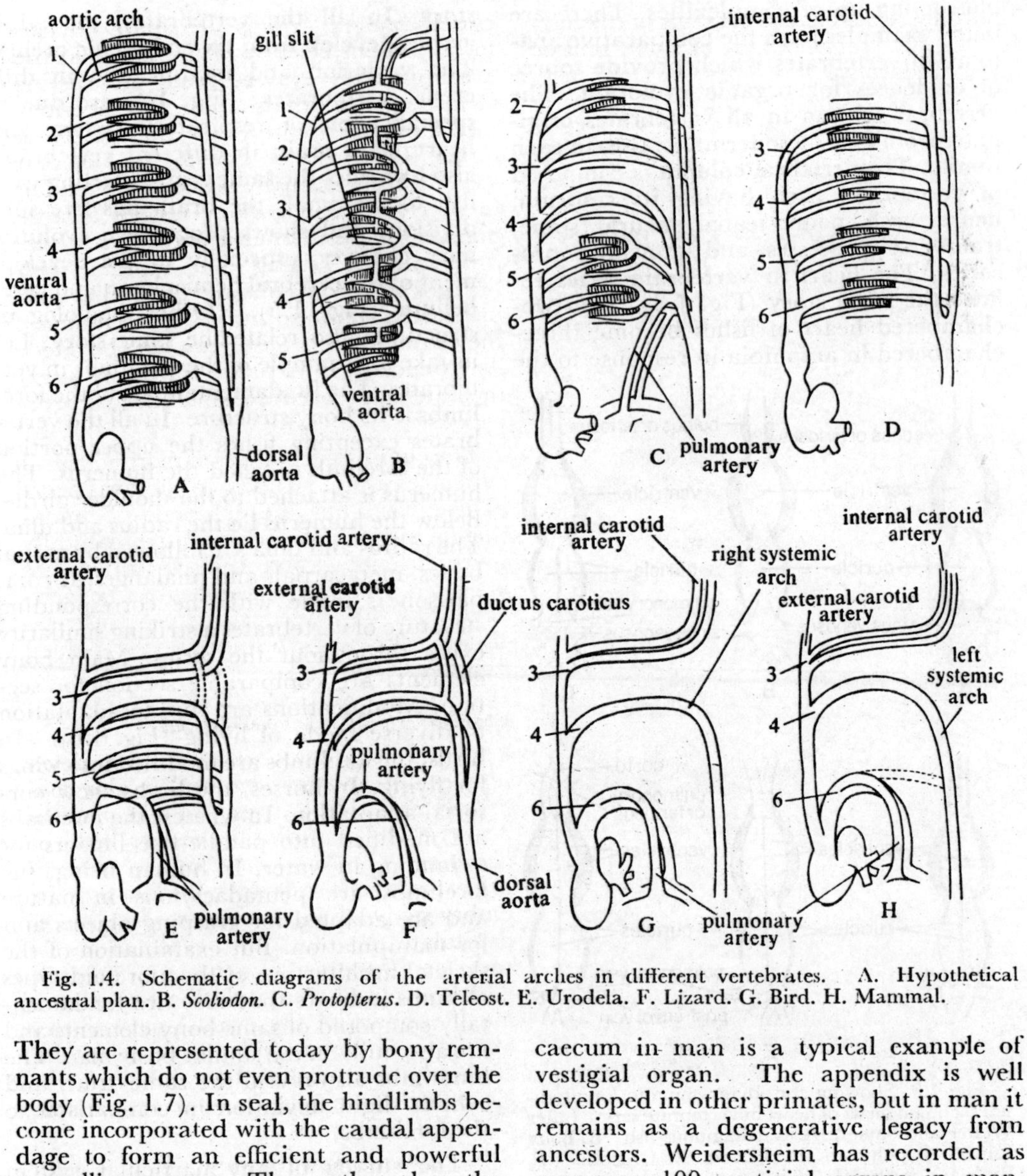

Fig. 1.4. Schematic diagrams of the arterial arches in different vertebrates. A. Hypothetical ancestral plan. B. *Scoliodon*. C. *Protopterus*. D. Teleost. E. Urodela. F. Lizard. G. Bird. H. Mammal.

They are represented today by bony remnants which do not even protrude over the body (Fig. 1.7). In seal, the hindlimbs become incorporated with the caudal appendage to form an efficient and powerful propelling organ. The python has the vestiges of the hindlimbs among the ventral scales. Practically in all vertebrates the remains of a membrane (nictitating membrane) is found in the inner corner of the eye. This membrane is complete and is perfectly functional in birds. The greatly reduced wings in some running birds, like Kiwi and Ostrich are also the examples of vestigial structures. Human body has a number of vestigial organs (Fig. 1.8). The muscles concerning the movement of the pinna in man are small and useless. Vermiform appendix in the caecum in man is a typical example of vestigial organ. The appendix is well developed in other primates, but in man it remains as a degenerative legacy from ancestors. Weidersheim has recorded as many as 100 vestigial organs in man. Wisdom teeth or third molars, vestigial caudal vertebrae (3–5 in number) of man are some of the typical examples.

Limitation of space will not permit to give an extensive survey of the vestigial organs in organisms. They are really immovable and there is hardly an organism whose body does not contain the tokens of the past. In human beings these vestigial organs are the roots of infection, because these are abortive and less vascular structures. It is extremely difficult to explain these useless structures, but we may assume that they represent the remains of the fully formed functional organs of some remote ancestral forms.

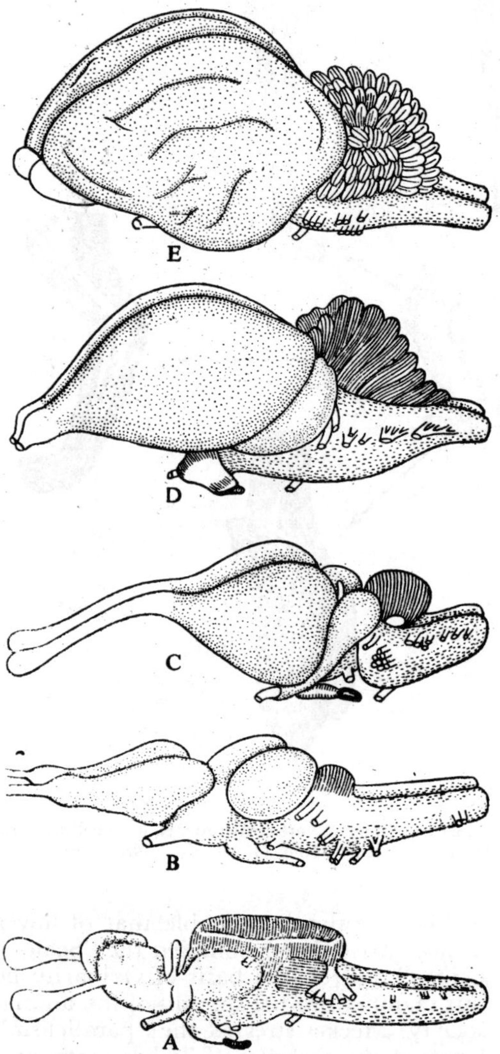

Fig. 1.5. Evolution of brain in vertebrates. A. Fish. B. Amphibian. C. Reptilia. D. Bird. E. Mammal. Note the similar divisions of the brain.

Intermediate forms. A survey of the animal kingdom reveals the existence of certain intermediate forms which bridge the gap in the sequence of evolution. Monotremes, as we see today, possess admixture of mammalian as well as reptilian features. They are actually 50% mammalian and 50% reptilian in their anatomical organisation. Monotremes form an intermediate bridge linking the reptiles to the mammals. Intermediate forms linking closely related groups are quite obvious to be present because of the fact that the existing forms evolved from pre-existing forms lower in organisation. But the intermediate forms are very rare amongst living forms. This may be due to the fact that such forms are usually weak and often fail to survive. Palaeontology reveals a large number of fossil intermediate forms, viz. *Archaeopteryx*, linking the reptiles with birds; Theriodonts linking the reptiles with mammals. Amongst living representatives, the example of *Peripatus* can also be cited. Anatomically it virtually forms an intermediate stage between annelids and arthropods.

The cumulative sources of morphological evidence cannot be taken as the final proof of evolution but the existence of fundamental similarities of structures naturally gives the indication that evolution has occurred. Homologous organs can be interpreted as the evidence of structural relationship existing throughout the animal kingdom.

EMBRYOLOGICAL EVIDENCES

Embryology deals with the early development of an organism and is also concerned with the changes that take place in an organism from fertilization to birth or hatching. The importance of the embryological evidence rests upon the fact that the embryogeny of higher forms seem to recapitulate the racial history of its ancestors or *Ontogony recapitulates Phylogeny*. This concept of recapitulation has given a lot of support to the evolutionary theory and gives indirect evidence of evolution. If evolution is taken to be a fact, embryological evidence appears to be very simple and straightforward.

Recapitulation theory. It is a celebrated fact that there exists a close similarity in the life history of different organisms, especially in different vertebrates. The development of an individual exhibits general conformities to the early creatures of evolution enroute to finality. This idea of recapitulation was first struck in the minds of the Embryologists and created a havoc in the minds of the Scientists and an epoch in the evolutionary history. This idea of recapitulation was criticised from many corners but it still occupies a prominent status in the discussion of evolution as a process.

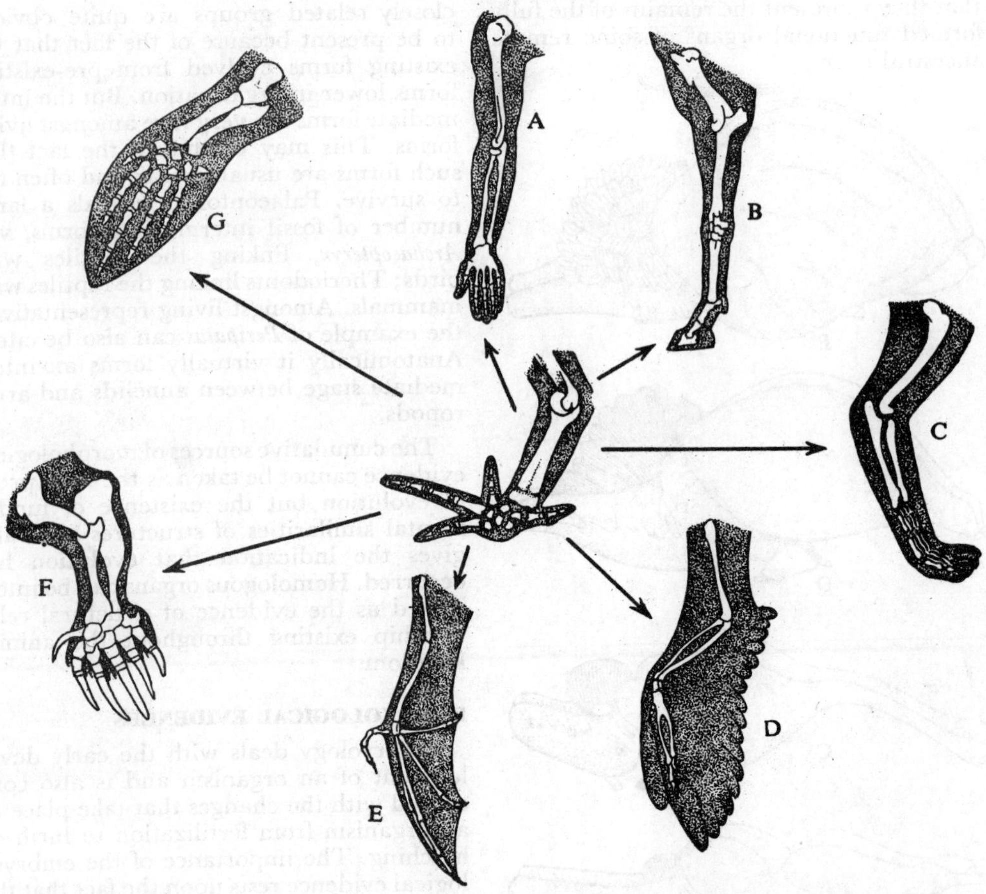

Fig. 1.6. Showing the modifications of the forelimb in different vertebrate forms. Forelimb of Toad (at the centre). A. Man. B. Horse. C. Cat. D. Bird. E. Bat. F. Mole. G. Whale. Note the similarity in bony structure.

Meaning of recapitulation. Embryos undergo development and their ancestors have undergone evolution. Recapitulation in biology may be regarded as a relationship existing between embryology and evolution on the assumption that the developmental stages of an individual (ontogeny) repeats the post racial history (phylogeny) in an abbreviated and/or accelerated way. In brief, ontogeny is a short resume of phylogeny or ontogeny repeats phylogeny.

Historical review. Men from early days tried to link ontogeny with phylogeny. The very term recapitulation finds its expression after much travel and difficulties. It is an wonderful blending of a number of ideas sponsored by the historic figures of Biology. Greek embryologists have established the fact that the embryos of higher animals resemble that of lower forms. Aristotle profounded the "Scale of beings" and on the basis of viviparity he put mammals at the summit of the animal society. Mecket stressed the "parallelism" between the embryos of higher forms and the adult stages of lower forms. Von Baer objected to Mecket's idea of parallelism and puts more weight to his views. He has expressed the results of his embryological studies in four points:

(1) During ontogeny the general characters appear prior to specialised characters. (2) From the more generalised characters, the less generalised and lastly the specialised characters are developed. (3) With the development, an animal gradually departs more and more from the forms of other animals. (4) The young stages in the development of an animal are not exactly

similar to the adult stages of other animals lower on the scale but resemble their young stages.

Müller, based on the concept of recapitulation, advocated that evolution might occur in two ways: (*i*) by divergence from the ancestral path during development and (*ii*) by the addition of new stages at the end of development.

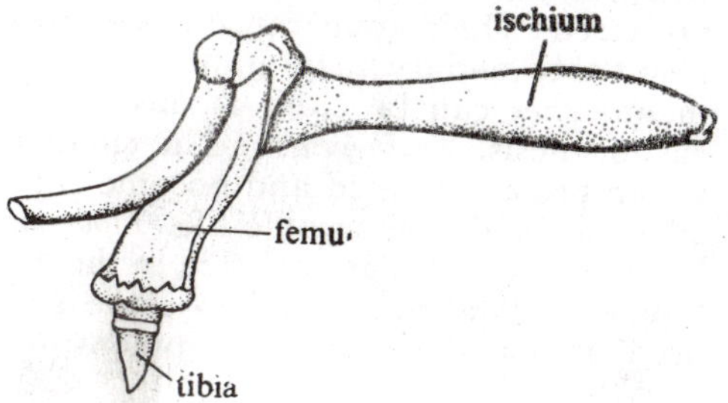

Fig. 1.7. Showing the vestigial hindlimb of whale.

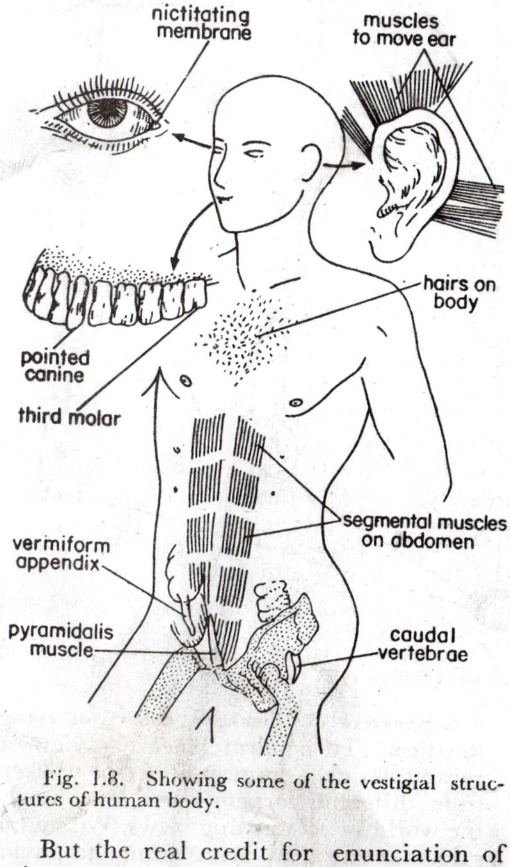

Fig. 1.8. Showing some of the vestigial structures of human body.

But the real credit for enunciation of the concept of recapitulation goes to the German scientist, Ernst Haeckel who set forth the "Biogenetic law" or the "Recapitulation theory". Haeckel placed the recapitulation concept on a solid basis.

Haeckel's theory of recapitulation. The spark of Haeckel's genius had lighted the darkness of embryology. The theory of recapitulation indicates that the entire phylogenetic processes occurred for billions of years are telescoped together in the short process of differentiation found in an individual sequence of development. To sum up the recapitulation theory of Haeckel, it can be stated that ontogeny is condensed as a recapitulation of the phylogeny. His theory has two aspects: (*i*) Ontogenesis is the recapitulation of phylogenesis and (*ii*) Phylogenesis is the mechanical cause of ontogenesis.

To an embryologist the striking similarity between developing embryos of different vertebrates is a well-known fact (Fig. 1.9.). Haeckel supposed that each animal during its development from the egg to adult, passes through a series of stages reflecting the evolutionary history of the species to which it belongs. Fig. 1.10 relates the relationship between the ontogenetic and phylogenetic relationship of frog. Presence of gill-slits in mammalian embryo may be a case of 'hang-over' as Waddington calls it, from the time when the ancestors of the mammals were fishes. The embryonic history gives sure information not only about the facts of evolution but also the general course it took in a species.

Evidences in support of recapitulation theory. Haeckel and many embryologists put forward many evidences in support of recapitulation theory. These are:

(1) Existence of close similarity between embryos of different vertebrates in early stages of development. (2) The heart in vertebrate series is built on common basic plan. It consists of two portions—receiving parts and forwarding parts. The receiving parts comprise of auricle and sinus venosus while the forwarding parts consist of ventricle and conus arteriosus. (3) In embryonic stages of all vertebrates, the arterial arches are alike. In all vertebrate embryos, there are six pairs of arterial arches. (4) Presence of gill-slits in all vertebrate embryos. (5) In vertebrates, nervous system originates by infolding of dorsal ectoderm. (6) Nearly all the crustaceans pass through *nauplius* stage. (7) Tadpole larva of frog resembles fish. (8) Mollusca and annelida

pass through Trochophore stage. (9) Covering of hair in human babies (*Lanugo*). (10) Antlers of some living deer resemble those of fossil forms. (11) Often tail is present in human being. (12) In the ontogenetic development of kidney in higher vertebrates there is a succession of Pro-, Meso- and Metanephros types. (13) Almost all the coelenterates have 'planula' larva. (14) The vertebrae develop from the same source and in a similar way in different vertebrates.

Haeckel was not so clear as to the anatomical construction of the later ancestral forms. An important part of his theory is development. Presence of gill-slits in vertebrate embryos is a case of palingenesis. *Caenogenesis*. The characters which are new adaptations. The young stages of a developing animal present certain structures which no adult ancestor could possibly have possessed. Their appearance is due to the result of secondary adaptation. The foetal membranes can be cited as instances of caenogenesis. *Trachygenesis*. The characters which are accelerated and become crowded back in the embryonic life. *Lipogenesis*. When some stages are omitted in the development. *Bradygenesis*. When some stages are lengthened during development.

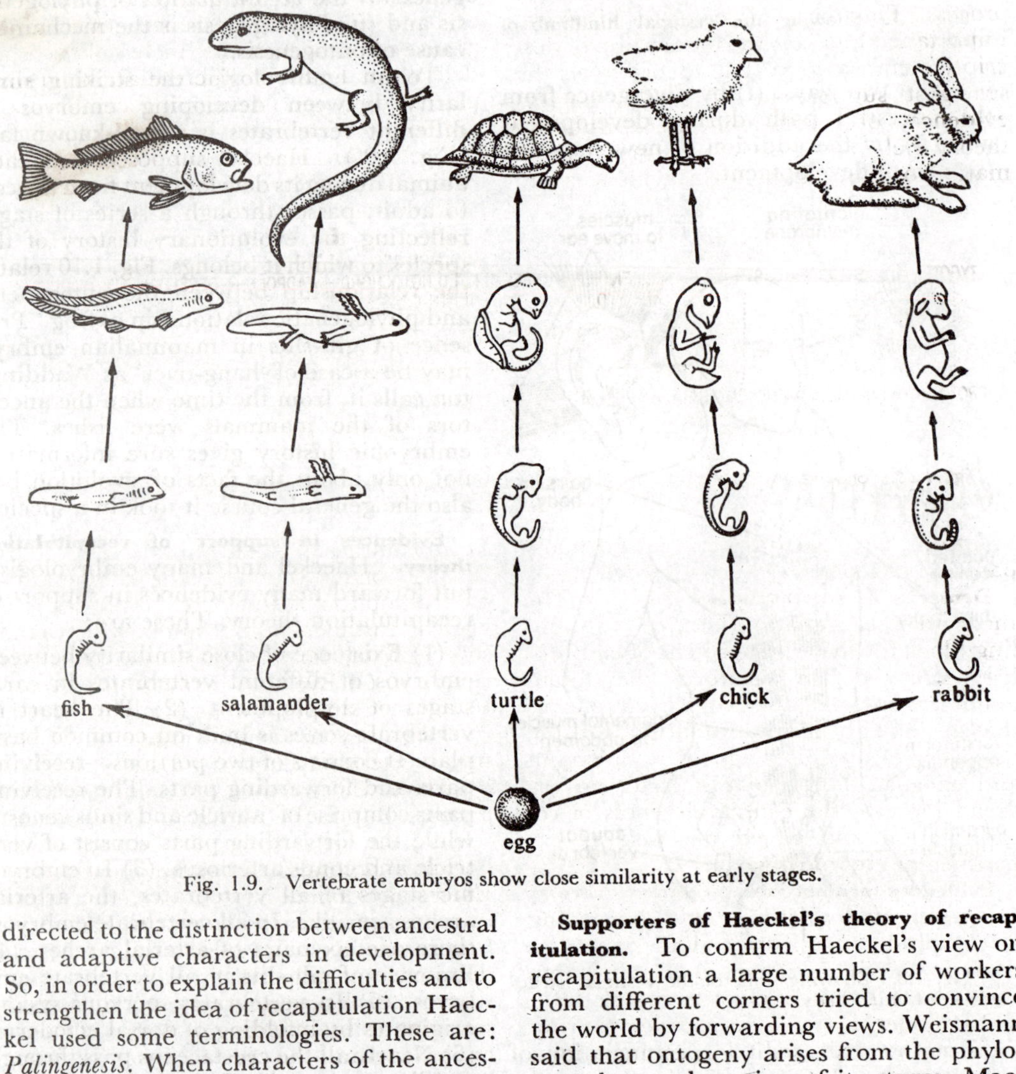

Fig. 1.9. Vertebrate embryos show close similarity at early stages.

directed to the distinction between ancestral and adaptive characters in development. So, in order to explain the difficulties and to strengthen the idea of recapitulation Haeckel used some terminologies. These are: *Palingenesis*. When characters of the ancestors are conserved by heredity and are faithfully recapitulated at the time of one's

Supporters of Haeckel's theory of recapitulation. To confirm Haeckel's view on recapitulation a large number of workers from different corners tried to convince the world by forwarding views. Weismann said that ontogeny arises from the phylogeny by condensation of its stages. Macbride advocated that the larval phase of development of an individual represents a former condition of the adult of the stalk to which it belongs. Graham Kerr advocated that the vertebrate larvae differ very little from that of the common ancestral type. Agassiz observed that the developmental stages of present echinoid show parallelism with the fossil genera.

Merits of recapitulation theory. After the establishment of the recapitulation theory by Haeckel the idea of recapitulation was accepted readily by most of the workers. But recently the idea of recapitulation is modified and is interpreted correctly. However, Haeckel's theory of recapitulation must be discussed while explaining evolution as a process. This theory is of great historical importance and lays the foundation of descriptive embryology. The theory gives the strongest support to the embryological evidence of evolution. Recapitulation theory helps to determine correct systematic position when other evidences are unknown, e.g., *Saculina*, *Ascidia* and many parasitic copepods.

Demerits of recapitulation theory. The theory of recapitulation became a stumbling block for the advancement of embryology. Every worker, whatever, they found about the facts of embryology tried to focus that on the light of recapitulation. But the discovery of organiser action by Spemann and exogastrulation of Holtfreter gave a dread blow to the conservativeness of recapitulation and paved the way of experimental embryology.

Evidences against recapitulation theory. 1. *Thymus gland in vertebrate series*. According to recapitulation theory similar structures must develop in the same way. But in Salmo the thymus gland develops from ectoderm and endoderm. In Talpa it develops from ectoderm and in man it develops from endoderm. So the diversities in the development of thymus gland in vertebrates goes against recapitulation concept. (2) *Tooth and tongue relationship*. Excepting man, in almost all vertebrates teeth develop before tongue, but in man tongue develops first then come the teeth. This is a case of deviation of recapitulation. (3) *Germ-layers and their potencies*. Haeckel regarded that the ectoderm, mesoderm and endoderm are the three fundamental germ layers from which similar differentiation occurs in the same way in all vertebrates. But modern experimental embryologists have changed the concept of rigidity of germ layers. (*a*) X-ray irradiation experiments on chick development shows degeneration or stoppage of differentiation of the embryo due to interference. (*b*) Transplantation experiments in embryos can defy normal path of differentiation. Transplantation experiment with eye on tail are some of the instances. Weiss has produced a tail in the neck region of the embryo by grafting. (*c*) If sugar is lacking in *in vitro* culture of chick embryo, it develops in an abortive way. (*d*) The famous exogastrulation experiment of Holtfreter is a typical instance of deviation, where the direction of morphogenetic movement is reversed. (*e*) Eyes can be made to develop to the median position by applying $MgCl_2$ in the embryonic condition of fish. (4) In most vertebrates, neural tube is formed as a groove whose dorsal edges fuse to form a tube. But in Petromyzon, Lepidosiren, teleost, etc., it arises as a solid rod which becomes hollowed out subsequently. (5) Emergence of pro-, meso- and metanephros in succession in the embryo of higher vertebrates is not a case of recapitulation. Waddington, by experimental means, has shown that the pronephros gives the necessary stimulus for the development of mesonephros, similarly mesonephros in turn provides the necessary stimulus for the development of metanephros. (6) Cetacea

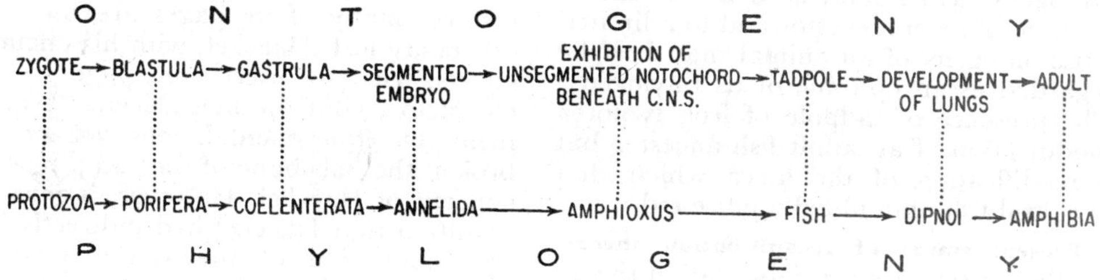

Fig. 1.10. Schematic representation showing the relationship between ontogenetic and phylogenetic history of frog.

originated from ancestors who had well-developed hindlimbs but no trace of hindlimbs is present in their embryo.

Recapitulation theory criticised. The concept of recapitulation of Haeckel is severally attacked and the modern Biologists have modified the original concept. W. His was the first man who challenged the recapitulation theory. At early stages, developing animals possess the characteristics of the class, order, genus, species and sex to which they belong as well as the individual characteristics. F. R. Lillie also pointed out that not only the final result but all the stages of ontogeny are modified in evolution. Roux altogether rejected the theory and deduced mechanical process by which different structures arise from particular area of germ. Garstrang gave a new interpretation and regarded that the phyletic line of succession of zygotes run more or less parallel with the adult sequences and ontogeny does not recapitulate phylogeny. Ontogeny, as he said, is to be the modifications of its predecessor and to a limited sense ontogeny of an animal may best be regarded as an epitome of its phylogeny. The presence of tadpole of frog is not a modification of an adult fish-ancestor but a modification of the larva which that ancestral fish undoubtedly possessed.

Present status of recapitulation theory. At the present time recapitulation theory is no longer regarded to be correct. In different embryos, there is no doubt, striking similarities exist but this is not the case of repetition of phylogenetic stages as visualised by Haeckel. Most of the parallelism existing in ontogenetic development may best be regarded as the reflection of characters not of the adults but of the embryos. Recapitulation theory of Haeckel may be said a theoretical explanation having no practical support. Presence of vestigial structures is not the fulfilment of phylogeny. Experimental embryologists have altogether denied the notion, "ontogeny repeats phylogeny". By suitable experimental studies, J. Needham in his book "Biochemistry and Morphogenesis" has regarded recapitulation as a phenomenon and not as a law. The recapitulation theory in its original sense is untenable and has many things to do with the embryos but has little to derive out of it.

Relationship between phylogeny and ontogeny has been explained by many Russian scientists who regarded that the relationship between the two processes does not follow any rule. They have coined several terms to explain the relationship existing between them. *Addition*. When development follows the racial pattern up to the last racial ontogenetic stage. To this, new stages are added to give rise to new adult. *Anchallaxis*. Ontogenetic development deviates from the racial pattern so that recapitulation is absent. *Deviation*. Ontogeny of the organism follows the ancestral pathway of development up to a certain stage and then deviates along a new line. *Abbreviation*. When the last stages of the racial ontogeny are omitted. *Acceleration*. During ontogenetic development the intermediate stages of the racial ontogeny are dropped, so that the later characteristics appear in the ontogeny earlier than they did in the intermediate racial ancestors.

The science of embryology has a long history, many of its pages are torn and others are lost. Haeckel, with his uncunny efforts, tried to insert some pages in that old history. But the magnificent advancement of experimental embryology has broken the backbone of the recapitulation concept of Haeckel. However, it must be admitted that Haeckel had indirectly laid the foundation of the development of embryology. Nowadays his original idea is not taken as such but has been modified. As the embryonic development is an epigenetic phenomenon, superficial similarities between the embryos are likely to be present. Haeckel's recapitulation theory can be interpreted by saying that the individual developmental stages (ontogeny) may at best repeat the ontogenetic stages of the racial forms (phylogeny) and not the phylogenetic stages as such.

GEOLOGICAL EVIDENCES

The science of geology furnishes many clues in understanding the process of evolution. The most important branch of geology which provides the greatest support is Palaeontology (Gr. *palaios*=*ancient*) which deals with the fossils. This particular branch of science links Geology with Zoology.

Estimation of time. Estimation of the geological time scale is necessary to determine the age of various fossils. There are several methods to estimate the geological time.

Sedimentation rate technique. The sedimentary rocks are formed by settling down of the eroded materials and other sediments. The time and rate of formation of sedimentary rock are fairly known at present. Geologists are of the opinion that the sedimentation rates are more or less constant and by examining the sedimentary rock deposits containing fossils the age of the strata can be estimated. But this method is not taken to be accurate, because the sedimentation rates are not always uniform.

Radiation of heat from earth. Physicists estimated the age of the earth from the radiation of heat from the earth. It is assumed that the earth was once a molten mass and has cooled down subsequently to its present state.

"Radioactive clocks"—to date the age of fossils: Scientists determine the age of the fossil deposits from the fact that the radioactive elements undergo decay at regular rates and thus form the 'radioactive clocks'. There are several 'radioactive clocks''. They are: uranium-lead 'clock', carbon 'clock' potassium and argon 'clocks', etc.

Uranium-lead method. The most accurate method of estimation the age of the geological strata comes from the principle of disintegration of radioactive element, uranium (uranium–238) into stable lead (lead–206), at a constant rate. It is calculated that through such natural disintegration one per cent. of uranium disintegrates in 66,000,000 years. By comparing the weights of uranium and lead in a particular strata, the actual age of the strata can be determined more accurately.

Radiocarbon method. The uranium lead method is proved to be very accurate, but uranium is very rare. So technique of dating by using other radioactive elements other than uranium becomes more convincing. The most commonly employed method is the radiocarbon (carbon–14) method. Carbon is utilised by all organisms. Of the total quantity of carbon used by the organisms, a constant and insignificant quantity is radioactive. It is known that a particular quantity of radioactive carbon loses 50% of its weight in 5,760 years.

Unfortunately materials of more than 40,000 years old cannot be determined by radiocarbon technique. But this method is highly accurate to determine the age of the recent fossils. In the fossil forms the quantity of the radioactive carbon may be estimated and by comparing it with the quantity present in fresh forms—the age of the particular test piece can be determined. Very recently dating of fossils has been based upon **isotopes of potassium and argon.**

Fossils form the basis of understanding the process of evolution. The ever-changing variabilities of life are incomplete if not for the contribution made by the palaeontological aspects of geology. The buried animals and plants in the earth's strata give us strongest support to the principles of organic evolution. The sole basis of support is the fossils—the nature's heiroglyphics. The vanished species imprinted in the pages of the earth furnish the documentary evidences of organic evolution and provide the alphabets with which the language of the biological history is written. Nature has preserved these relics to carry the tale of evolution. They are actually the milestones in the path of evolutionary progression.

This history of the study of fossils dates back to a very ancient time. Fossil was discovered first by Empedocles who collected a fossil of Hippo in Sicily and regarded that as the remnant of God. Aristotle thought the fossil forms as the evidences of an attempt of the inorganic materials to organise the shape and form of organic life. Henrion (1718) regarded fossils as casts and molds left over in the creation of plants and animals. Leonardo da Vinci (1452-1519) first recognised the fossils as the evidences of animal life of ancient ages. Georges Cuvier (1769-1832) made thorough study of the fossils. He published an account of fossil elephants in the year 1800 and contributed much to the new approach towards the study of fossils and their significance in understanding evolutionary process.

Fossils are the remains of organic life present long long ago preserved by natural process in the strata of the earth. Fossils are of diverse forms. Nature has performed many physico-chemical experiments to preserve them to carry the thread of evolution. There are certain preconditions for fossilisation. Fossilisation is a very slow and gradual process. Fossilisation is only possible if the living organisms are *imme-*

diately buried under earth to escape the contact of atmospheric air to prevent oxidation and must also *overcome other destructions.* Commonly it is regarded that fossil means petrifaction, meaning thereby turning to stone. Several types of fossils are recorded which can be grouped under four heads:

Petrifaction or pseudomorph. In this form, original structural pattern is more or less preserved and it has to undergo mineralisation. In this category of fossil usually the hard skeletal parts of the animal body are preserved. Petrification is a gradual process where replacement of molecule for molecule occurs. The petrified fossils retain not only the external features, but the original histological picture also. Fig. 1.11 shows the petrified fossils of Trilobites.

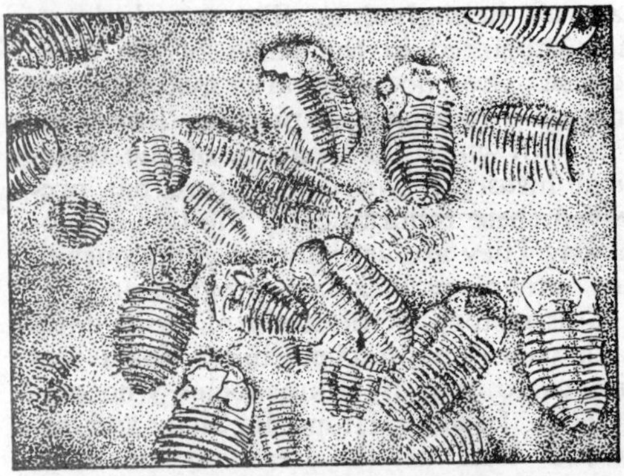

Fig. 1.11. Petrified fossils of Trilobites.

Preservation intact. Certain animals are preserved by nature with original substance more or less intact. Many pre-historic animals are preserved in nature's cold storage, specially in the snowy bed of Siberia. The notable example is the Mammoth preserved in the frozen bed of Siberia. Similar remains have also been discovered in Alaska. Certain animals, particularly the insects are preserved in crystal clear amber—a fossil resin of pines (Fig. 1.12). At the time of exudation, the resin is soft so as to engulf small insects. The resin becomes subsequently hard and changes over to amber. The amber preserves the delicate part of the insect without causing slightest injury.

Moulds or casts. Natural casts or moulds are formed when the surrounding region which encircles the organism becomes hard and is converted to stone. The organism, thus enclosed is disintegrated and subsequently removed by natural process. The surrounding stone leaves a cavity which gives the exact contour of the body of the vanished organisms. In this category of fossils no internal structure is preserved.

Foot-prints or imprints. The impression of the foot-prints or other parts of many organisms are kept preserved on the stone.

Fossilisation occurs when a sector of the earth containing the buried animals are converted into stone by natural processes. In such formation the oldest layer must be the deepest and the newly formed strata will be topmost. This original orientation of the earth's strata may be greatly changed by natural catastrophe but the fossils embedded in the strata help to determine the exact age of the strata. Because along with the fossils, certain radioactive substances like uranium which undergoes change into stable lead are also present.

The actual remains (such as mammoths in ice and insects in amber), minute replacements (petrifaction), coarse replacements (molds and casts) and impressions or prints gave direct evidences of organic

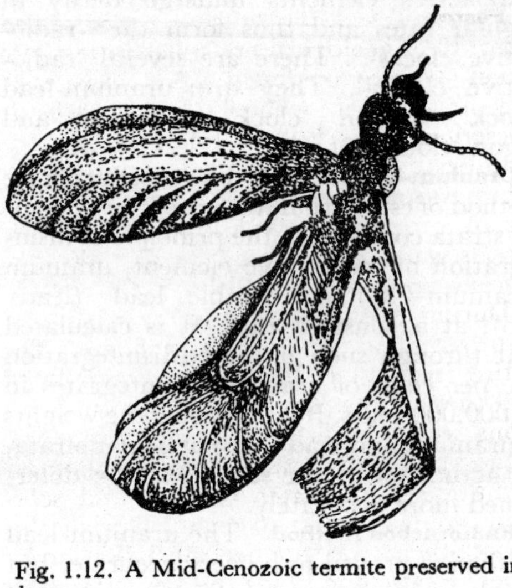

Fig. 1.12. A Mid-Cenozoic termite preserved in amber.

evolution. Besides, there are many fossils which furnish indirect evidences of organic evolution. They are:

(A) **Coprolites.** Solidified excreta or the casts of the same.

(B) **Artifacts.** Prehistoric fashioned flints or ant-hills.

(C) **Burrows, tracks, trails,** etc., of living animals.

(D) **Geologic** formations from organic sources like graphite, flint, limestone, coal, petroleum, etc.

Uses of fossils

Fossils, the "medallions of creation" help to analyse (1) racial history of plants and animals, (2) past climatic conditions of the earth and (3) to measure the geologic time.

Geological time scale. Based on the study of rocks with its contained fossils geologists have divided the earth's past history into a number of eras.

Like human history, the earth's history is also divided into a number of major *eras*. The eras are divided into *periods*. The periods are subdivided into *epochs* (Fig. 1.13).

Geological chronology. Down to the unimaginable corridors of the geological history, the thread of life has passed from generations to generations ever-varying but unbroken.

Precambrian—the darkest period of geology. Archeozoic and Proterozoic are the first known eras of the earth's history. Palaeontology cannot lay its hand on these eras. It was the darkest period of geology when nature was nursing the new-born life.

Postcambrian—the brighter period of geology. Real geological era starts from Cambrian, when true fossils were discovered. The first creature whose body was perfectly preserved was that of Trilobites. In the Cambrian period almost all major invertebrate groups were evolved. In the Ordovician, progressive evolution of the Cephalopods was observed and the first fossils of fishes were also recorded.

During Devonian period animals first came to land from aquatic home. The entire sea-water became overpopulated and there was struggle for existence. Scorpions were the first to leave the aquatic home and came to land. This was a transformation from water to land and scorpions were the first to come. During this period there was further specialisation of fishes. *Ichthyostega* holds the key of amphibian ancestry. The bodily emergence was all right but there was a hindrance in the reproductive system. They had to go back to the primal aquatic home for the purpose of reproduction. During Carboniferous a peculiar amphibia, *Eryops* evolved where an initial adjustment of reproductive phenomenon occurred. These amphibians became very sensitive and due to lack of physiological adjustments, they had to bid good-bye to this world. The last amphibia which could stand was the *Seymouria*. Seymouria carried the burden of evolution and gave the reptilian line of evolution—the *Limnoscelis* which in turn diverged into two lines—Ophiacodont and Sphenacodont.

Mesozoic—the unique and interesting era. The entire era is regarded as the *Age of Reptiles*. There were fluctuating geological conditions. Triassic was the age of turtles, snakes and Saltophosuchus. This period also observed the first sign of birds and mammals. From Saltophosuchus "explosive evolution" occurred and it gave rise to the two branches of Dinosaurs—the Saurischians and Orintheschians.

But during this era geological changes made environmental conditions unfavourable and inhospitable for the huge dinosaurs to continue. They had to leave the world keeping behind a sad story. The key of dominance of life fell on some humble creatures. Small in size but unique in their endowments, they constituted a group of animals—the mammals. Hugeness of size has nothing to do in evolution but it was the brain faculty which was of paramount importance.

Cenozoic—the age of mammals. The Cenozoic era was the age of mammals and birds, on which geology has a lot to say. The fossil remains of the horse, camel and elephant furnish vivid instances of unmistakable transformations of life through geological history.

Palaeontological discoveries provide the strongest evidence of evolution. Fossils in the earth's crust in almost all cases bridge the gaps left in the march of life through the geological time. Palaeontology provides many 'missing links'. The most famous transitional form linking the reptiles and the birds is the fossilised bird, *Archaeopteryx*. It had feathered wings like bird, but possessed reptilian teeth and lizard-like tail. Palaeontology also records the complete history of the origin and evolution of many modern mammalian forms. A comprehensive documentary record of the said animals has been vividly discussed. It must be recorded that the fossils buried in the earth's strata give an admirable support to the principles of organic evolution. It helps us to understand the basic trends of evolutionary processes. It makes us convinced that the living organisms change in course of time from simple to complex. This progressive trend is amply recorded in the successive geological strata.

TAXONOMICAL EVIDENCES

Taxonomy, the science of classification of animals, gives a convincing evidence in favour of organic evolution. Modern classificatory scheme of animals is based on structural similarities amongst organisms and such structural resemblances also indicate their relationships. In the realm of taxonomy a phylum is regarded as the largest group of animals having a common

Fig. 1.13. Diagrammatic representation of the major geological eras of the earth. 4,700 million years of the earth's history is compressed into one year. Each day of this calendar is equivalent to more than 12 million years of earth.

TABLE EVOLUTION–I

Percentage of the year of earth	Eras	Major divisions or periods	Advances or marches of life	Dominant form	Climatic condition	Remarks
	Psychozoic	Recent or Post-glacial period	Rise of human civilisation and the era of mental life	Age of Man	Stable geological condition	Dominancy of man over land, water and air
Cascadian Revolution or Ice-age						
4%	Cenozoic (Modern life)	Pleistocene	Extinction of great mammals.	Age of modern mammals.	During Oligocene aridity of the climate was increased. Approaching of glaciation in Pleistocene. Occurrence of periodic glaciation in the later part of this era.	During this era extinction of great and ancient mammals occurred.
		Pliocene	Origin of man.			
		Miocene	Culmination of mammals.			
		Oligocene	Evolution of higher mammals.			
		Eocene	Vanishing of the archaic mammals.			
		Paleocene	Rise of archaic mammals.			

Laramide Revolution

Percentage of the year of earth	Eras	Major divisions or periods	Advances or marches of life	Dominant form	Climatic condition	Remarks
11%	Mesozoic (Medieval life)	Createaceous	Specialisation of reptiles and extinction of great reptiles.	Age of reptiles	Earthquake, flood-volcanic eruptions cause the crustal change of the earth.	Land reptiles achieved specialisation and perfection. Extinction of giant reptiles due to hostile geological conditions and origin of birds and mammals.
		Jurassic	Rise of flying reptiles and birds.			
		Triassic	Rise of dinosaurs, birds, egg-laying mammals.			

Appalachian Revolution

Percentage of the year of earth	Eras	Major divisions or periods	Advances or marches of life	Dominant form	Climatic condition	Remarks
30%	Palaeozoic (Ancient life)	Permian	Rise of land vertebrates; modern insects; amnictes, etc. Many ancient forms became extinct.			During these periods all the invertebrates and vertebrates were evolved. During late Silurian and early Devonian there was aridity—the impelling cause for the rise of air-breathing fishes and amphibia. Carboniferous had warm and moist climate which is necessary for the metamorphosis of insects and amphibia. Periodic glaciations caused the extinction of many ancient forms.
		Carboniferous	Rise of insects, echinoderms, ancient sharks, primitive reptiles.	Age of Amphibians	Great revolutionary geological impacts; periodic glaciation. Changes of the face of earth.	
		Devonian	Rise of amphibia and widespreading of fishes.	Age of fishes	First emergence of mountains, submergence of lands, formation of low land, formation of green mantle.	
		Silurian	Rise of scorpion, fishes, lung-fishes, etc.	Age of higher invertebrates		

Percentage of the year of earth	Eras	Major divisions or periods	Advances or marches of life	Dominant form	Climatic condition	Remarks
		Ordovician	Rise of corals, nautilids, armoured fishes, etc.		Stable geological-conditions, submergence of land.	
		Cambian	First known marine faunas—rise of shelled animals—dominance of Trilobites.			
Grand Canyon Revolution						
55%	Proterozoic and Archeozoic		Scanty evidences of living things. Organic deposition at the bottom of the sea indicated the existence of life in simpler forms.	Protozoa	Unstable geological condition, extensive volcanic activities.	These eras are called the age of primitive invertebrates.

ground plan of organisation. Possession of common characteristics by the members of phylum makes it reasonable to conclude that all of them have evolved from a common ancestral stock. By this way it is possible to arrange the phyla according to their complexities beginning with the Phylum Protozoa and culminating in the Phylum Chordata. In the line of evolution there are many unbridged gaps. To connect such gaps we may have to go back to the past history of the animal forms and much of the information is kept hidden amongst extinct forms. The existing animals represent only the terminal twigs of the evolutionary tree. Direct connection between terminal twigs may sometimes be absent, but the phylogenetic relationship can be established if we follow the branches and stem of the phylogenetic tree. The interrelationship amongst different phyla gives a clue to the progressive changes in evolution. The taxonomic evidence of evolution stands on morphological and physiological similarities amongst organisms.

EVIDENCES FROM COMPARATIVE PHYSIOLOGY AND BIOCHEMISTRY

In course of evolution, the change of structure is always accompanied by physiological and biochemical changes. Evolution is basically a summation of different biochemical phenomena. All living organisms are basically built on one substance —protoplasm which varies slightly from species to species. The chemical composition and functions of protoplasm can be described in a similar way with few exceptions throughout the organic realm. This impressive fact strongly suggests community of origin. The fundamental properties of living ,things remain rather constant, while variation has produced immensely varied forms in the living world. The chemistry of chromosomes—the physical basis of heredity also relates the same story. The chromosomes, in all living cells consist of basic proteins in combination with nucleic acids. Histone and protamine are the simplest types of proteins present in chromosomes, but globulin and some other proteins are also identified. The nucleic acids are rather similar. The nucleic acids differ mainly in the sequence of base pairs which unite the nucleotide chains together. The specificity of genes upon which the traits depend is largely due to the sequential arrangement of base pairs joining the nucleotides. The constitution of chromosomes, like that of protoplasm, also indicates the general uniformity of the fundamental units of life.

A survey of the evolution of living organisms reveals that the primitive members of the animal phyla inhabited the palaeozoic seas while their terrestrial relatives became adapted to the terrestrial condition. Physiologically the terrestrial vertebrates have retained many characteristic traces of their descend from the aquatic ancestors. In both aquatic and terrestrial animals, the ionic composition is more or less same.

It is recorded that parasitism is an ancient phenomenon and goes far back into the geological time. The existence of parasites has always been a part of the general pattern of vertebrate life. The parasites, indirectly or directly, furnish evidences of evolution. Certain parasites are recorded to attack only organisms that bear a close resemblance to each other. This indicates uniformity of body chemistry. Malarial parasites occurring in reptiles, birds, insectivores, some primates, suggest their origin in Mesozoic era before the origin of reptiles. Pinworms occur in amphibia, reptiles, some rodents, primates and indicate their early origin at the time when vertebrates arose. Similarly parasites like Sclerostomes, Tapeworms that occur in isolated animals, such as South African *Strutheo* and South American *Rhea* suggest common origin and their subsequent diversification.

Enzymes and Hormones. The chemical composition and physiological functions of the enzymes and hormones are closely identical. A protein-splitting enzyme, trypsin, is present from the protozoans to mammals. The starch-splitting enzyme, amylase, is found from sponges to man. The hormones in different mammals are fundamentally alike. Insulin extracted from other mammals is used in diabetic human beings. The thyroid hormone is similar in vertebrates. Beef thyroid is used biochemically for the treatment of human thyroid deficiencies. Thyroid hormone regulates the metamorphosis of frogs. If thyroid glands are removed surgically the metamorphosis is stopped. But administration of human thyroid tissue extract into such frogs will rectify the deficiency and accelerates metamorphosis.

Similarity in chemistry and physiological role of enzymes and hormones, indicate the fact that some identical genes controlling the activities of the specific glands are present amongst the vertebrates.

Phosphagens. The chemistry of muscle contraction relates that ATP (adenosine triphosphate) breaks down during contraction to release energy. Phosphagen, an energy-rich compound, breaks down and liberates energy for the resynthesis of ATP. Phosphagen is a specific compound in the muscles of vertebrates. Phosphagen is present in echinoderms and hemichordates—a fact suggesting the phylogenetic relationship between them.

Serological studies. The comparative study of the body fluid, especially the blood in different forms according to their chemical nature suggests resemblance between different organisms. If foreign protein (antigen) is introduced into the blood of an organism, counteracting substances (antibody) are produced. This antigen-antibody reaction is very significant in our body-system. Nowadays immunisation against particular diseases may be artificially produced by causing antibody production. Such antigen-antibody reactions

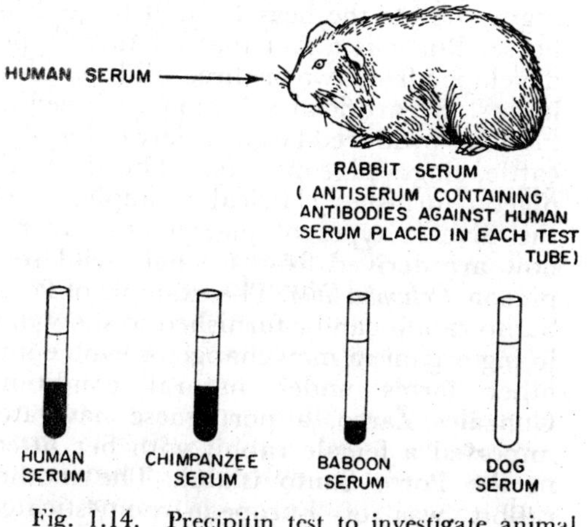

Fig. 1.14. Precipitin test to investigate animal relationship.

are highly specific. These antigen-antibody reactions are used as tools to determine the genetic relationship. It is seen that antibodies against the blood of an animal react vigorously with the blood of closely related forms and less vigorously with the blood of distant related forms. Serological tests have been invaluable in taxonomy, because they help to establish natural relationships amongst animals.

Serological tests depend on the property possessed by the living body to protect itself against any foreign intruders. An entry of bacteria and viruses into the body system results into the production of defensive substances called the *antibodies*. A substance of protein nature which induces the formation of antibodies is designated an *antigen*. For an example, if the serum of horse's blood is inoculated into a rabbit, the latter will form antibodies against the serum of horse. The antibodies are protein substances and an antibody-containing serum is called the *antiserum*. If the blood serum from a rabbit containing antiserum is removed and mixed it with the serum of horse, the antibodies in serum of rabbit will react with that of horse resulting in the formation of a white precipitate. These kinds of antibodies are called the *precipitins* and these serological tests are known as the *precipitin test*. The precipitin tests have recently been extensively employed to study the evolution of animals.

Similarity in serum proteins. Domestic rabbits have been extensively used as experimental antibody producers. Inoculation of human serum into the blood of rabbits results in the formation of antibodies against the human serum. The antiserum of rabbit will contain antibodies against human serum. If a little quantity of antiserum is mixed with human serum in a test tube, a white precipitate will form and will settle at the bottom of the test tube. Mixing of antiserum containing antibodies against human serum with the serum of chimpanzee, baboon and dog in three separate test tubes shows:

(*i*) The test tube with chimpanzee serum shows similar amount of precipitate like that of human serum. This fact signifies that the serum of chimpanzee and the serum of human are exactly alike in chemical substance.

(*ii*) The test tube containing baboon serum results formation of a smaller amount of precipitate. These results indicate that the baboon serum contains a few proteins seems identical to those in the serum of human, while the major proportion of proteins in the serum of baboon are different to those found in human serum.

(*iii*) The test tube containing dog serum does not form any precipitate. This is because of the fact that the proteins of dog serum are completely different from those of human serum, because the antibodies formed against the human serum fail to react. Sometimes a small reaction might occur if the antiserum is very potent. This is because of the fact that all mammalian forms have some chemical similarity in their serum proteins.

The above-mentioned precipitin tests show that the serum of chimpanzee and human is indistinguishable. This similarity is due to common chemical organisation of blood proteins which is due to inheritance from a common ancestry of chimpanzee and man. The baboon serum is less like human serum than the serum of chimpanzee. This can be explained that in distant past, possibly in the oligocene period, the baboon, old world monkeys and man shared a common ancestry. They inherited similar type of serum structure from that common ancestor. Subsequent evolutionary divergence from the common ancestral stock has led to the modification of the pattern. Man and chimpanzee shared a common ancestry more recently than the baboon and man. That is why the serological tests show close similarity in serum proteins in man and chimpanzee than the baboon and man.

Similarity in serum structure explains evolutionary process and the serological tests afford a device to detect the degree of phylogenetic relationship between animals. Serological tests advocate that the degree of chemical similarity is always proportional to the degree of phylogenetic relationship. Closely related animals have similar serum proteins while distantly related forms possess serum proteins which are less alike.

CYTOGENETICAL EVIDENCES

The similarities and dissimilarities amongst organisms depend upon the similarities and differences in their genetic constitution. Various types of relationships occur at gene level. Cytogenetical studies on *Drosophila* exhibit close genotypic similarities amongst different species. The chromosomal configuration of various forms of *Drosophila* shows close similarity. Cytological and genetical differences and resemblances are used as tools to determine their ancestry. Cytogenetic resemblances amongst several species are indicative of the fact that they share a common ancestry.

In recent years genetics have contributed much in understanding how the heritable changes, the basis of evolution of a particular type of organism, have operated. Heredity is a conservative process and in rare cases allelomorphic genes may mutate which causes the alteration of the structures dependent upon them. Such changes or variations are also heritable. In the laboratory and in nature, origin and continuation of such changes can be explained. Like all other evidences, cytogenetic studies provide good evidence of evolutionary process.

EVIDENCES FROM DOMESTICATION

Different types of animals can be produced artificially under domestication. The great range of size and form of the different breeds of horses is a most striking feature. The small long-haired pony appears as pigmy beside the heavily built large draft horse. But it is a fact that all these types developed from two or three wild species of horses. By artificial selection in breeding man has produced large variety of forms in cattle, dogs, pigeons, etc. The domestic pigeon furnishes classical example. All the diverse types of pigeon encountered now are derived from a single wild rock pigeon, *Columba livia*. The example of Porto Santo rabbit can be furnished to show that living organism may change or evolve into other forms under natural condition. Gonzales Zarco, a portuguese navigator imported a female rabbit with her litters in the Porto Santo island. The mother rabbit was of European domesticated variety. The climate of the said island was very favourable for the rabbit and due to the absence of carnivores the rabbit family grew excellently and multiplied at a prolific rate. After a few hundred years when the representative of the island rabbits are compared with the original European stock it was noted with surprise that the two types are quite different which lead some biologists to describe them as distinct species.

The foregoing discussion makes it crystal clear that the living organisms can be changed within a very brief span of time under Artificial Selection. If this be the fact, it will not be unscientific to hold that natural forces are instrumental in producing various new types in millions and millions of years.

EVIDENCES FROM SPECIFIC ADAPTATIONS

Evolutionary changes are possible because living organisms can adapt successfully to existing environmental conditions. Geological history reveals two categories of evolutionary changes in living organisms. Some organisms are extremely plastic and can successfully adapt themselves to escape any eventuality that they confront, while others cannot respond suitably and as a result they become extinct. The evolutionary history of elephants furnishes both the facts. About 350 species of fossil elephants are recorded in the geological history, but most of them disappeared from the globe excepting the two surviving genera living in two widely separated countries of India and Africa. The cause of overall extinction is due to the failure of response to the changing environmental conditions. Few ancestral forms were plastic enough to evolve into the existing types. Of the two surviving types, the African elephant, *Loxodonta* is fast disappearing because of the ivory-hunters. The Asian elephant, *Elephas*, has comparatively better chance of survival, because of two possible reasons—the tusks do not yield high-grade ivory and they can be domesticated.

Animals do change in a most striking way to meet specific needs. As a result structural and functional adaptations occur in living organisms. All the adaptive changes furnish the most convincing evidence that living things change in course of time by making new adaptations.

Multitudes of striking adaptations are observed in animals. The protective adaptations of mountain Ptarmigan can be cited as a single instance. They are protected in a wonderful way by changing the colour of the plumage seasonwise. In winter the body is covered over by white feathers to match with the snowy covered surrounding while in summer the colour of the feathers is more or less brown to correspond the background of rocks, earth, leaves, etc. The instance of an Indian butterfly, *Kallima*, may be presented here as a remarkable instance of protective mimicry.

The adaptations in animals are the outcome of interaction of the organisms with the environmental dynamics. This phenomenon is usually a very slow and gradual process and it requires sufficient time. But there are examples where the adaptive changes can be recorded within a very short time.

The evidences discussed so far in understanding the process of evolution are cumulative. The claim of evolution as a theory rests upon the cumulative effects of all the evidences. No one of the evidences is able to explain the evolution alone but in conjunction with others it gives the final and positive support. Morphological evidence can only explain that living organisms have changed in course of time and are phylogenetically related with one another. The resemblances in ontogenic development establishes the hereditary relationship. The geographical evidence reveals that the variation in geographically isolated forms are the adaptive feature to different environmental condition. Palaentological evidences as fossils preserved in the earth's crust show gradual evolution of the organic forms. One amongst them cannot prove with finality but all of them jointly point to the same conclusion. Evolutionary changes are very slow and far-reaching and such changes cannot be observed in the brief life span of a human being. But the evidences force us to believe in unprejudiced manner that evolution has actually occurred.

ORGANIC EVOLUTION—THEORIES OF EVOLUTION

Organic evolution is a process of "descend with modifications" which signifies that diverse animals and plants of the present day have been evolved from earlier forms by the inheritance of variations induced either by extrinsic or intrinsic factors. There are numerous positive evidences in favour of the concept of organic evolution and many theories explaining the mechanics of evolution. The important theories of organic evolution are given below.

LAMARCK AND THE INHERITANCE OF ACQUIRED CHARACTERS

Jean Baptiste de Lamarck (1744–1829), a French Scientist, was undoubtedly the first biologist to propose a definite theory explaining the evolution of living organisms. He published his theory in the year 1809 in *Philosophie Zoologique*. Lamarck, as a philosophical zoologist, attempted to explain the forces operating in evolutionary process. Lamarck's idea makes a compromise of Buffon's and Erasmus Darwin's views on evolution. Lamarck supposes that changes in organisms in course of time are imperative but these changes are, according to him, not the outcome of direct action of the environment, but it acts on the internal structures in an indirect way. His theory is best known today as the theory of *Inheritance of Acquired Characters*. Lamarckian analyses of evolution are unacceptable to the present day biologists, because of obvious reasons to be discussed later on.

Lamarckian theory. Lamarck based his theory of evolution on certain biological principles. He believed that the living organisms change in course of time to become more complex. From amoeba to man there is a march of evolutionary progression caused by the transmutability of species. This particular aspect of his theory is universally accepted to be true. According to Lamarck, nature works to provide incentive for the development of new structures out of an outcome of new function. Lamarck holds that variations in organisms arise either through conscious efforts or through reaction to a change in the environment or through the effects of *use* and *disuse*. It is supposed by him that the continual use makes a structure greatly developed and disuse, on the other hand, makes the structure atrophied. The variations thus acquired during the lifetime of the organisms are heritable and by this way a new or different species may eventually be produced.

The main theme of Lamarckian doctrine is the inheritance of acquired characteristics. A thorough analysis of his theory reveals some truth of evolution, but most of his interpretations are subjected to violent criticism. His attempt to explain all the phenomena in evolution by his concept of transmission of acquired characters appears inadequate and his arguments in most cases are solely deductive. Acquired characters originate by change of function in somatoplasm which transmit the characteristics in successive generations. Lamarck regarded the acquired characters as bodily characters initiated in one generation and these are heritable.

The essence of Lamarckian concept

The essence of Lamarckian concept of progressive evolution thus can be categorised in the following way:

1. Spontaneous generation of simple forms of life was followed by development of more complex organisms. Living organisms and their component parts tend continually to increase in size.

2. The environment governs the natural tendency of an organism to show structural changes.

3. Use and disuse cause variation in a structure.

4. Characters acquired in one's lifetime are transmitted to one's subsequent generation.

The first part of Lamarck's first law states that the lower organisms were capable of spontaneous generation from matters of inorganic nature and as such there might be some validity in it. But the second line of the law is the statement of a supposed fact which unfortunately is far from truth. Many groups of organisms show no tendency whatever to produce strains leading to gigantism, though in some restricted groups of terrestrial animals, viz., elephant, horses and camels the increase in size is quite established. Conversely the reduction in size is a prominent feature of evolution in a number of groups.

The second law states that environment acts directly on organisms and causes the production of new characters. Lamarck here believes in the fact that in animals the environment acts through the nervous system or in other words the desire of the animal brings forth the formation of new structures. To put the idea in its simplest form Dodson (1960) has written "the man who mused—Birds can fly, so why can't I? should have sprouted wings and taken to the air. Lamarck never presented such crude examples but the fact remains that the second law is false.

The first two laws of Lamarck display a vitalistic outlook. These two laws never blossomed further and have been disregarded by later biologists.

Since there is some mechanistic truth in the third and fourth laws of Lamarck they are considered together as a principle of evolution though greater emphasis has always been on the fourth law.

Evidences advanced by Lamarck in favour of his third law

The supposition made by Lamarck that the continual use of a particular structure in the body of an individual causes the enlargement and strengthening of the structure has certain factual bases. Lamarck cited the case of the webbed toes of the aquatic birds. The development of web in them is caused by constant stretching of the skin between the toes. The length of rabbit's pinna, that of the neck of giraffe, the long talons of eagle, more development of right hand than the left hand in man are examples of use. The long-necked modern giraffe had short-necked ancestor. As a response to stretch up the neck further up to take the leaves from high trees, the neck increased in length and this continued for generations until the modern state is reached. Lack of use or disuse causes the degeneration and ultimate elimination of a particular structure. The instance of modern snakes can be placed here. The ancestor of the existing snakes was lizard-like forms with two pairs of well-developed limbs. By constant disuse of the limbs as a response to fossorial adaptation the limbs completely disappeared in all snakes excepting their vestiges in pythons. Lack of pigment and degeneration of visual apparatus in cave dwellers due to lack of light in caves are the instances of disuse. Degeneration of lateral toes in the evolution of horse is another example of disuse.

The fourth law: The final proposition arrived at by Lamarck states that the modifications produced by the other three laws during the lifetime of an individual will be inherited by its offsprings, with the result that changes are cumulative over a period of time.

The trial of Lamarckian concept

Lamarck defended his theses vigorously until his death though he failed to convince his contemporary scientists. But the apparent simplicity of his theory drew the attention of many workers in the line of evolution and an array of experiments followed.

Inheritance of instinct. William Beebe showed the inheritance of instinct in birds. Birds show many instinctive characters. Instinct is usually defined as the natural impulses by which organisms are guided to perform certain activities. The instinctive characteristic is based upon inherited knowledge. Selection of bird to demonstrate the inheritance of instinctive traits has certain advantages. These features are attributed to the offsprings by the parents. Beebe showed by experimentation that the birds, raised from incubated eggs, without having parental care, manifest all the characteristic instinctive features of the race.

Kammerer's experiment on salamanders. Kammerer performed experiments with salamanders to give support to Lamarckism. He reared salamanders in different coloured boxes, such as red, yellow, blue, etc. The salamanders according to him afterwards took the hue of the surroundings.

Thus he supported the doctrine of inheritance of acquired characters. Similar experiments were repeated but it could not be proved. Noble and Przibram reported that Kammerer's experiment was a fraud. It came to light that Kammerer was let down by his associates and as a result he had to commit suicide.

Weisman (1834–1914) started performing experiments by cutting the tails of mice for successive generations (35). He found that even in the last generation the mice grew tails as long as their first ancestor. Crude did long term but similar experiments with the tails of sheep but could not prove the truth of Lamarckian theory.

McDougall (1938) performed experiments on learning in rats. Rats were dropped into a tank of water from which there were two exits, one lighted and other dark, but not always the same one. The experiment was so devised that a rat leaving by the lighted exit received an electric shock, while one leaving by the dark exit received no shock. Thus the number of trials required by a particular rat to learn to escape by the dark exit constituted a measure of the speed of learning. These rats were then bred and their descendants were similarly subjected to training. It appeared that the speed of learning increased from generation to generation. McDougall concluded that learning is an acquired trait and is inherited. He himself demonstrated that the speed of learning varies directly with the intensity of the shock.

Repetition of similar experiments by Crew (1936) for 18 successive generations and that of Agar and his colleagues (1954) for 50 generations failed to produce the results obtained by McDougall. Thus it became clear that effects of training in mice are not inherited. Crew is inclined to ascribe the discrepancy to an insufficiency of controls and an inadequate attention to genetic method on McDougall's part.

C. C. Guthrie exchanged the ovaries of black and white coloured hen and claimed that a change came about in them. Davenport showed that no such effects are produced.

Gayer and Smith (1918–1924)—It is known that germ cells of the ovary of the female and testes of the male are responsible for inheritance. It is also a fact that germ cells are supplied by blood and the same blood is supplied to other body parts. It, then, may be presumed that the substances carried by blood might serve as *go between* conveying to the germ cells the effects of bodily changes. Gayer thought that the role is played by *Antibodies*. He together with Smith performed an experiment. A solution of lens substance obtained from rabbits was inoculated into fowls. These fowls produced antibodies against this foreign lens protein. The fowls' serum bearing these antibodies was now injected into pregnant rabbits. Some of the offsprings thus born were with degenerated or malformed eyes. When these offsprings were raised and bred, it was noted that the eye defects were inherited in subsequent generations. Though the experiment gave a positive support, it is questionable whether the case reflects inheritance of acquired characters or not. In the experiment it is more likely that the antibodies acted on the eyes of the embryo or on the genes responsible for eye structures of the embryos directly. If this be correct then it is never a case of inheritance of acquired characters. Pavlov (1923) withdrew his earlier statement that the conditioned reflex in mice is inheritable.

Lamarckism, however, persisted in U. S. S. R. Karl Marx and Engels were strong supporters of Lamarck and opined that Lamarckian way of inheritance would guarantee the future improvement of human race. After the arrival of T. D. Lysenko on the Soviet scene with the vernalization technique Lamarckism got a strong foothold in U.S.S.R.

Lysenko drew big conclusions from a single plant. He and his school were not above falsifying data or censoring classic references, and they systematically opposed opponents at the most overt level of xenophobia and political partisanship. "It is clear that in the absence of public criticism, with a monopoly of publication and support, surrounded by an atmosphere of repression and fear, the provocative, self-praising Lysenko could prosper" (The rise and fall of Lysenko. Translated by L. M. Lerner, Columbia University Press).

Lysenko's strange view had cost Russian science and agriculture a generation of progress. Lysenkoist geneticist is a most derogatory compliment.

The Jews and Muslims practice circumcision for centuries and yet there is no reduction of prepuce in them. The feet of the Chinese woman is another example. For generations the feet of the Chinese women were kept bound yet, the practice has failed to bring any modification of the feet of present-day Chinese women.

The work of Metalnikov (1924) on immunity of wax-moths and of Sladden and Hewer (1938) on food preference in stick insects seems to demand *prima facie* a Lamarckian explanation. Huxley, however, is of opinion that in view of the fate of other claims and of the theoretical difficulties too much weight must not be attached to such isolated ones.

Darwinian concept of pangenesis—an attempt to tide over the Lamarckian Crisis

Charles Darwin was supporter of Lamarckian concept of inheritance of acquired characteristics. These are acquired anew within the lifetime of an individual. Darwin postulated a theory of heredity to explain the inheritance of acquired characters. His theory is called the *theory of Pangenesis* which explains how characters are transmitted from parents to the offsprings. He assumed the existence of imaginary hereditary particles called the *pangenes* or *gemmules*. These particles are produced in the life-time of an individual by each and every part of the body. The pangenes possess all the distinctive characteristics of the particular parts from which they were produced, together with modifications acquired in the life-time of the individual. The pangenes are discharged into the circulating blood-stream, from there these are all collected together to form the germ cells. The germ cells eventually unite to form the zygote which gives rise to the new individual. This phenomenon insures the recurrence and development of the parental characteristics in the offsprings together with handing over of the acquired characteristics.

Weismann's theory of germplasm. Weismann was not convinced by the concept of pangenesis of Darwin. He advocated that living organisms are composed of two types of materials, the somatoplasm and the germplasm which are quite different. The foundation of the germplasm theory of Weismann in 1892 directly gave a dread blow to the Lamarckian concept of inheritance of acquired characters. He regarded that every individual in bisexual forms starts its life from the zygote formed by the fusion of male and female gametes. The zygote constitutes the marvellous bit of hereditary capital which forms the bridge of continuity between generations. So the inheritance of the parental characteristics by the offsprings must pass through the zygote. In organisms, germplasm becomes isolated in the very early phase of ontogenic development. Boveri has shown the splitting off of germplasm in *Ascaris* as early as at the 16-celled stage. Of the 16 cells, 15 cells form the somatoplasm and the rest one is set aside as the germplasm.

To strengthen his contention, Weismann performed extensive experiments with white mice. Repeated artificial mutilations, such as cutting the tail of parents fail to produce abnormalities in the progenies. Variations due to mutilations are not inherited. Any characteristic which is to be inherited, must certainly be stamped in the germplasm. Somatoplasmic changes are not inherited. Castle and Philips have transplanted ovaries of a black Guineapig to a white female whose ovaries have been replaced. The youngs produced were all black.

Objections

1. Not a single experiment has so far been able to provide positive evidence favouring the inheritance of acquired characters.

2. Lamarckian concept is not applicable to higher forms. However, it may be true for lower unicellular organisms where there is little or no separation of soma and germplasm.

3. The conscious effort as has been advocated in case of giraffee can hardly be expected in the plant kingdom.

4. Some of the laws speak in favour of vitalism which is antagonistic to materialism in scientific thought.

5. Lamarckism suggests a theory without evidence.

6. It has been made clear that new organisms develop from the germ cells of parents and not from somatic cells. The germ cells are set aside earlier in the growth of an individual and are not affected by body cells.

7. There are numerous cases where Lamarckian inheritance is either impossible or exceedingly restricted. In higher mammals the internal environment is regulated with an extraordinary degree of constancy. The temperature of blood, the salt composition and acidity of blood are kept constant by elaborate and special mechanism. The reproductive cells like all other cells in the body, are exposed to the internal environment supplied by blood-stream. How then can changes in the external environment be transmitted to them? The regulation of the internal environment provides an effective shock-absorber for alterations which could occur in the external environment. Yet higher mammals have evolved as rapidly and in as obviously adapted ways as any lower type in which this buffering does not exist.

8. Insects provide a number of hard nuts for Lamarckian cracking. Social

Hymenoptera is such an insect. The bulk of the work in the Hymenoptera colony is done by neuter females, while reproduction is entrusted to full females and the males which are less abundant. How is it then possible on any Lamarckian view to discover a mechanism by which the special instincts and structures of the workers have been evolved? The workers cannot transmit them because they do not reproduce.

9. Modern biology, taken by and large repudiates Lamarckism.

Modern concept of Lamarckism. Lamarckism fails as an all-embracing cause of evolutionary process, because it has many shortcomings. It cannot explain the origin of new species and the origin of adaptations remains unsolved. It also cannot give any clue as to how the acquired characters are inherited. The proofs advanced by Lamarck and some followers of Lamarck are proved to be based not on scientific fact. It will be needless to advance all the proofs of Lamarckian inheritance. Kammerer's experiment with salamanders are not confirmed by subsequent workers. Other experimental demonstrations have not been confirmed by subsequent investigations. Certain Lamarckians have advanced the view that inherited effects of function or environmental modification are so slight that they cannot be detected experimentally and require cumulative action through thousands of generation to become obvious. Payne (1911) has shown an excellent example to demonstrate the failure of production of Lamarckian effects by disuse in *Drosophila*. Disuse of eyes in *Drosophila* cannot produce the effect of disuse even up to sixty-nine generations. Thus to plead the impossibility of detection is a counsel of despair which is totally unscientific.

Modern biologists do not recognise Lamarckism as a causative factor to explain the origin of adaptations during evolution. The discovery of genes and their role in heredity throw much light on the origin and inheritance of hereditary variations. Lamarckian acquired characters are phenotypic variations and they cannot affect the genotypes. The converse is actually the fact. The phenotypic variations are the result of interactions of the genotypes and the environment. Although some characters are well explained by Lamarckism, there are numerous cases which Lamarckism cannot explain properly. Lamarckism is not sufficient enough to explain the origin of species. Modern evolutionists regard the importance of Natural Selection as the only agent in evolution. Natural Selection works at the gene level. The units of inheritance are the genes and all sorts of changes or characteristics must pass through the pathway of genotypes. Waddington has developed a new evolutionary point of view which tries to make a compromise between Lamarckian theory and the selection theory. Based on his work on Drosophila, he has advocated that the adaptive characters in Drosophila are genetically assimilated. If this particular concept is substantiated in future, this will revolutionise the entire concept of evolutionary progression.

Lamarckism as such is incomplete in itself to account for evolution. In the opinion of Ray Lankester, Lamarckism is "self-contradictory". Although Lamarckian doctrine is quite inadequate to explain evolutionary process, it cannot be denied that Lamarck was the foremost figure to contribute much to focus the diverse attention of the biologists upon the central problem of organic evolution. Modern genetics makes the fact crystal clear that whatever changes are induced in the phenotype, the change is not necessarily in the genotype. The failure of Lamarckism is mainly due to the failure of recognisation that the phenotype is the by-product of gene-reproduction.

DARWINISM AND THE THEORY OF NATURAL SELECTION

With the rise of scientific methods of inquiry, the concept of organic evolution began to occupy a prominent place in the minds of the thinking men of the world. Although the germ of evolutionary idea has its inception with the Greeks, Charles Darwin was undoubtedly the foremost figure of the evolutionary thought. He paved the way for the general acceptance of the truth of evolution by his reasonable and acceptable theory of evolution. Gradual change is the rule in living world and new species are constantly being produced by the evolutionary dynamics. Darwin advanced the concept of **Natural Selection** as a force to account for evolution. Natural Selection has been the central strand on which all the main facts of evolution rest. Darwin's monumental work in evolution is one of the most efficient chapters in Biology and every fact of biology, however, diverse they appear to be, fall into one perspective, the Natural Selection.

Natural Selection is regarded as a causo-mechanical process of transmutating of species ended in inductions and deductions. It is a sort of selective process eternally operating in Nature to improve and maintain the adjustment of living organisms to their surroundings and the way of living, whereby misfits, unfits and weaklings are eliminated. To explain Natural Selection as a force in evolution, Darwin called attention to certain facts of nature. With the remarkable advancement of modern genetics, Darwin's original concept of Natural Selection has taken rebirth and the whole of Darwinian concept is embarked upon a phase of modern synthesis. Evolutionists of today, with the growing scientific advancement, are trying to penetrate deeply into the operations of the biological systems. Although Darwin's idea of evolution has undergone extensive modifications, his concept is regarded as the most dominating feature in biology.

Foundation of Darwin's theory. Charles Darwin (1809–1882) profounded the concept of Natural Selection as the principal cause of evolution and this is universally taken to be true. He created a sensation throughout the world with the magic form of a single idea of the mechanism of evolution. The publication of his famous book and the Bible of evolution, "On the Origin of Species by means of Natural Selection" in the year 1859 is the crowning event in the rise of evolutionary thought. Darwin came to the conclusion after about twenty years' of thinking and researches. He visualised living things in a new light during his memorable voyage in *H. M. S. Beagle* in the Atlantic and Pacific islands including Galapagos. His voyage was a voyage of discovery. Darwin with unquenchable intellectual curiosity gave shape to his observations and researches into a reasonable theory of evolution. The work of Malthus on population stimulated him very much in arriving at this conclusion.

Alfred Russel Wallace—co-discoverer of Darwinism

The name of Alfred Russel Wallace (1823–1913) in the Scientific literature will always be associated with that of Charles Darwin so far as the publication of the theory of evolution is concerned. He was a contemporary and collaborator of Darwin and contributed a paper "On the tendency of varieties to depart indefinitely from the original type" in a symposium of the Linnean Society in the year 1858. Wallace came from social origins so poor that he could not aspire to life in college. He left school at 14 and apprenticed to a railway surveyor. Later he became a school master. The books in the library that were his Oxford and Cambridge. Wallace spent a wanderer's life. He and Bates decided jointly to see the world. The British Museum assured them that almost any specimen they might collect could be sold to pay their expenses. They aimed to keep duplicates for their own study as a path towards solving the problem of origin of species.

Off they went to the Amazon. Wallace returned from the Amazon after 4 years. Forced to flee a fire at sea, he spent 10 days in an open boat and lost nearly all his collections. After a year and half writing and talking in London, he concluded that "the very finest field for an exploring naturalist was to be found in the great Malayan Archipelago". He spent 8 years in these islands, making 60 to 70 separate excursions. He made his base in the island town of Ternate in a rather ruinous house. Here he found peace to write the famous brief paper on Natural Selection that he sent out to Darwin to initiate public knowledge of the great theory. Like Darwin, Wallace also gave emphasis on "Natural Selection" as the main agent in evolution. But the insufficiency of data and his theological inclination in later years brought no honour for him. Wallace also went on an expedition of nature and came to a theory, the working principles of which are exactly the same as that of Darwinism. Owing to the honesty and generosity of Darwin, a warm friendship grew between Wallace and Darwin. Darwin always recognised the originality of Wallace's conclusion and their two research papers were combined and published under the joint authorship in 1859. The whole credit went to Charles Darwin, but it should always be remembered that Wallace's work was almost a paraphrase of Darwin's work.

Credentials of the Darwin's theory. Before Charles Darwin, his grandfather, Erasmus Darwin and Lamarck brought the scientific idea that the present-day living organisms have evolved from simpler forms of the past. Darwin presented large body of evidences to convince the world about the truth of evolution. Heterogeneous collection of facts are marshalled into some deductions which are known as Darwinism. As stated earlier Darwin discussed his theory of evolution with Natural Selection as the principal factor and supported his theory by elaborate evidences derived from the published research work and personal observations. It is a piece of doctrine elevated to biological accuracy and a sort of assemblage of diverse biological phenomena. The deductions brought by Darwin from certain biological principles are shown in Fig. 1.15.

INDUCTIONS	DEDUCTIONS
PRODIGALITY OF PRODUCTION – CONSTANCY OF FOOD AND SPACE –	STRUGGLE FOR EXISTENCE
STRUGGLE FOR EXISTENCE – OMNIPOTENCE OF VARIATIONS AMONGST THE INDIVIDUALS –	SURVIVAL OF THE FITTEST
SURVIVAL OF THE FITTEST – OPERATION OF NATURAL SELECTION AMONGST THE SURVIVORS –	SPECIATION

Fig. 1.15. Showing the inductions and deductions of Darwin.

Darwin attacked the process of evolution from all corners. He tried to establish the fact of evolution in one hand and also tried to discover the mechanics through which it operates. Darwin based his Natural Selection concept on certain observable facts of nature and he has drawn the deductions from them. Darwin was convinced by the fact that an enormous discrepancy existed between the number of individuals born and the number of individual survive. The balance of nature is maintained by Natural Selection. To illustrate this fact he has brought a large number of examples.

Prodigality of production. Power of reproduction is one of the basic characteristics of living organisms. The living creatures reproduce at a terrific rate. Phenomenon of reproduction is a matter of fact and is a power bestowed on all living things. Darwin was acquainted with the famous doctrine of Malthus and found it applicable to the animal world. Organisms increase in a geometrical ratio, but the food production and space almost remain constant. Without some checking mechanism, organisms will exhaust the available food and space. The following examples will show the rate of egg production in some animals. A female Salmon produces 28,000,000 eggs in a season, an *Ascaris lumbricoides* passes 700,000 eggs in 24-hours, an oyster liberates 114,000,000 eggs at a spawning and *Rana catesbeiana* produces 20,000 eggs annually. Darwin has calculated that in oyster, if all the eggs produced by a single female, survive and multiply, the shells will heap up to eight times the size of the earth. He has also calculated that starting with a single pair of elephants (the slowest breeder amongst mammals) about nineteen millions of descendants would be produced within 750 years.

Constancy of food and space. Without artificial means of increasing, the food and space can hardly be changed. Extraordinary power of productivity on the part of all living organisms will result overcrowdedness. Darwin believed that the prodigality of reproduction is very important because it results into a crowded condition and to the sharp competition for the necessities of life. Darwin recognised that this overcrowdedness results into the struggle for existence.

Struggle for existence. As a result of the prodigality of production and the constancy of food and space a competition for survival amongst the animals results, because numerous youngs are always being produced than can be sustained. The struggle for existence is a sort of competition and is threefold. It may be (*a*) *Intraspecific struggle*, (2) *Interspecific struggle* and (3) *Environmental struggle*.

The intraspecific struggle is the competition against the organisms of their own kinds. The interspecific struggle is the competition between members of different species. Instances of such type of struggle is universal in living world. Almost every living form depends directly or indirectly upon other living forms. The environmental struggle is usually against the physical environments such as against excess of moisture or of drought, against extreme temperature (cold or hot), against other geological conditions.

Variations. No living organisms are alike and they vary appreciably. Darwin observed slight variations occurring in all parts of the organisms and these variations are the primary cause for which the children do not resemble exactly their parents. Geology also furnishes innumerable instances of changes in nature during the phylogenetic development of a particular form of animals in course of time. As the struggle for existence is universal and the variations become the rule in nature, some variants become advantageous in the struggle for survival and others are disfavoured. Favourable variants will have great chances for survival and the unfavourable variants will fail to survive.

Survival of the fittest. In the struggle for existence the organisms with variation in structure, habits or instincts may be better adapted to new conditions and will have better chance of survival. Consequently their offsprings will inherit those variations. In course of time, these organisms will inevitably acquire superiority over the rest and less adapted forms. Variations bring organic diversity in response to environmental dynamics and result into adaptation.

Natural selection. Species are plastic and dynamic entities that have been moulded over long periods of time through Natural Selection. Species do not come *de novo*. Natural Selection enforces adaptation amongst organisms. Nature will select good varieties and others are crushed out. Natural Selection operates amongst the fittest and the new forms are stabilised and thus leads into speciation.

Darwinian concept of Natural Selection. Darwin conceived that the consequence of potential increase in number will result merciless competition amongst and between the species. Any quality possessed by an individual which offers an advantage over others in the struggle for existence would be insured to the survivors. In subsequent generations a large number of individuals with advantageous characteristics will be produced. With the repetition of selection in each generation all the survivors would eventually be of the favoured types. Darwin had the idea that the improvement of organisms is actually the work of Selection. The ultimate result of Natural Selection is that each kind of organism tends to become more and more improved in relation to its condition. Such improvement inevitably leads to the gradual advancement in the organisation of a greater number of living beings. The struggle for existence and the Natural Selection are highly metaphorical terms. The struggle for existence merely signifies that a number of individuals in each generation is bound to die before they can reproduce and the Natural Selection is responsible for the differential survival and the reproduction of the variants.

A hundred years' history of Darwinism. Darwinism has stood the test of more than one hundred years and it underwent through differential phases—until its pervading influence makes its status high in the intellectual thought. Its hundred years' history is very dynamic in nature.

Phase of ready acceptance. After the enunciation of the theory of Natural Selection by Charles Darwin in 1859, it was accorded a warm reception. It was proclaimed as a masterpiece and was readily accepted by all the people of the world. Before him, no one had forwarded any concrete suggestion as to how evolution could have occurred. Weismann was the greatest supporter of Darwinism and regarded Natural Selection as the principal cause of evolution.

Eclipse of Darwinism. In late nineteenth century, the evolutionary studies became stereotyped and resembled the concept of

Natural Theologian school. Gradually people began to lose confidence on Darwin and the evolutionary studies became merely routine work. Whatever results they obtained out of their researches, Scientists wanted to focus it in the light of Natural Selection and the evolutionary studies became merely "case-books of adaptation" in the language of Julian Huxley. Bateson first opposed the Darwinian concept of continuous variations in nature. Many workers were convinced that Natural Selection was not able to do things claimed for and did not accept evolution as a slow and continuous accumulation of minute advantageous mutations.

The discovery of mutation in the evening primrose, *Oenothera*, by de Vries diverted the attention of prominent biologists. He believed in the origin of hereditary variations through mutation and disproved the role of Natural Selection by supplementing it by his *theory of Mutation*. Entire Darwinism was eclipsed in the period from 1914 to 1924. At this time Biology was fragmented into different branches. Rediscovery of Mendelian principles in 1900 by Correns, de Vries and Tschermark wanted to put forward that mutation is sharp and discontinuous while the continuous variations widespread in nature are not heritable. The geneticists (Goldschmidt, Muller, Morgan, Fisher) did not believe in Natural Selection because they were able to discover the mutation and the genes. Pearson and Weldon established a branch, Biometry as a science of mathematical applications to understand evolution and did not recognise the genetic facts as well as the idea of Natural Selection. The biometricians adopted hypothetical modes of inheritance and genetic variation. This mathematical theory was of no service to the evolutionists. The experimental embryologists started under the leadership of Roux and Spemann, contradicted the classical concept of recapitulation. Experimental physiologists lost all connections with the sister sciences and developed into its own. This period was really a chaotic phase and the real cause of losing the prestige of Darwinism was the sudden division of different branches of Biology within a very short time.

Phase of unification. After a period of eclipse, Darwin's mechanistic generalisation raised its head and regained its vitality with full honour and prestige. The opposing factors reconciled themselves as the younger branches of biology and achieved a sort of biological synthesis within themselves. With the classical disciplines, the reconciliation was converged upon the Darwinian centre. Many realised that mutation has no value if there is no Natural Selection. Biology for the last thirty years has become a unified science and has embarked upon a phase of synthesis. All disciplines are analysed on the platform of harmony. A new compromise school sprang up to be called the Neo-Darwinism group. The school placed the theory of Natural Selection on a solid basis and Darwinism took a rebirth. T. H. Morgan was originally a great critic of Selection Doctrine, but later on with his genetical researches became an upholder of Darwinism. The Darwinism what we know today is a modified Darwinism and operates with facts unknown to Charles Darwin.

Shortcomings of Darwinian doctrine. The evolutionary theory proposed by Darwin has many shortcomings. Darwin believed in blending inheritance but modern genetics has proved beyond doubt that blended inheritance does not occur. Modern genetics establishes that the inheritance is particulate and the particles are borne on the chromosomes. Darwin has left many obscurities in his theory and modern genetics provides a particulate theory of inheritance to solve the obscurities which Darwin had left in his theory. In addition to the concept of blending inheritance, Darwinism has many weaknesses. These are as follows: Darwin accepted the occurrence of variations, but he did not explain the actual mechanism of variation. The title, Natural Selection, coined by Darwin seems to be misnomer, the real title would have been *'Natural Rejection'*. Darwin's idea of Natural Selection is unidirectional which will inevitably lead to overspecialisation. Natural Selection cannot account for the beginning of organs. Natural selection of Darwin cannot explain degeneration. Darwin was a believer of Lamarckian inheritance of acquired characters which has weakened his theory. Darwin's concept of Natural Selection may select favourable individuals but that does not explain the origin of species. Objection to Natural Selection is quite obvious because it is essentially a materialistic concept.

Efficacy of Neo-Darwinism. Neo-Darwinism has enabled us to analyse the working of Natural Selection in a way that was unimaginable in Darwin's time. Modern concept of selection not only includes the idea of rejection but also the maintenance of the *status quo* of the species. Natural Selection operates in a most contradictory fashion. It is actually a biological duality and has got two aspects. These are: **Stabilising aspects.** This aspect of Natural Selection keeps down the deleterious mutant genes as well as the gene combinations. By this way Natural Selection protects and stabilises the developmental pattern of the species. It results homeostasis. **Dynamic aspects.** This aspect of Natural Selection allows the species to maintain its hold on the changing ecological condition and also to control and overcome new ecological opportunities. It establishes new genotypes quite suited to the new conditions. The dynamic aspects of Natural selection operate in two ways—one is *directional* and the other is *disruptive*. This aspect is dealt separately in the last part of this topic.

The Neo-Darwinian concept of selection gets the greatest support from Mendelian laws of heredity. There is no organism without genotype and the genes constituting the genotype are the fundamental particles of life. It is the corporate genotype that determines the course of development in an environment. The genotypes should be changed to give an efficient functionating organism. A variety environment is open to the genotypes and the genotype will choose that in which their fitness is highest and will eventually maintain a condition of equilibrium. The selection mechanism gives shape and order to these variabilities. Neo-Darwinian concept of selection maintains that selection not only includes the idea of rejection of 'unfits' but also of the maintenance of 'fits'. The fitness is due to the selection of genes. The efficacy of Neo-Darwinism is represented in Fig. 1.16.

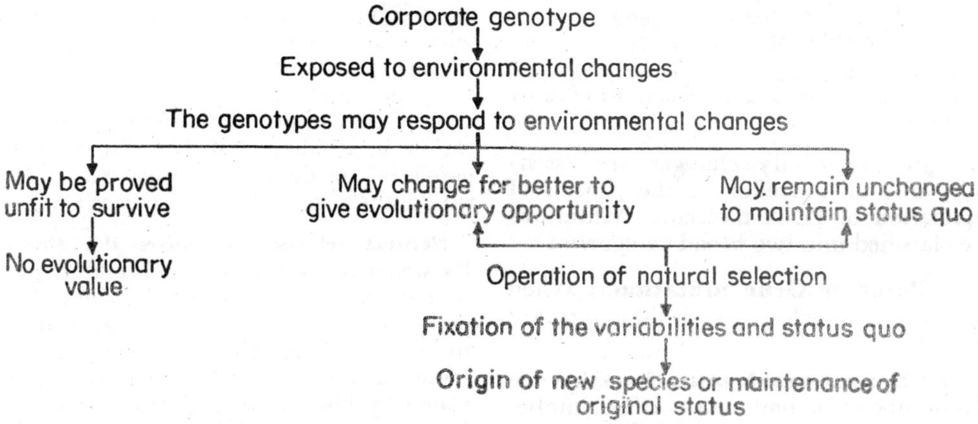

Fig. 1.16. Showing the efficacy of Natural Selection

Environment—the perennial source of genetic variabilities. Every organism exists at the expense of an environment. New environments are constantly being produced. To survive, an organism must react to the environmental changes by homeostatic modification of the developmental pattern. Environment is a directive force in evolution. Environment does not change the organism as such, but it offers challenges to which organism may respond by its genotype.

Reproduction produces variety in the genotype. The main consequence of reproduction is to produce an endless variety of genotypes. Some of them may have different longevity and viability and thus Natural Selection always gets new materials to work with. Fig. 1.17 shows the schematic representation of the relationship between reproduction, mutation and Natural Selection.

Mutations—source of hereditary variation and provide the material basis. Any change in the hereditary materials is known as *mutation*. Evolution of diversified living organisms in this world is due to hereditary

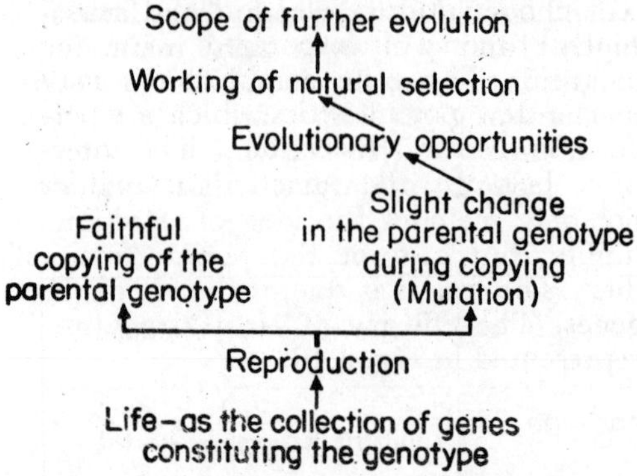

Fig. 1.17. Showing the relationship between reproduction, mutation and natural selection in evolution.

changes. Hereditary changes are essentially due to 'mistakes' in the process of replication of DNA molecule. Mutations are classified into two broad categories:

A. **Point or Gene mutations:** When one or a few nucleotides within a gene are affected.

B. **Chromosomal mutations:** When the number of chomosomes or the number or the sequence of genes in a chromosome are affected. Chromosomal mutations are of different types, *viz.*

1. **Changes in the number of chromosomes:** Changes in the number of chromosomes are due to (i) *Fusion*—involving the fusion of two non-homologous chromosomes into one resulting in loss of a centromere. (ii) *Fission*—involves splitting of one chromosome into two with the acquisition of an additional centromere, (iii) *Aneuploidy*—When one or more chromosomes of the normal set may be lost or present in excess and (iv) *Haploidy and polyploidy*—Most organisms are diploid having two sets of chromosomes in somatic cells. Some organisms are normally haploid possessing only one set of chromosomes. Polyploidy arises when more than two sets of chromosomes are present.

2. **Changes in the number of genes in chromosomes:** Changes in the number of genes are due to: (i) *Deletion:* when a segment of DNA containing gene/genes is lost from a chromosome. and (ii) *Duplication:* when a segment of DNA containing gene/genes is present more than once in a chromosome.

3. **Changes in the sequence of genes in the chromosomes.** These are due to: (i) *Inversion*—When the sequence of a block of genes is inverted in a chromosome and (ii) *Translocation*—Change of the sequence of a block of genes in chromosomes.

Heredity is a conservative process while the genotype has the inherent property of stability. But the genotype of an organism may be changed by mutation and recombination. Mutation provides the raw materials for Natural Selection to work with. Evolution is impossible without mutation, because mutation introduces novelty into the genotype and evolution is impossible without new organisation. Mutation is accident and Natural Selection converts accident into apparent design, randomness into organised pattern.

Natural selection—stabilises the species. Existence of a species depends mainly upon its genetic quality and equilibrium. Random mutations are likely to cause damage to the living organisms. But the improvements as well as the adjustments are established by Natural Selection by perpetuating the useful ones. Selection mechanism always tries to fix the frequencies of gene-mutation at values that are most advantageous under the environment the organisms live.

Selection is essentially a process of conservation. Mutations provide new evolutionary opportunities and selection mechanism stabilises those mutations so that continuous productions of new species will occur in nature. It must be regarded that mutations are not the final offer to a species, it must be subjected to the action of Selection.

The researches on industrial melanisation reflect the operation of directional natural selection and also the instance of the origin of adaptiveness in biological organisation. The development of dark melanic species of moths in the industrial regions of Western Europe furnishes the most striking example of evolutionary changes. The case of a peppered moth, *Biston* sp. of Manchester can be stated below. The genus *Biston* has two species: (*i*) *Biston betularia* is the lighter variety and (*ii*) *Biston carbonaria* is the darker variety. Before industrialisation in Manchester, *B. betularia* was very prevalent and *B. carbonaria* was absent. But with the advent of industrialisation in Manchester, the lighter variety vanished and the darker variety became dominant. The sequence of changes is stated in Table —Evolution-2.

TABLE EVOLUTION—2

Year	% of *B. betularia*	% of *B. carbonaria*	Environmental condition
1848	100%	First appearance	Industrialisation just started
1910	50%	50%	Industrialisation fast progressing
Today		100%	Fully industrialised

The two species of moth differ in single Mendelian gene. The darker variety is usually dominant, but the ligher varieties are the normal ones. Industrialisation causes the removal of lighter types and the darker varieties become prevalent. The darker varieties are actually the mutants of the lighter forms caused by the deposition of melanin. The most possible interpretation of the spread of melanic variety is that the darker forms are more vigorous than lighter varieties. Abundance of lighter varieties in lighter environment is due to the fact that they are camouflaged by being protectively coloured to match the hues of surroundings and thus escape the eyes of the enemies. In the dark surrounding caused by industrialisation, the darker varieties nearly match the colour of their surroundings. Industrialisation removes the disadvantages of the dark varieties which become widespread and eventually suppress the lighter varieties. In this particular case the change in environmental condition causes the change of gene responsible for colour production. The darker forms are the mutated forms of the lighter varieties. Full industrialisation makes the surrounding darker and the darker varieties are stabilised by Natural Selection because of their greater survival value in that particular environment. The lighter varieties vanished due to their non-adaptiveness.

Adaptation—the outcome of Selection

Mutations that improve the adaptive value of the species in an environment are tied up by Natural Selection and subsequently become incorporated with the genotype. The gene combinations may produce new potentialities which are perfected by Natural Selection. The interaction between the genetic variabilities and Natural Selection produces greater adaptive plasticity. The dynamic aspect of Natural Selection establishes new genotypes adapted to new conditions.

Darwinism has a long checkered history of success and eclipse, until very recently it is fully vindicated and shines high in the intellectual fabric of modern thought. Darwinism has undergone evolutionary transformation but the essential cognancy remains still undisputed and operates with new facts and factors unknown to Charles Darwin. Neo-Darwinism is the lawful product of modern researches which has largely retained the selective implication as shown by Darwin. Neo-Darwinian school establishes the intimate relationship between Mutation and Natural Selection. The importance of genotype in the realm of evolution is most vital. Mutation and Natural Selection work hand in hand in causing evolution. Mutations provide the raw material and Natural Selection directs the way. Mutations, in wide sense, are the changes in the genotypes and evolution is possible because of the plasticity of the genotype. The genotypes give the ground-plan on which selection mechanism composes the various themes of evolution. In the realm of evolutionary dynamics, Natural Selection establishes new genotypes adapted to new condition and thus leading to progressive evolution. The advancement of physiological genetics modifies the original concept of selection to a great extent. Selection mechanism is universal but it works at the level of genes. The present *status quo* of organisms rests on the balance between environmental flux and adaptive intensity where selection mechanism directs the way of evolution.

THE SYNTHETIC THEORY OF EVOLUTION

All modern Biologists are unanimous to accept evolution as a fact. But no Biologist has actually observed the origin of major groups of organisms. Because major evolutionary steps have taken millions of years to be completed. The evolutionary processes which gave rise to major evolutionary groups, took place in the remote past—long before the origin of man. So there were no people to observe them. But the process of evolution can be studied in two ways:—**Indirect method** by observing the comparative sequence of events in the past and **Direct method**—based on the experimental causes of evolution from the events in the world at present.

Efficacy of Modern synthetic theory

Since the publication of the theory of Evolution by Charles Darwin more than a century ago, biologists studied the subject in two ways. Most of the earlier biologists were interested in the course of evolution. But the modern biologists tried to explain the process in a different way. They wanted to explain the processes and causes of evolution. The synthetic theory of evolution tries to explain the processes and mechanisms of evolution. This theory is based on five basic processes. They are:

(i) **Gene mutation,** (ii) **Changes in the structure and number of chromosomes,** (iii) **Genetic recombination,** (iv) **Natural selection** and (v) **Reproductive isolation.** The first three agencies cause genetic variability while the remaining two lead a population into adaptive channels.

Charles Darwin provided the evidences to establish that evolution has actually taken place. He put forward the concept of Natural Selection to explain it. His ignorance of hereditary mechanism caused the rise of conflicting group—the early Mendelian geneticists. The synthetic evolutionists in the years 1920-30 made a compromise view on the whole problem. They removed all the obstructions left between the Darwinian naturalists and early Mendelian geneticists. The strongest support on this issue comes from the researches in population genetics.

Natural Selection establishes the link between environmental change and organic evolution. Darwin first recognised that the link between variability or constancy of the environment and evolutionary change or stability is established by Natural Selection. This process is the result of the following characteristics of population: (i) genetic variability, (ii) genetic recombination, (iii) hereditary continuity, (iv) ability for mutation, (v) excess reproductive ability, (vi) integration of the genotype, and (vii) limitations of genepool of a population. Genetic variability and cross fertilization may produce enormous number of genetically different kinds of individuals. A new generation may possess only a sample of the aforesaid combinations. The interaction between living organisms and their environment may or may not involve struggle for existence between different organisms. But it will affect the genetic composition of the sample. The pathway in which natural selection guides the population depends on the environmental change and the content of the genepool. The diversity of population is due to the fact that, natural selection can either promote constancy, direct continuous change or promote diversification depending upon environmental changes.

Natural Selection—its three kinds

Based on the organism-environment relationships three types of selection have been recorded. They are:

Normalising or Stabilising Selection. This selection promotes constancy in a population. If environmental interactions remain constant (or unchanged) through time, normalising selection prevails and thus evolutionary change is arrested. The interaction between mutation and natural selection is essential. Mutation is essentially an accidental change in genetic system. Mutations are usually of no use to a species living in a 'normal' environment. Most of the mutations range from neutral to deleterious to lethal. Useful mutations are incorporated in the genepool of a species. Some mutations become beneficial in heterozygous condition but become deleterious in homozygous state. Normalising selection checks the accumulation of such mutations that would lower the fitness of a species in an environment. Mongoloidism (Down's syndrome), chondrodystrophy (Achondroplasia), Phenylketonuria, Albinism arise repeatedly by mutation in human population. Mongoloidism is a dominant lethal mutation in man which results due to duplication of chromosome-21. Mongoloid individuals have 47 chromosomes instead of the normal number of 46. Most of the mongoloid individuals survive, but seldom possess reproductive ability. Chondrodystrophic dwarfism in man is one of the many dominant mutations in man. Phenylketonuria and albinism are two instances of numerous deleterious recessive mutants in human populations. Both of them occur very rarely (1 : 10,000 and 1 : 20,000). Phenylketonuria is a disorder in the metabolism of phenylalanine (an amino acid). Albinism causes inability to synthesise pigment in skin, hair and iris of eyes. In human population all these disorders arise by mutation but normalising selection checks their undue accumulation by lowering Darwinian fitness from minimum to zero.

Genetic variability and normalising selection are common in the living world. Normalising natural selection is a conservative force which purges the genepool of a population of deleterious genetic variants and thus tries to keep the species unchanged or constant. Genetic variability is one of the prerequisites for evolutionary change and mutation is the source of variabilities. Mutations are mostly deleterious. Deleterious mutation in one environment may be useful in another. So deleterious and useful mutations are not fixed categories. Use of insecticide causes resistance to insect population. The insecticide resistance is imperative for survival when an insect population is being regularly treated with insecticide. This results change in genes, i.e. emergence of genetic variability. Genetic variability enables a population to adapt to changing environments.

Directional Selection: This selective force directs changes. Directional selection occurs when environment interactions change in one direction. When specific interactions have been established (such as predator-prey, grazing animal-forage etc.) between two different types of organisms this type selection force is likely to operate. Directional selection operates by repeated feedback interactions between predator and prey. This is well illustrated in evolution of horses on the plains of North America. The grasses having hardest leaves possess Darwinian fitness as they were least likely to be damaged by grazing. The horses having hardest teeth with complex patterns of enamel were well equipped to feed maximum number of such grasses. Instances of such co-evolution stimulated by feedback interactions are plenty in the evolutionary history of vertebrates.

Directional natural selection can be demonstrated in industrial melanisms in many species of moths. This has been extensively studied by H. B. D. Kettlewell, 1961; Ford, 1971, and many other workers in England.

The peppered moth, *Biston betularia*, is a normal lightly pigmented form and its melanic (darkly pigmented variant) form is *Biston carbonaria*. These two species of the genus, *Biston*, are different due to a single gene. The 'dark gene' is dominant. Since the middle of last century the melanic forms have become widespread in industrial areas where the vegetation became blackened due to pollution. In polluted regions due to shoot and other wastes, the normal light forms have been replaced by melanic forms. But in non-polluted regions the normal light forms are still present. Predation of the moths by birds is the chief selective force which promoted industrial melanism. The melanic forms are protectively coloured on polluted blackened vegetation. They become prominent in non-polluted regions where the light forms are well protected.

Diversifying or Disruptive Selection: Diversifying natural selection promotes diversification in living organism. If a previously homogeneous habitat becomes diversified, the interactions between biotic populations and their environments diverge from one other. This initiates the process of adaptive radiation. Evolution of reproductive barriers to make the diverging lines genetically separate will ultimately lead to the permanancy of such divergence.

A species or a population lives in an environment. Environment is hardly completely uniform. Natural environment is actually mosaics of more or less similar or different habitats or subenvironments. The habitats or subenvironments may be suitable for some genotypes while others for competing genotypes. Diversifying natural selection enhances the adaptedness of populations living in heterogeneous environments. Genetic diversity is an advantage in changing environment.

Relationship between Reproductive isolation and origin of species

It has already been mentioned that the changes in the organism-environment relationship cause the evolution of new characteristics in a population. This is caused through the agency of natural selection. Environment is diverse in nature, hence organism-environment relationships are bound to be different for related populations inhabiting different zones of the earth's surface.

From a homogeneous population some individuals adapted to new conditions may be sorted out by natural selection and can establish themselves in new habitats. Availability of new habitats causes the evolutionary divergence from the initial population—thus resulting adaptive radiation. Individuals entering in the new areas will form subpopulations which will differ significantly from the ancestral population—resulting in the emergence of polytypic species. The polytypic species containing many races are adapted to different environmental conditions. Sometimes races can diverge from the ancestral populations which may prevent exchange of genes with other races. Such barriers to gene exchange is called isolating mechanism. When the isolating mechanisms become strongly developed the populations are regarded to be reproductively isolated from each other and they then rank as distinct species. Sympatric species (i.e. different species living together in the same area) may maintain their identity because they do not ordinarily exchange genes. Mutation, genetic recombination and natural selection help in the origin of isolating mechanisms. Isolating mechanisms are less effective when the habitats are unstable and rapidly changeable. In regions where the environment is relatively stable or remained stable foı a long sequence of time isolating mechanisms are comparatively strongly developed.

DE VRIES—THE THEORY OF MUTATION

A Dutch Botanist, Hugo de Vries (1848-1935) published the mutation theory in the year 1901 to explain the process of evolution. He based his theory on the results obtained from the breeding experiments on a plant called the evening primrose, *Oenothera lamarckina*. In his experiment, de Vries observed certain striking differences in the forms appeared suddenly amongst a population of normal evening primrose. de Vries regarded this phenomenon as **mutation** (L. *mutare*=to change) or **saltation** (L. *Saltare*=to leap) and these changed forms as the **mutants.** The conservative process of heredity causes a mutant to breed true and the progenies are produced like the parents. The term mutations or saltations means sudden large changes or discontinuous variations in the organisms and that are heritable. Saltationist school is of opinion that National Selection has nothing to do in species formation and mutation is the only causative agency in Speciation. Saltationists regard mutation as the sole agency of evolution. Mutations are best regarded as the initiating force in evolution and Natural Selection as the limiting force.

Mutation is actually the change in gene. The genes are highly stable and faithfully duplicate with exactness in the process of reproduction. Despite such phenomenon, genes may mutate. Mutation results through some fault in gene reproduction. Muller established mutation on a solid foundation. Mutation can be caused by X-radiation. Auerbach has produced mutation by chemical mutagens. Mutation in non-reproductive cells of an organism has no evolutionary significance, but such changes in the reproductive cells produce significant effects on the species. In fact, most of the mutations produce detrimental effects on the organism but few may have favourable effects on the organism in combination with other genes. The genes that contribute to form the genetic make-up of an organism form a highly co-ordinated system.

Mutation is a random affair and takes place in all directions. Genes are giant molecules and slight alteration in their structural organisation results mutation. Selection mechanism plays the most vital role in evolution. Selection incorporates favourable mutations into the hereditary constitution. Thus it is the genes in the chromosomes that are to be selected. Mutations would be meaningless until these are selected by Natural Selection. The co-ordinating action of Natural Selection and mutation gives rise to evolutionary directiveness. As already discussed mutations provide raw materials for Natural Selection to work with. Usually the favourable mutations appearing in any species are incorporated in the stable genotype by the action of Natural Selection. By this way a species may change for the better by accumulating advantageous mutations.

ORGANIC EVOLUTION—ORIGIN OF ANIMALS

The evidences and theories of organic evolution discussed in the preceding portions have indicated the fact that all the species inhabiting the earth today, are the products of evolution from simple pre-existing forms. The evolutionary history of most of the species is far from being complete, because of the non-availability of specific documentary records to construct the pathway of evolution. But in some cases the line of descent is amply recorded by a series of fossils leading stepwise from the ancestor to the present-day forms. This portion deals with the origin and evolution of some animals to strengthen the concept of organic evolution.

ORIGIN OF MAMMALS

Mammals stand at the pinnacle of evolutionary success amongst the living organisms. The question of mammalian ancestry is of special interest to man, because their history includes ours in a broader sense, as man is included under this great class, *Mammalia*. Mammals constitute a very unique group and separate themselves from others by having many morphological and physiological characteristics which help them to overcome inhospital environmental hurdles. A survey of the past geological records reveals that in Mesozoic era, there was a storm of evolution. This era is regarded as the age of reptiles. The majestic and dominant forms were the Dinosaurs, the then giant rulers of the horizon. When the Dinosaurs ruled the earth, a group of small and insignificant creatures evolved under their shadows. Though lowly in origin, but unique in their endowments, they constituted the mammals. The cause of extinction of Dinosaurs is quite obvious. Of other factors, hugeness in size, lowly developed nervous system, inability to retain constant body temperature, can be regarded as the prime causes of extinction. Mammals, with efficient mechanism of regulating body temperature, well-developed nervous system, efficient power of swift movement confronted the challenges of evolution. From such a lowly start arose the mammals—which became diversified and ruled the earth.

Definition of mammal. Mammals are warm-blooded vertebrates. The body is covered more or less with hairs. Young mammals are born alive (excepting the egg-laying monotremes) and are nourished by mothers with milk in infancy secreted by the mammary glands. Separate opening of the anus and the urino-genital system (excepting monotremes) is characteristic.

Biological organisation of mammals. For a better understanding of the mammalian ancestry, a generalised survey of the mammalian organisation of existing as well as of the extinct forms is desirable.

Morphological peculiarities. Mammals possess (1) Heterodont dentition, with differentiation into incisors, canines, premolars and molars. (2) Thecodont teeth, i.e., each tooth is embedded in an alveolar pocket on the jaw bone. (3) Diphyodont teeth, i.e., milk set is replaced by permanent set of teeth. (4) Normal dental battery is: i $\frac{3.3.}{3.3.}$, c $\frac{1.1.}{1.1.}$, pm $\frac{4.4.}{4.4.}$, m $\frac{3.3.}{3.3.}$. (5) Presence of two occipital condyles. (6) Lower jaw is made up of one pair of bones—the dentaries. (7) Articular and quadrate bones are transformed into malleus and incus respectively. (8) Secondary palate is complete and functional. (9) Cervical vertebrae are seven in number. (10) Ribs are double headed with capitulum and tuberculum. (11) Vertebrae are gastrocentrous. (12) Coracoid of the pectoral girdle is absent as a separate bone and is represented as the coracoid process of the scapula. (13) Heart is fully four-chambered with only left aortic arch. (14) Mature erythrocytes are enucleated, biconcave and circular. (15) A muscular diaphragm separates the thoracic and abdominal cavities. (16) Presence of sweat and sebaceous glands. (17) Highest development of brain, great development of neopallium and presence of corpus callosum (absent in monotremes, rudimentary in marsupials). (18) Pinna is present excepting some aquatic mammals.

Physiological peculiarities. (1) Complete separation of oxygenated and deoxygenated blood with a free scheme of circulation. (2) Homoiothermous condition is a new attribute which keeps the body temperature constant and is independent of environmental fluctuation. (3) Viviparous excepting monotremes and youngs are nourished in the uterine cavity by placenta in embryonic condition.

A brief note on some archaic mammals. A short survey of the archaic mammals will be informative to gain a clear idea regarding the ancestry of mammals. Mammals became well flourished at the beginning of Cenozoic era but have started their true career in the Mesozoic era. So the description of some Mesozoic and Cenozoic mammals will serve the purpose.

Mesozoic mammals. The mammal-like reptiles were quite abundant in Permian period. But at the early Triassic time, those forms became gradually extinct and ultimately they became almost entirely absent by the end of Triassic. But again in the Jurassic and subsequent periods fossil mammals became more abundant. Mesozoic mammals possessed some basic characteristics : (1) small size, (2) sharp-pointed teeth, (3) teeth were generally heterodont and (4) habits were varied.

Pantodonta. The fossil forms were discovered in upper Triassic period. Dentition was more like that of reptile than mammalian. But the simple jaw consisting of a single bone established their mammalian affinity. The fossil records of this group are rather incomplete. *Allotheria*. They were very important group amongst the extinct mammals and their fossils were recorded in the upper Triassic rocks of Germany. The most important mammalian characteristic was the nature of the molar teeth which had two to three longitudinal rows of tubercles. The presence of these tubercles gave the name *Multituberculata* to this group. *Triconodonta*. These fossils were recorded in Jurassic period. The molars had three cusps arranged in a single longitudinal row. *Symmetrodonta*. This group was confined to the upper Jurassic period of England and North America. The molars had three main cusps arranged in a triangular fashion. *Pantotheria*. The symmetrodonta was distantly related to this group. The molars had four or more cusps. The three principal cusps assumed the form of a trigon. Their fossils were recorded throughout the Jurassic period.

Cenozoic mammals. Besides the archaic mammals, a few important Cenozoic mammals need mentioning in connection with the ancestry of mammals. They belong to three groups, *Creodonta*, *Condylarthra* and *Amblypoda*. *Creodonta* had small reptile-like brain, carnivore-like teeth and terminal phalanges with claws. Condylarthra represented the primitive ungulates and structurally formed a sort of connecting link between the hoofed and clawed mammals. The cranial portion was located behind the orbit. Teeth were suited for herbivorous habit. Molars were low crowned. *Amblypoda* was short footed and hoofed forms of almost elephantine size.

Fate of archaic mammals. With the close of reptilian age, the Mesozoic mammals became well flourished in the Mesozoic era and exhibited wide adaptive radiation. But very quickly they became extinct and were replaced by modernised mammals which were fully equipped to face all eventualities of nature. The cause of extinction of archaic mammals is primarily due to their incapabilities to respond to new environmental conditions. The archaic mammals had a brief evolutionary career and they were subjected to racial extinction. The challenges of life were received by modernised mammals who flourished well in time and space. Some of the archaic forms transmutated into higher types while some became emigrant. Marsupials passed beyond the limits of South America and crossed the

Antarctic land-bridge to come to Australia where they still persist.

In search of ancestry. In tracing out the folds of life it is often said that the higher forms have originated from lower forms. So the ancestor of mammals must be some lower vertebrate forms. The chances for the fishes and the birds are out of question, because one is too low and the other is highly specialised. Ancestry from them is in no case possible. Now remains the possibilities for the amphibia and the reptiles to hold the respected position of mammalian ancestor.

Amphibian ancestry. The amphibian ancestry of mammals was emphasised by T. H. Huxley in the year 1880. He advocated that mammals arose from amphibia. He assumed that *Hypotheria* is a sub-mammalian evolutionary stage, which arose from amphibia and holds the key of mammalian evolution. In support of his contention, Huxley put forward some arguments. First, amphibia as well as mammamals possess double occipital condyles. Another argument in this context is that in mammals only the left aortic arch is present but in amphibians and reptiles both the arches are present. Of the two arches, the left one is weakly developed in reptiles which indirectly supports amphibian ancestry of mammals. But this theory was subjected to vigorous criticism. The points of resemblance are not rigid enough to support Huxley's assumption excepting the occurrence of double occipital condyles. The occipital condyles in amphibians and mammals are not strictly homologous structures but differ in development. In amphibians the occipital condyles are the derivatives of the exoccipitals, but in mammals the source is the basioccipital. This assumption is further weakened by the lack of any intermediate forms and palaeontological data.

Huxley's view of mammalian origin through Hypotheria is untenable today and does not offer any tangible evolutionary significance. In spite of that this view is still kept as a piece of historical document in the discussion of mammalian ancestry. The very concept of mammalian ancestry directly from amphibia without the intervention of reptiles is most unscientific and absurd.

Reptilian ancestry. The reptilian ancestry of mammals is full of evolutionary significance and gives a vivid indication of biological transformation towards the mammalian organisation. The reptilian ancestry of mammal can be established from three angles, (1) evidences furnished by some living mammalian groups, (2) from serological tests and (3) from palaeontology.

Evidences from living mammals. Some of the primitive living mammals give the strongest support of reptilian affinity of mammals. The monotremes and the marsupials occupy intermediate status between reptiles and mammals.

Monotremes offer strongest support. Monotremes constitute a sort of connecting link between the mammals and their reptilian ancestor and serve to bridge the gap between the two. Monotremes are undoubtedly mammalian, but possess many reptilian features too. The reptilian features are:

(1) Pectoral girdle is typically reptilian with a large coracoid, an interclavicle and without scapular spine. (2) Pelvic girdle possesses an epipubic bone. (3) Ribs excepting the cervical ones are single headed. (4) Presence of cervical ribs. (5) Bones of the skull are typically reptilian. Lacrymal bone is absent, jugal is reduced or absent, auditory bulla is absent, presence of reptilian provomer. (6) Corpus callosum is absent. (7) Reproductive system is based on reptilian plan. (8) Oviparous. (9) Presence of shallow cloaca. (10) Meroblastic segmentation. (11) They are not fully homiothermous.

It appears that the monotremes are roughly fifty per cent. mammal and fifty per cent. reptile in their organisation. They are very primitive and represent the end product of an independent evolutionary line of mammal-like reptilian stock.

Marsupials—also offer support. The marsupials also show many reptilian features. These features are:

(1) Brain lacks corpus callosum and the olfactory part is comparatively large. (2) Presence of epipubic bone. (3) Palate is deficient in ossification. (4) Epicondylar foramina occur frequently.

Evidences from serology. Serological tests have revealed that the reptilian blood shows close affinity with that of mammals which are warm blooded.

Palaeontological evidences in support of reptilian ancestry. The fossil remains of the mammal-like reptiles offer a sound evolutionary basis of mammalian origin through reptiles. The mammal-like reptiles are:

Synapsida. Synapsida comprises of a very interesting group of reptiles that appeared in the rocks of late Carboniferous time. They form a bridge between the primitive reptiles and the primitive mammals. The latest genera approached so close towards mammalian stage that it is really questionable whether they should be included amongst mammals or reptiles. They showed the tendency of the loss of bones in the skull. Teeth showed differentiation. The lateral temporal opening was situated behind the eye and it became enlarged gradually. The synapsida had a large number of genera. The early members were very close to the ancestral Cotylosaurs, while some of the latest genera approached very close to the mammals.

Pelycosaurs—the first synapsid. They initiated a line of evolutionary development which is most completely recorded in geological history. The first Pelycosaurs are the Ophiacodon of which *Varanosaurus* was a primitive member. In Varanosaurus, the skull was narrow, marginal teeth were socketed and showed little regional specialisation. *Bifurcation of Pelycosaurs.* The Pelycosaurs evolved along two lines—the large plant eating *Edaphosaurs* and the large aggressive carnivorous *Sphenacodonts. Edaphosaurs.* Springing from an Ophiacodon ancestry in Permian age, the Edaphosaurs diverged into a different line. The skull was small, stout and shallow. The teeth were not differentiated. *Sphenacodonts.* The second evolutionary line from the Pelycosaurs was the Sphenacodonts which led the channel of evolution. Teeth were strongly differentiated which made the refinement of the skull.

Therapsids. The fossils of these forms were recorded in the middle and late Permian and in Triassic age. They were the forms that took the road leading towards mammalian organisation. Possession of the following characters certainly supports the contention. These are:

(1) There was a strong trend of lateral temporal fossa to enlarge.

(2) Presence of a secondary palate.

(3) Dentary was enlarged at the expense of other bones.

(4) Double occipital condyles were present.

(5) Teeth were differentiated.

(6) Quadrate and quadratojugal were reduced and loosely attached with the skull.

The therapsids followed two lines of adaptive radiations. (A) *Anomodonts.* They were archaic therapsids who retained many primitive features of the Pelycosaurs. They lack secondary palate and the pineal opening was very large. The dentary was moderately developed. The typical representative of this group is exemplified by *Tapinocephalid* (*Moschops*). (B) *Dinocephalians.* They were the most successful therapsids of phylogenetic longevity. They appeared in the middle of Permian to lower Triassic. Bones of the temporal region of the skull were thin due to enlargement of temporal opening. Acromion process was present on the scapula. The phalanges were reduced to mammalian count (2–3–3–3–3).

Theriodonts. When the Dinocephalians were evolving in their own way, the theriodonts developed rapidly in a direction that was led directly to the mammals. The theriodonts were varied in shape. *Cynognathus* was a typical genus of the theriodonts. *Cynodontia* had many mammalian features. They possessed the following mammalian characters: (1) Double occipital condyles. (2) Lower jaw was almost made up of a single bone—dentary. (3) Heterodont dentition. (4) Presence of hard palate. (5) Double headed ribs. (6) Scapular spine was present.

Ictidosauria formed a bridge between reptiles and mammals. They had almost crossed the threshold line that separates the reptiles from mammals. They possess: (1) Skull was highly organised and mammal-like. The cranium was spacious. (2) Quadrate was greatly reduced. (3) Lower jaw was made up of dentary. (4) Presence of double occipital condyles. (5) Heterodont dentition.

Molar hypotheses support reptilian ancestry

Mammals possess characteristic molar teeth. The origin of molar teeth in the phylogenetic history of mammal has given rise to a number of theories. All the theories indirectly support reptilian ancestry of mammals.

Tritubercular theory. Cope and Osborn hold that mammalian molar teeth passed through a trituberculate stage found in Cynognathus. **Multitubercular theory** of Woodward and Goodrich advocates that mammalian molar teeth arose from a multituberculate stage found in *Cynodontia* (*Diademodon*) and the modern mammals are the evolutionary derivatives of the *Diademodon*. **Compromise theory of diphyletic origin** of Mivart and Beddard advances the logical idea of mammalian evolution from Cynodontia having both tritubercular and multitubercular molar teeth.

Judging all the theories the inevitable conclusion that can be drawn is that reptilian ancestry is full of evolutionary significance. Cumulative evidences from palaeontology, comparative morphology and serology thus prove that the mammals sprang from reptiles. Very likely the source is best represented by either Cynodonts or Ictidosauria.

Probable ancestry of mammals

According to most evolutionists of today, the Theriodont reptiles hold the key

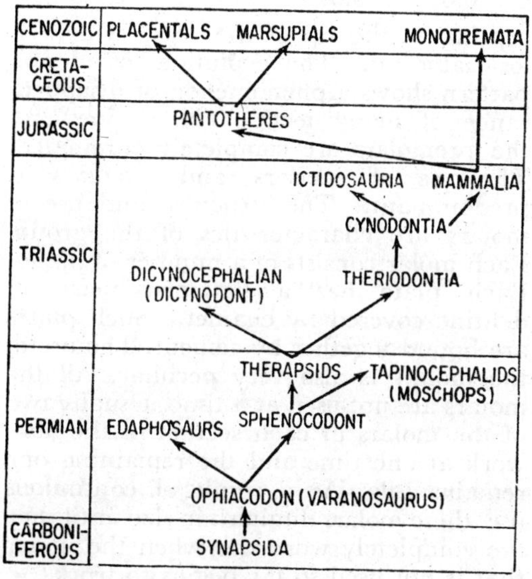

Fig. 1.18. Schematic phylogenetic tree of mammalian evolution (after Colbert).

of mammalian ancestry. From Cynodontia, the Ictidosauria and the mammalia arose and underwent parallel evolution. The probable line of ancestry is shown in Fig. 1.18.

Biological trends in mammalian evolution

A review of the fossil mammal like reptiles reveals that the evolution of mammals is preceded through the following biological trends.

(1) The gradual growth of brain which is related to the increment of the cranial cavity. (2) Gradual acquisition of hard palate. (3) Loss and condensation of some bones of the skull and lower jaw. (4) Gradual reduction of elements of the pectoral girdle. (5) Gradual condensation, differentiation and compactness of the general body system to increase the efficiency.

Time of origin. The mammal like reptiles were abundant in Permian and in early Triassic periods. The probable time is assigned either to be late Permian or early Triassic.

Place of origin. There are two views regarding the place of origin of mammals. First view is that the mammals arose in South Africa as evidenced from the abundance of Therapsid fossils. Based from the history of reptilian disappearance the second view holds that the place of origin was Central Asia.

Cause of origin. From Permian to Jurassic age, there was a great climatic fluctuation. There had been periodic aridity alternating with glacial periods. So nature has to provide a new mechanism of temperature control. The achievement of temperature control probably saved the insignificant mammals from the onstroke of geological changes.

At the present state of knowledge it will be better not to go too far regarding the origin of mammals. Further researches on palaeontology, comparative anatomy are necessary to picture out the completeness of mammalian ancestry. However, with all certainties and probabilities, the insignificant Cynodonts whose existence was at the mercy of the mighty Dinosaurs, survived the perilous existence and gave origin to mammalian form of life in time and space.

ORIGIN OF PROBOSCIDEANS

Proboscideans or the elephants constitute one of the largest and majestic groups in the animal kingdom. They represent uniqueness in their anatomical organisation. They possess an admixture of a large number of primitive as well as specialised characters and show the most spectacular success in course of phylogenetic development. Evidences advanced by fossil elephants furnish an unmistakable transformation from simple to complex. The elephants have reached the peak of evolutionary perfection that indirectly led them into a blind alley. Though they were quite abundant in their geological career, the kingly elephants are now represented by only two genera threatened with extinction.

The most striking feature in the elephants is the hugeness in size, pillar-like legs matching the proboscis and tusks. Because of the presence of peculiar anatomical features, its systematic position is quite problematic. Excepting remote phylogenetic connection with the Sirenia and Hyracoidea, the elephants stand as an isolated group. Starting with their early evolutionary career, they acquired numerous peculiar features which separate them from others. Most of the features are adaptive in nature.

Biological organisation of Modern Elephants. Before going to the actual process of evolution of proboscidea, a brief account of the anatomical organisation is necessary for better understanding. In addition to the possession of a large number of specialised adaptive features, the surviving elephants have retained many primitive characters. The primitive features are largely confined to the soft parts, as they are more conservative in evolutionary dynamics than the hard parts.

Primitive features. To enumerate briefly, the primitive features are: (1) The condition of cerebellum. It is well convoluted, but is uncovered by the cerebral hemispheres. (2) The stomach is simple. The liver has two distinct lobes, but lacks gall bladder. (3) Presence of two superior venae cavae carrying blood to the heart. (4) The lungs are simple and not so much lobated. (5) The testes are permanently abdominal. (6) Presence of primitive placenta. (7) There is a tendency towards the reduction of lateral digits, especially in African elephants. The feet are five-toed. The carpal bones are serially placed one above the other. (8) Although there is a general tendency towards reduction of ulna in cursorial animals, the ulna forms a most dominant bone in elephants.

Specialised features. The skeleton shows many specialised features. The hugeness of size is a paramount feature in elephants. The development of proboscis by the elongation of snout with the support of upper jaw is a unique feature to note in elephants. Transformation of the incisors into tusks is also notable. The limbs are pillar-like and are modified to carry the enormous weight of the body. The neck becomes shortened to carry the unusual weight of head with the proboscis. The skull shows many specialised characteristics. It is smaller in comparison to the body, but is amply modified to offer leverage for the enormous weight of the proboscis. To minimise the unwanted weight, special formation of air spaces or *diploe* is observed in skull. The development of the diploe causes obliteration of most of the sutures.

The dentition is highly characteristic and the dental pattern is : i $\frac{1.1.}{0.0.}$, c $\frac{0.0.}{0.0.}$, pm $\frac{0.0.}{0.0.}$, m $\frac{3.3.}{3.3.}$. There is no enamel excepting at the tip of the tusk, which is lost later on. The evolution of dental pattern shows a phenomenon of disappearance of many teeth. The canines and the premolars are completely eliminated. The use of incisors and molars is predominant. The structure and use of molars are characteristics of the group. Each molar consists of a number of plates. Each plate has a flattened mass of dentine covered by enamel. Such plates are bound together by cement. The use of the molars is also very peculiar. All the molars are not used at a time. Usually two of the molars in each section of the jaws work at one time and the remaining one remains idle. As a result of continuous use, these molars diminish in size until they are completely worn out when the molar that is not used so far begins to work.

The brain is very large in size and its specialisation lies in it. Though the brain shows extensive convolutions, it possesses many primitive features. Elephants have a remarkable memory and mental development. They possess relatively dull vision, but olfactory and auditory senses are very well developed.

Though palaeontology has shown a variety of forms of elephants, they are represented today by two genera. Other elephants are all lost in evolutionary struggle. The living elephants are: *Loxodonta africana* of Africa and *Elephas maximus* of India or Asia. Both of them possess common structural organisation, but differ slightly. In *Loxodonta* the molars are comparatively narrow and low crowned. The ridges are usually fewer in number and assume rhomboidal pattern, They are greater in size and possess big pinna. The tusks are longer and curved. Occurrence of several subspecies of the genus *Loxodonta* is claimed by many, but definite evidences to establish them are lacking. They are most probably the geographical races rather than the subspecies. In contrast to *Loxodonta*, the Indian form, *Elephas maximus* possesses short skull with a rounded vault. The tusks are short and more or less straight. The tusks are directed

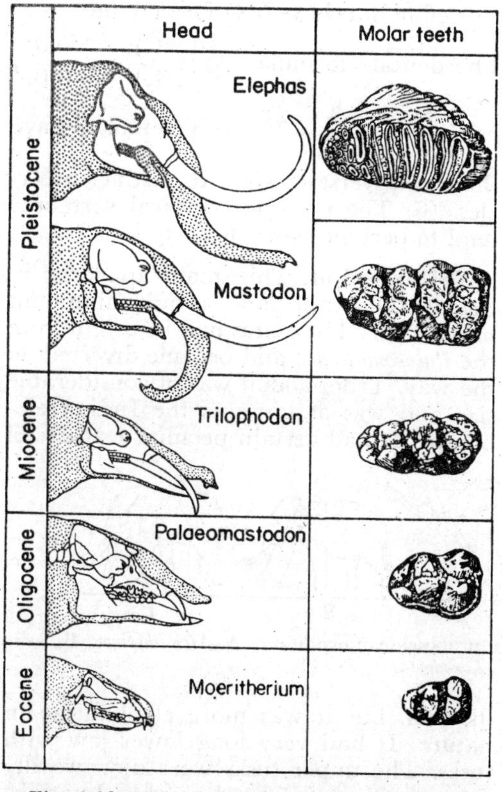

Fig. 1.19. Diagrammatic sketches showing the evolution of skull and molar teeth in proboscidean evolution. Note the gradual development of proboscis and tusk.

more downward than forward. The molars are broad and high crowned. The ridge plates are many and are closely appressed.

In search of ancestry. The phylogenetic changes of the elephants are completely recorded by series of fossil forms (Fig. 1.19). But great controversies exist as regard the exact starting point of the proboscidean evolution. Authorities differ in their views regarding the earliest proboscidean. Lull regards *Moeritherium* as the earliest form directing the channel of proboscidean evolution. But Scott assumes that this stage is not in the main line but represents a side line originating from the common ancestor. According to him, *Palaeomastodon*, another fossil proboscidean, is the main source for proboscidean evolution and holds the pivotal position from which all forms of elephants evolved in time and space. Modern researchers in this line also think that *Palaeomastodon* is the primal source.

Moeritherium. The fossil forms of the *Moeritherium* have been discovered in the upper Eocene and early Oligecene bed of Egypt. Moeritherium possesses many elephantine features. Although they were much smaller in size than the latter forms, the height of the animals was estimated to be about three feet. The skull was fairly long with eyes placed farforward. Initiation of development of air-cells was observed especially in the back of the skull. There was no proboscis, but its development was initiated in the form of a short flexible tapir-like snout. Dentition shows the sign of proboscidean evolution in which there is reduction in the normal number of teeth. The dental formula of the *Moeritherium* is represented as: $i\,\frac{3.3.}{2.2.}$, $c\,\frac{1.1.}{0.0.}$, $pm\,\frac{3.3.}{3.3.}$, $m\,\frac{3.3.}{3.3.}=36$. The molars had not yet been specialised. These were small and short crowned. Transverse crests were simple and filled with dentine. The formation of tusks has already begun. The second pair of upper incisors became elongated and sharply pointed. The second incisors of the lower jaw assumed the appearance of procumbent tusks and were horizontally directed.

Palaeomastodon. The fossil forms of the *Palaeomastodon* were recorded in the --

Oligocene bed of Egypt. They were comparatively larger in size than Moeritherium. They attained appreciable size. The proboscis was better developed and was probably extensible. The bones of the skull had air-spaces (diploe). Occurrence of diploe is a clear and definite indication of proboscidean evolution. The dental battery had more progressed towards proboscidean evolution in the reduction of teeth. The dental formula is represented as: i $\frac{1.1.}{1.1.}$, c $\frac{0.0.}{0.0.}$, pm $\frac{3.3.}{3.3.}$, m $\frac{3.3.}{3.3.}$=28. The molars possessed three crests.

Biological trends of Proboscidean Evolution. Linking the earliest fossil proboscideans and the modern surviving genera, series of fossil forms have been discovered. Fossil records of elephants furnish a complete evolutionary history. The most remarkable feature to note in proboscidean evolution is that their evolution has progressed towards certain basic biological trends. The trends are as follows:

(1) Reduction in the number of teeth and change in the disposition and sequence of arrangement.

(2) Gradual transformation of incisors into tusks.

(3) Acquisition of specialisation of molars.

(4) Gradual loss of enamel.

(5) Gradual development of proboscis or trunk.

(6) Development of air-spaces (diploe) in the skull bones.

(7) Gradual increase in size (Fig. 1.20).

Fig. 1.20. Showing the gradual increase in size in proboscidean evolution. A. *Moeritherium*. B. *Palaeomastodon*. C. *Stegodon*. D. Mammoth. E. *Elephas*. F. *Loxodonta*.

(8) Mechanical readjustment of the limbs to bear enormous weight of the body.

The above-mentioned biological trends are largely concerned with the head of the elephants and other parts have undergone little modifications with the progression of time.

Palaeontological procession of the elephants.

Phiomia. The fossils of this genus were also discovered in the lower Oligocene bed of Egypt. Exact position of *Phiomia* in the evolutionary series is controversial. Lull regards that *Phiomia*, originating from *Moeritherium* directed the channel of evolution. But authorities like Scott inclines to think that Phiomia, differentiating from *Palaeomastodon* in late Oligocene represented a side line and showed the tendency towards deviation. Phiomia is undoubtedly a proboscidean of comparatively larger size than the previous forms. This genus possesses the following features: (1) The height of skull is increased considerably. (2) The development of diploe in the skull is noticed in the skull bones. (3) Indication of development of an extensible, short proboscis is observed. (4) The dental formula shows loss of entire canines. The second pair of incisors have transformed into tusks. The dental formula is: i $\frac{1.1.}{1.1.}$, c $\frac{0.0.}{0.0.}$, pm $\frac{3.3.}{3.3.}$, m $\frac{3.3.}{3.3.}$=28. (5) The molars have three transverse crests with distinct tubercles. (6) The posterior cervical vertebrae tend to become shortened.

Trilophodon. This genus appeared in the late Oligocene and was quite abundant in Miocene. The genus had its origin from the *Palaeomastodon* and became diverted on the way. Trilophodon was of considerable size and was as large as the Indian elephants. It had certain peculiar features of its own, but it was more elephantine in nature. It had very long lower jaw with tusks. The upper tusks were downwardly curved and were enamel-banded. In adults the molars were of enormous size and contained three crests. A marked increase of the diploe in the skull was observed. The nostrils were placed high on the skull. Trilophodon had a great power of migration and reached North America for the first time from Asia in Miocene period and constituted an important member of the North American fauna for a short time.

Mastodon. Arising from the *Palaeomastodon*, *Mastodon* became a very dominant group in Miocene time. This group constituted the best known American elephants. The lower jaw was very much shortened. The molars were simpler and were devoid of intervening cusps. Vestigial and functionless tusks may occur in lower jaw of male. The upper incisors lack enamel. The dental formula of Mastodon is:

$$\text{deciduous—i}\frac{1.1.}{1.1.}, \text{c}\frac{0.0}{0.0}, \text{pm}\frac{0.0.}{0.0.}, \text{c}\frac{3.3.}{3.3.} = 16.$$

$$\text{permanent—i}\frac{1.1.}{0(1).0(1)}; \text{c}\frac{0.0.}{0.0.}; \text{pm}\frac{1.1.}{1.1.}, \text{m}\frac{3.3.}{3.3.}$$

$= 18$ and 20.

The massiveness of the limb bones and broadness of the pelvis were the two noticeable features in *Mastodon*. The development of diploe was more pronounced. Only two fully developed molars were present in the jaws at one time.

Exact relationship of Mastodon is not precisely known. They were plenty in Pliocene and Pleistocene time and ranged from Europe to Alaska and the United States. Romer inclines to think that the *Palaeomastodon* and the *Phiomia* constituted the group known as the Mastodon. The Mastodons, though extinct today, were quite numerous and diversified group in tertiary age.

Stegodon—the transitional Elephants. The *Mastodons* were replaced by the *Stegodon*. Possession of transitional characters by the Stegodon made the issue confusing. They arose in Pliocene time and were placed by the palaeontologists in the direct line of proboscidean evolution. They were considered to be the successful arrow line in the way of the evolution of elephants and hold the pivotal position from which the modern elephants evolved. Stegodon enjoyed a wide range of distribution and the members were well flourished in the Eastern hemisphere of the world. They were known as the transitional elephants. They differ from the modern elephants by the possession of large skull and peculiar dental battery. The dental battery resembled the modern elephants excepting the condition of the molars. The vallies of the molar crests were dificiently laid with dentine.

Stegodon gave origin to the modern elephants. One of these forms persisted along with the early Pliocene Mastodon in North America, which later on invaded South America. This form became further advanced in jaw structure and molar structure where the cusps increased to the maximum of seven or eight. Romer named this form as the Stegomastodon. There occurred a shortening of the facial portion. The fleshy nose above presumably became free to form a long proboscis like that of modern forms. The molars were high crowned and had gained cement.

True Stegodon fossils were abundant in Southern and Southeastern parts of Asia, which indicates the places as the original home of true proboscideans. Stegodon gave origin to the true elephants.

Mammoth. During late Pliocene and early Pleistocene, in Asia and Europe, the first typical elephant took its origin possibly from the Stegodon or from the Stegomastodon. They were popularly known as the mammoth. Mammoths were adapted to cold climate which resulted in the development of a coat of long, coarse black hairs with a coat of brown wool beneath. They had short head and tuskless lower jaw. The teeth exhibited maximum degree of complexities. The molars had a large number of cusps. During pleistocene period various other elephants were recorded which were referred to as the Mammoth. The remarkable frozen mammoth of the Siberia tundras preserved in nature's cold storage presents a typical representative of mammoth. They became all extinct in Pleistocene.

Living Elephants. Two well-defined groups had their origin from the Stegodon source and are surviving today in two widely separated parts of the earth. They are the Indian or Asiatic elephants, *Elephas maximus* and the African elephant, *Loxodonta africana*. Biological organisation of these two forms have already been discussed.

Fossil elephants have left behind a

documentary record of their evolution. Fig. 1.21 shows the scheme of the phylogenetic development of elephants. The earliest recorded fossil forms of the elephants were present in the Eocene bed of Egypt. Great controversies exist as regards the exact starting point of proboscidean evolution. Both the *Moeritherium* and the *Palaeomastodon*, whoever be the primal source, were discovered in the Eocene and the Oligocene bed of Egypt respectively. From this point it is natural to think the place of origin to be Egypt. Another opposite view exists which advocates the place of origin to be Central Asia.

The, study of proboscidean evolution forces the attention to the evolutionary afford a sound basis of descend with modifications of the train of life. For better understanding of the evolution of horses, a brief review of the biological organisation of the modern horses, and their diversifications will be discussed first, then we will analyse the fossil records to ascertain what can be demonstrated about evolution.

Biological organisation of modern Horses and their diversity

Modern horses have reached the highest grade of cursorial adaptation. The whole of structural organisation of horses is due to the perfection of food-getting mechanism and the attainment of speed. The modern horses show considerable genetical

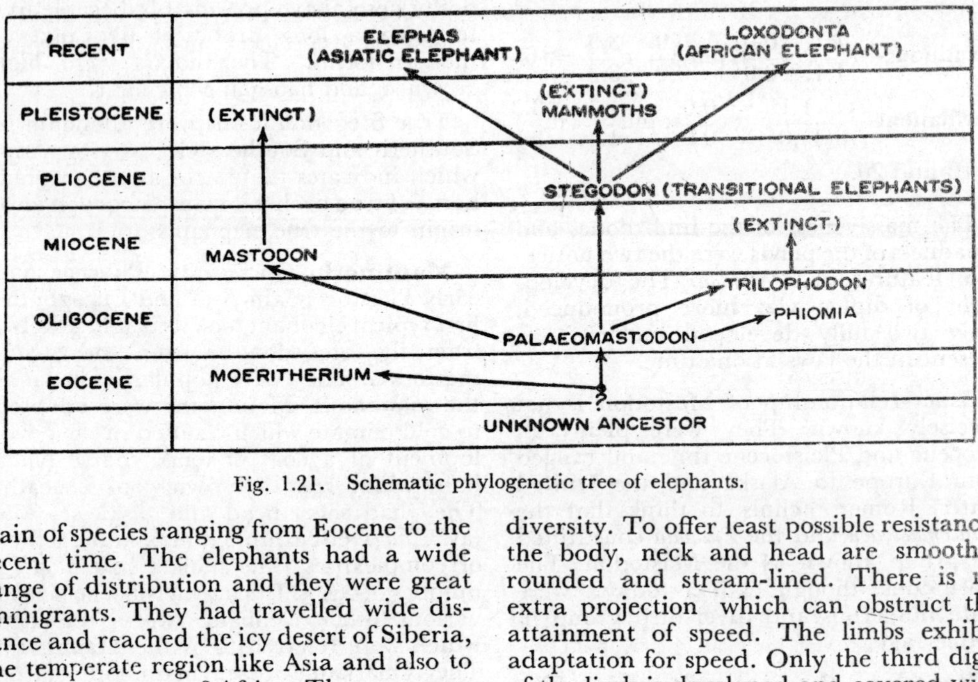

Fig. 1.21. Schematic phylogenetic tree of elephants.

train of species ranging from Eocene to the recent time. The elephants had a wide range of distribution and they were great immigrants. They had travelled wide distance and reached the icy desert of Siberia, the temperate region like Asia and also to the subcontinent of Africa. They are now represented by two genera—*Elephas* and *Loxodonta* in two widely separated regions of the earth.

ORIGIN OF HORSES

The origin and evolution of horses signify the most speculative success of a race in course of its phylogenetic development. Evidences advanced by the fossils in the pages of the earth's history give a complete and convincing documentary record of the evolution of horses. The records of the horses go back all the way with successive links from Eocene to the recent time and

diversity. To offer least possible resistance, the body, neck and head are smoothly rounded and stream-lined. There is no extra projection which can obstruct the attainment of speed. The limbs exhibit adaptation for speed. Only the third digit of the limb is developed and covered with hoof. The metapodials of second and fourth digits are present as small splint-bones. The distal segment of the limb becomes lengthened by the elongation of the bones. Presence of a cannon bone is a characteristic feature. The limb joints are of the tongue and groove variety and the movement is restricted to the force-and-aft plane. With the lengthening of limbs, the neck and skull have become very much elongated. The skull has a spacious and well-developed brain-case. The facial portion is elongated. The orbits are completely surrounded by bone. There is a general tendency towards reduction of teeth. The canines are rarely developed in females. The premolars are becoming molarised. The molars are deep-crowned.

In search of ancestry. Piercing the gloomy curtain of geological history many Eocene horses have been recorded which show the dawn of equine evolution.

Eohippus stamps the most primitive horse-like form. In the lower Eocene bed of North America, *Eohippus* was the first recorded fossil form which showed equine adaptation. It was a four-toed form. The forelimb had four complete toes and each terminates in a hoof-like nail. There was no trace of pollex. But the hindlimb had three digits with vestigial remnants of the first and fifth digits. The ulna and fibula were slender bones. The size was smaller and attained about twelve inches in height, but several species may exceed the limit. The premolars were becoming molarised. Modern researches tried to establish the fact that the Eohippus is the direct source from which different forms of horses evolved in time and space.

Gradual loss of digits from five to one. (5) Attainment and perfection of hoof. (6) Gradual reduction of ulna and fibula, formation of cannon bone. (7) Limitation of movement of limbs. (8) The teeth become elongated and attain complexities. Gradual loss in the number of teeth. (9) Pre-molars are gradually becoming molariform.

Palaeontological procession of horses. The fossil remains of horses have been recorded with considerable accuracy and the series can be constructed with completeness from Eohippus to Equus (Fig. 1.23). The pathway of evolution sometimes becomes confusing due to the presence of certain divergent lines. The side-lines are so numerous that they cast doubt on the concept of single and uniform trends in equine evolution. The evolutionary sequence of equine evolution becomes further complicated by the migratory power of the race. It is believed that the main course of the evolution of horses occurred in North America with migration at various times to the Old world and South America.

Fig. 1.22. Showing the gradual increase in size in the evolution of horse. A. *Eohippus*. B. *Miohippus*. C. *Merychippus*. D. *Pliohippus*. E. *Equus*.

Biological trends of equine evolution. Starting from the lower Eocene Eohippus upto the modern horses, a great number of fossil stages have been recorded. A survey of the palaeontological record shows the following changes that may be listed as:

(1) Gradual increase in size from cat-like Eohippus to modern horses (Fig. 1.22). (2) Concurrently with the lengthening of limbs, the neck and the skull become elongated. (3) Conversion of foot posture from semi-plantigrade to unguligrade. (4)

Eocene—The epoch of four-toed horse. Several forms of horses have been recorded in Eocene period. Amongst the Eocene horses Eohippus is regarded as the dawn horse. **Eohippus—the lower Eocene form.** The equine characters of the Eohippus have been discussed previously and this form represents a typical Eocene form. The other forms of the horses recorded in the Eocene period are the *Orohippus* and the *Epihippus*. These two forms show little difference from the Eohippus excepting certain features in the nature of dentition. **Orohippus—the middle Eocene form.** The next stage of the equine evolution beyond the Eohippus is the *Orohippus*. This form showed some advancement over the Eohippus as indicated by the loss of the splint of the fifth digit in the foot. The middle finger of the forelimb becomes slightly increased in size and the outer fingers become much shortened. The third and fourth premolars are becoming molariform. **Epihippus—the upper Eocene form.** This form is further developed than the previous forms where the third and the fourth premolars become completely molariform. According to many, *Epihippus* may not stand as the direct ancestor of Oligocene horses. The outermost digits in the forelimb become much diminished and in the hindlimb the middle digit becomes dominant. The Eocene horses are small in size but they show a gradual increase in size. Distribution of Eocene horses from Europe to New Mexico reveals their power of migration.

Oligocene—the epoch of three-toed horses. Beyond the Eocene forms, the Oligocene horses show further advancement towards equine evolution. The Oligocene horses had to confront many adverse conditions and they have given origin to some divergent forms. Two distinct forms are recognised. **Mesohippus—the lower and middle Oligocene form.** This form attains a size of about 24 inches in height and possesses three functional digits both in the fore- and hindlimbs. The middle digit becomes much elongated. The ulna and fibula are slender bones. All the premolars excepting the first one become molariform. **Miohippus—the upper Oligocene form.** This form basically resembles the Mesohippus in all respects except that the size is larger.

Miocene—the epoch of diversification. With the advent of Miocene period there was extensive branching out of horses towards several lines. Miocene horses were of several forms and exhibit at least three adaptive lines. Of the three evolutionary lines, two overcome the evolutionary hurdles and the other line became extinct. **Parahippus—the lower Miocene horse.** This form represents the transitional stage. In this stage the valleys between the crests in the teeth started to become filled up with cement. The lateral toes are much reduced. **Merychippus—the middle Miocene horse.** This form holds the key of the direct line of equine evolution. The teeth showed a transition from the uncemented short-crowned forms to the fully cemented long-crowned forms. Though this form is morphologically three-toed, but functionally they are one-toed. The orbit, for the first time, was almost complete. In Merychippus the milk set was uncemented and short-crowned but the permanent set was long-crowned and cemented. **Protohippus—the upper Miocene form.** This form evolved from the Merychippus but showed certain advancement in that both the milk set as well as the permanent set were long-crowned and cemented. The limbs were still three-toed. *Evolutionary sidelines.* During Miocene there were two evolutionary sidelines. *Hypohippus.* This form had low-crowned teeth suited for browsing on succulent herbs. They were called as the *browsing horses.* Like all other Miocene horses the limbs were three-toed. They attained the size of pony and were doomed to racial extinction. *Hipparion.* They showed certain progressive features than the previous forms. Perfection of teeth and development of special features in the skull were progressive in nature. But they were conservative in retaining three toes in the limbs. They continued up to the Pliocene time when they became all extinct excepting few forms who struggled upto the early Pliestocene period.

Pliocene—witnessed the rise of one-toed true horses. Pliocene period was geologically of great unrest. *Pliohippus* originated from the Protohippus in the upper Miocene period. They were very progressive forms. The development of the skull characteristics and the perfection of the teeth were like modern horses. They were single toed forms and there was almost no trace of the lateral toes. *Plesippus.* This form arose in the late Pliocene period and attained the size of the Arab horse but the limbs were much smaller. They were one-toed forms with no trace of lateral digits. The skull was Equine-like. The dental pattern was more advanced than that of Pliohippus and the teeth were as in Equus.

During the Pliocene period another notable form was the South American form—the *Hippidion*, a derivative of the Protohippus. The limbs were one-toed, but were very stout rather than slender. The Hippidion continued in the Pleistocene period and transformed into the *Orohippidium*.

Triumph of modern horses. The modern horses belong to the genus *Equus*. Both in North and South America a large number of extinct species of Equus belonging to Pleistocene period were recorded. Several forms of horse-like animals are still surviving to-day in Africa and Asia. The forms present in Europe and America are either feral or domesticated. The modern forms of horses belonging to the genus *Equus* are the descendant from the *Plesippus*.

The original site of the origin and evolution of the horses is North America, where they were plenty but became extinct in Pleistocene time. At the beginning of Pleistocene time, Equus had migrated from the original North American home to the other continents to become world-wide. In the modern time a number of forms commonly designated as the horses, Zebras and asses are present in the old world.

Phylogeny of horses. The phylogenetic sequences of horses are mostly unidirectional with few side-lines. Fig. 1.24 shows the phylogenetic tree of horses.

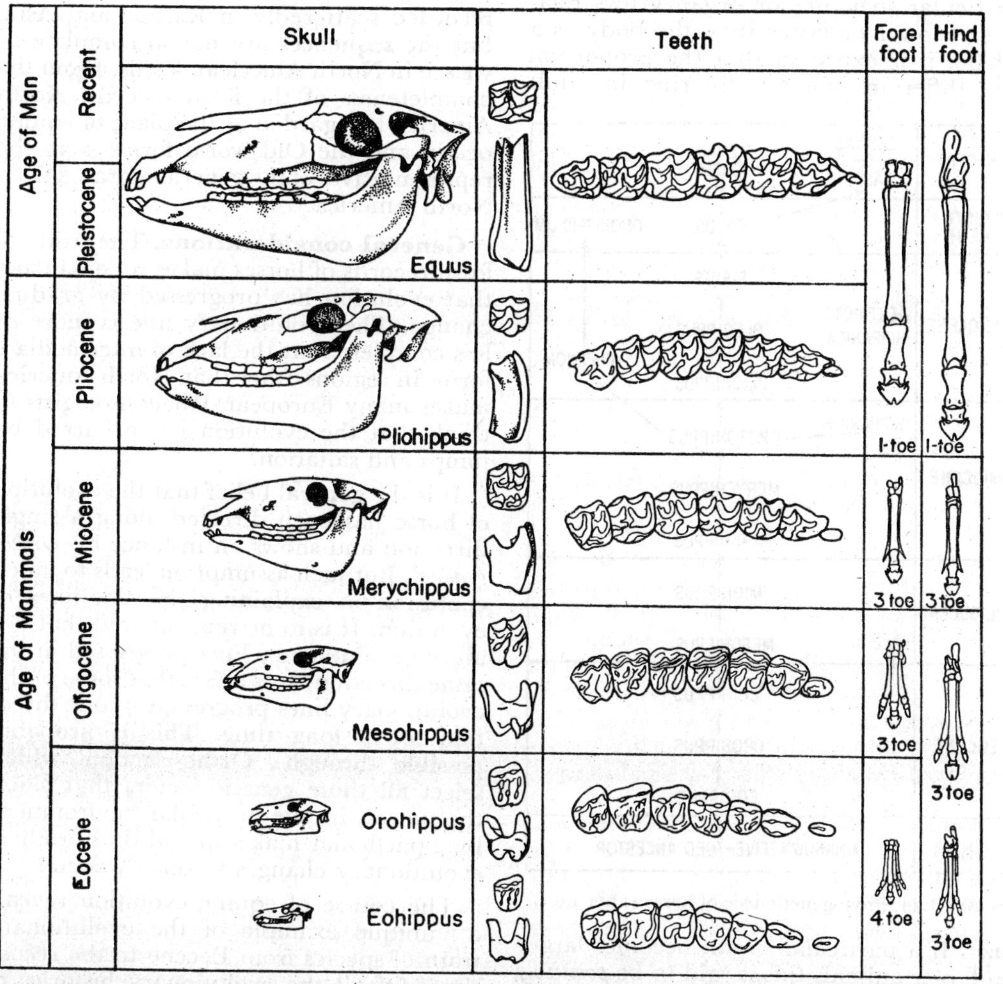

Fig. 1.23. Diagrammatic pictures showing the gradual evolution of skull, teeth and limbs in the phylogenetic development of horse. Note the gradual reduction of digits and the conversion of premolars into molars.

Study of allometry in the evolution of horses. Serious attempts have recently been made to estimate the rate of evolutionary changes occurred in the phylogenetic history of an animal. Many of the evolutionary changes encountered in an animal are measurable changes in proportions. This phenomenon is vividly observed in the evolution of horses. The present-day horse, *Equus* is certainly a different creature from *Eohippus*. Despite this fact, the differences observed in the evolutionary steps from *Eohippus* to *Equus* are the consequences of the increase in size. So if a particular structure or organ grows relatively slower or faster than the body as a whole, it is expected that the proportion will differ in animals differing in adult

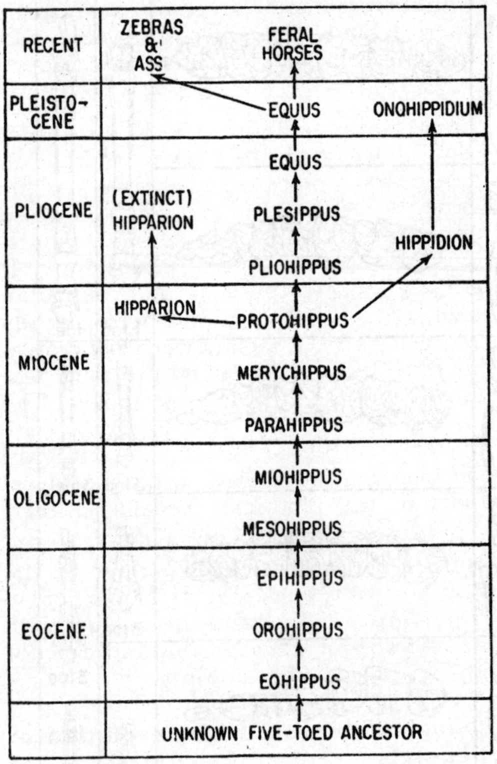

Fig. 1.24. Phylogenetic tree of horse (after Lull).

size. If a particular organ becomes larger in larger animals this is said to be *positively allometric*. Increase in the length of the facial portion of the skull furnishes an instance of positive allometry. It has also been advocated that the increase in the length of facial portion between embryo to adult bears a close parallelism of the changes occurred from Eohippus to Equus. If an organ or structure becomes smaller in larger animals, this is called as the *negatively allometric*. Reduction of lateral toes in the limbs in the evolution of horses shows the example of negative allometry, because the lateral toes are relatively smaller in larger animals. Although the underlying principles on such laws have been questioned by many, this demonstration showed that growth actually follows certain laws, especially in the evolution of horses it is true.

Place of origin. The earliest known fossil horse, Eohippus, has been recorded in North America. Fossil horses have beeen recorded scatteredly in Europe and Asia, but the sequences are not so complete as we see in North American fossils. From the completeness of the fossil records, North America is regarded as the place of equine origin and the Old world forms were the representatives of the genera migrated from North America.

General considerations. The study of fossil records of horses makes it convincing that evolution has progressed by gradual change. The evolutionary line is more or less complete, but the lack of intermediate forms in regions other than North America makes many European Palaeontologists to think that the evolution has occurred by jumps and saltation.

It is the general belief that the evolution of horse has been directed along a single direction and shows an instance of *Orthogenesis*'. But such assumption leads to much confusion in explaining the side-lines of evolution. It is to be remembered that not all the evolutionary lines progressed in the same direction. It is a fact that some of the evolutionary lines progressed in one direction for a long time. This is probably possible through 'Orthoselection' which select all those genetic factors that cause the success in one particular environment by a particular make-up and thus produce evolutionary changes in one direction.

The course of equine evolution reveals a unique example of the evolutionary train of species from Eocene to the recent time. Of all the evolutionary histories of animals, none is so vividly known as that of the horses. The excellent array of fossils give a remarkable progressive record in time and space.

ORIGIN OF CAMELS

The camels, the ships of desert, are of special interest to us because of their unique adaptation to the harsh and inhospital conditions in sandy deserts. Next to horses, the evolutionary history of camel is vividly known from the excellent collection of fossils. The remarkable fossils in the history of the earth give an instance of progressive evolution of camel in course of phylogenetic development. Despite these facts, the entire evolutionary history of camels is recorded in North America. They were exclusively North American upto Pliocene whence they migrated to South America, Africa and Asia where they are surviving to-day. Occurrence of many evolutionary side-lines in other continents does not make the phylogeny of camel so confusing as that of horses.

The camels are the representatives of the even-toed Artiodactyla. The camels occupy a transitional status and hold an isolated position. The biological relationship of camels with others is not yet clearly established.

Biological organisation of Camel

Two genera of camels are surviving today—*Camelus* and *Llama* or *Auchenia*. They are characterised by the possession of certain features. The limbs are two-toed and long. They are secondarily digitigrade showing an adaptation to sandy desert. The hoofs are reduced to nail-like forms. The metapodials are fused to form the cannon bone. The lateral movement of limbs is restricted to a limited range. One or two cushion-like pads are present in the feet. The number of teeth is reduced to 34. Only one upper incisor is present on each side and it is more canine-like. All the lower incisors are present. They are spatulate and precumbent. Canines are similar in structure and function as the incisors. The molars are high crowned, elongated and each has four crescentic cusps. The stomach is three-chambered and 'water-cells' are present in the wall of the rumen. The red blood corpuscles are unique amongst the mammals. They are elliptical and not circular in outline. The placenta is of primitive type. Presence of hump is a very characteristic feature in camel. It consists of a gelatinous fat and assumes a conical shape. It provides nourishment in time of scarcity of food. The hump is single in Arabian camel (*Camelus dromedarius*) and is double in central Asian camel (*Camelus bactrianus*).

Surviving genera

Two living genera are encountered—one, *Camelus* and the other, *Llama* or *Auchenia*. Two species of the genus *Camelus* are present. Llama is the South American genus and includes two wild species——*Llama huanacus* and *Llama vicunia*. The domestic derivatives of these wild forms are the Llamas and Alpacas of South America. They are built on the same structural plan as the camels excepting the absence of hump and of much smaller stature. Two pads, instead of one are present.

In search of ancestry. The primal evolutionary home of camels was North America and the ancestral source can clearly be traced back to the upper Eocene time.

Protylopus—holds the ancestral key. The first possible ancestral stage in the evolution of camel is observed in the Upper Eocene time. *Protylopus* possessed many features which are certainly camel-like. This creature had a small stature. It had the normal number of mammalian teeth, i.e., 44. The canines were slightly enlarged and the molars were low crowned. The facial portion of the skull was narrow and the orbit was incomplete. The limbs were subequal, the forelimbs were shorter than the hindlimbs. The forelimbs had four functional digits while the lateral digits of the hindlimbs became slender.

It is unanimously accepted that Protylopus is the earliest known stage in the evolution of camel (Fig. 1.25). Prior to Protylopus the ancestral line merges with the ancestry of other groups and can not not be traced with certainty.

Biological trends in the evolution of camels

Between Protylopus and the surviving genera of this day, a large number of fossils is recorded. Throughout the evolutionary line of camel the following trends are recorded. Excepting the retrogression of the limbs, the evolutionary changes show close parallelism to the changes undergone by the horses. The changes may be summarised in the following way:

(1) Gradual increase in size.

(2) Subsequent loss of lateral digits.

(3) Elongation and fusion of metapodials to form the cannon bone.

(4) Change over from unguligrade to digitigrade form.

(5) Reduction in the number of teeth.

(6) Gradual elongation of teeth.

Palaeontological procession of camels. From the Protylopus stage to the present-day form, a large number of fossil forms have been recorded in the successive strata of the earth.

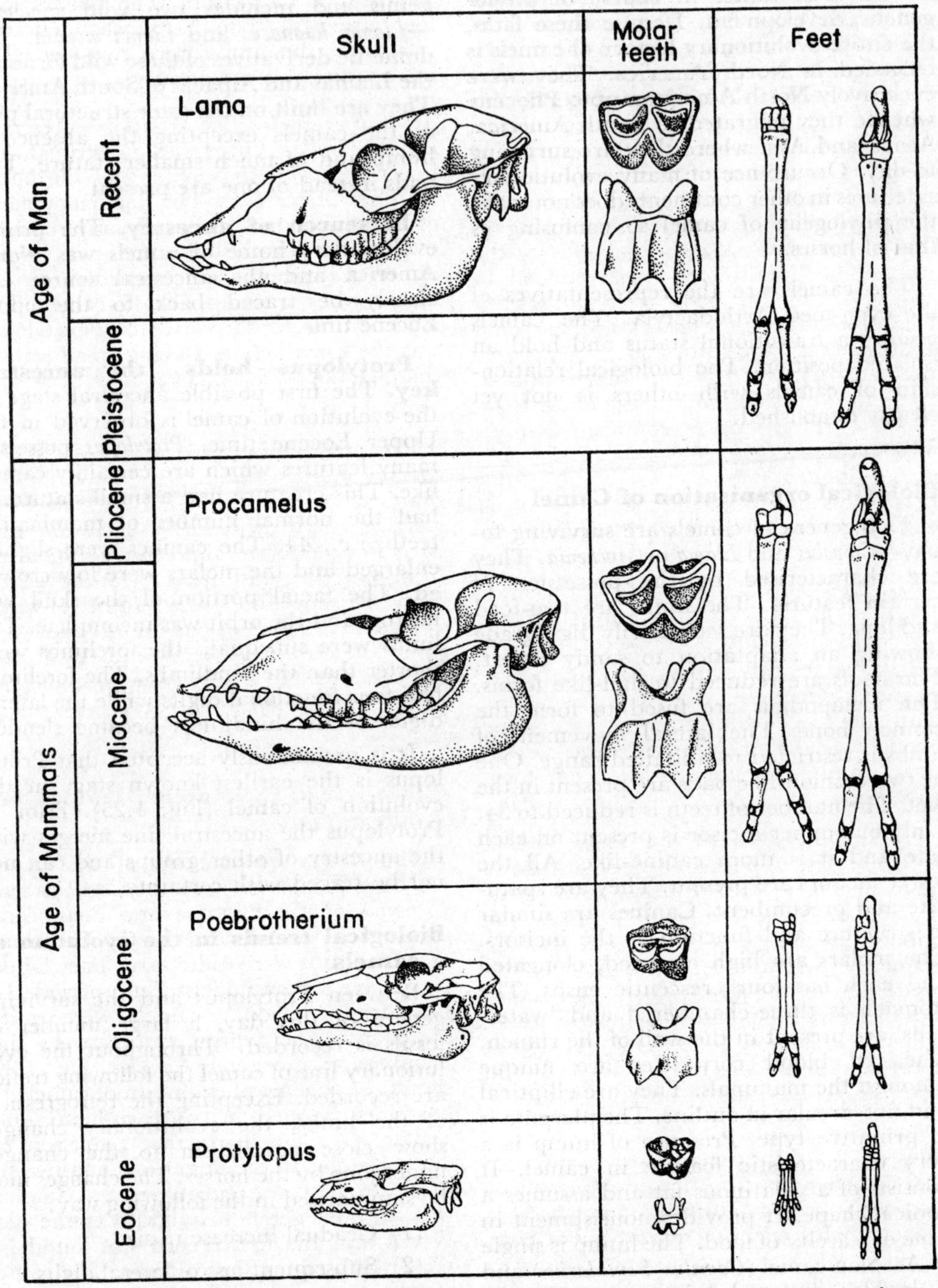

Fig. 1.25. Diagrammatic presentation of the evolution of skull, molars and limbs in the evolution of camel.

Oligocene—the influx of camel-like animals. During Oligocene period a number of fossil forms were recorded. Camel-like animals were quite abundant during Oligocene. They constituted a characteristic fauna of the time. **Poebrotherium—the middle Oligocene form.** The remains of the *Poebrotherium* were abundant in the middle Oligocene time. Poebrotherium attained the size of a sheep having longer limbs and neck. The facial portion of the skull was elongated and tapering. In the upper jaw, the grinding teeth were short-crowned and in the lower jaw they tend to become elongated. The number of teeth was 44. Digital reduction was observed in the limbs. The lateral toes were reduced to small vestiges. The hoofs were like that of deer. **Protomeryx—the upper Oligocene form.** Arising from Poebrotherium, *Protomeryx* directed the main evolutionary line of camel and continued up to the lower Miocene time. Protomeryx had all the features of Poebrotherium excepting the orbit which was completely encircled by bone. Protomeryx possessed full mammalian quota of teeth and the grinding teeth exhibit deepening tendency. Two digits were present and the hoofs were pointed.

Miocene—the period of divergence. Like that of horses, a number of well-defined divergent groups of camels was recorded. Of these divergent lines, the main line of evolution was directed through the lower Miocene form (*Protomeryx*) and the upper Miocene form (*Procamelus*).

Procamelus—directed the main line of evolution. The upper Miocene strata shows the fossil camel—the *Procamelus* showing for the first time the tendency towards the reduction of the number of teeth and the loss of the first and second incisors of the upper jaw in the adult stage. The metapodials showed the tendency towards fusion to form the cannon bone. The first sign of desert adaptation was quite evident by the presence of pads on the limbs.

Side-line evolution. *Gazelle-camels*. A large number of fossil remains of *Stenomylus* was recorded in the lower Miocene of Western Nebraska. The Oligocene ancestry of Stenomylus was not known but it continued as *Rakomylus* and became extinct in the lower Pliocene. Stenomylus was designated as the Gazelle-camel and had a very brief evolutionary career. Stenomylus had many anomalous characteristics, especially in dentition. The lower jaw had ten incisors like teeth. Six true incisors and the canines and first premolars assume the appearance of incisors. The molars were low crowned. The body was delicately built and the limbs were two-toed. The metapodials were separate.

Giraffe-camels. The group is represented by the upper Oligocene *Paratylopus*, the lower Miocene *Oxydactylus* and the upper Miocene *Alticamelus*. The particular name of the group does not imply any relationship with Giraffes. Paratylopus in the upper Oligocene was regarded as the ancestor of the group, Giraffe-camels. The lower Miocene derivative of Paratylopus was the Oxydactylus having small stature than its successor. It had shorter limbs and neck.

The metapodials were separate and had sharply pointed hoofs. It retained the full quota of mammalian teeth and the teeth were short crowned. The late Miocene and early Pliocene form, *Alticamelus*, originating from Oxydactylus became far advanced than its predecessors in a number of ways. Presence of cannon-bone and pads was some of the most remarkable features. Alticamelus had a very large size and the teeth were low crowned.

Pliocene and Pleistocene: Periods for true camels

The direct descendants of the upper Miocene *Procamelus* were the *Camelops* and *Eschatius* of the Pliocene and Pleistocene times. The fossil records of the Camelops and Eschatius were quite rare. They failed to survive beyond Pleistocene period. In Pleistocene period, camels were quite abundant. The typical camels were the members of the genus *Camelus*. Fig. 1.26 shows the scheme of evolution of camel. It is well developed in comparison to *Procamelus*. In Camelus further reduction of one premolar in the upper jaw and two in the lower jaw was encountered. Camelus was the first genus to migrate to the old World and the first record was found in the Siwalik formation of India. The cause of disappearance of North American camels is not known.

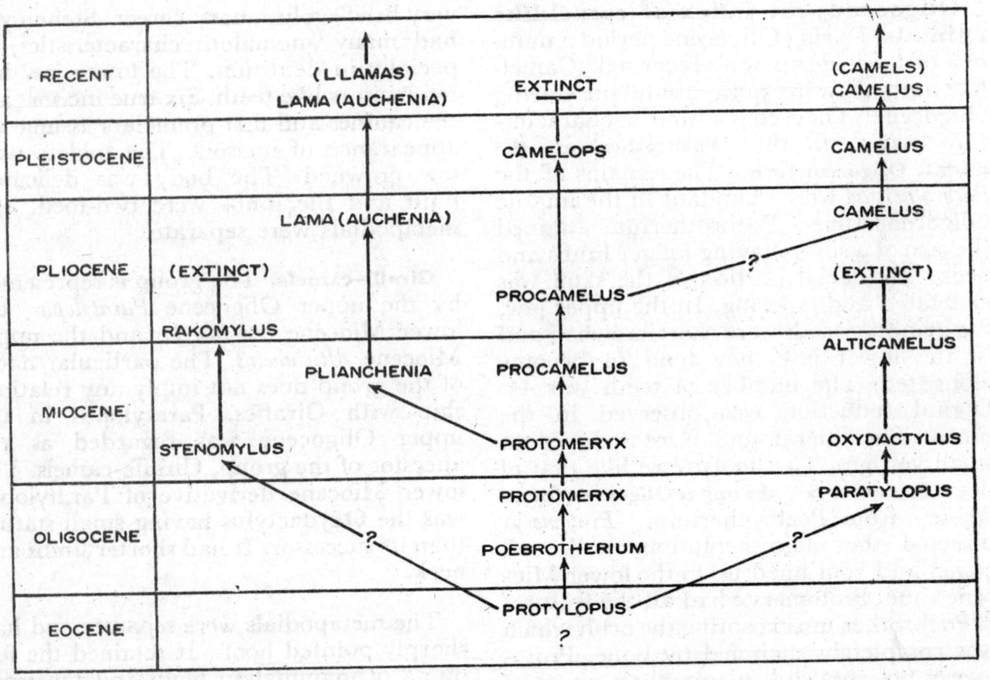

Fig. 1.26. Phylogenetic tree of camel (after Lull).

ORIGIN OF MAN

The origin and evolution of man, *Homo sapiens*, have been a topic of great biological interest since time immemorial. The idea that man is a creation of a supernatural power prevailed for long time in earlier days. But the Biologists view the origin of man using knowledge on morphology, physiology, embryology and fossil records. Man evolved from some unknown mammalian ancestor and reached the pinnacle of evolutionary fabric. Man is placed under the family Hominidae of the order Primate and differs from other anthropoid apes by having: *Large size of brain with greater functional ability* (*Maximum in Gorilla* =650 *c.c.*, *Minimum in Man*=1000 *c.c.*) *The brain case is larger than face region. The face is flatter with less protruding lower jaw. Continuous growth of long hair on head which are spare and short on body. Generalized hands with better developed thumbs and long leg with non-opposable big toe. Man is terrestrial in habit and walks erect on two feet.* They surpass all other animals by possessing the 'human features' which are exclusive for them.

Human features

In contrast to that of the anthropoids, the human line showed a large number of progressive features. The features are: (1) The face becomes flattened and is devoid of a nuzzle (Fig. 1.27). (2) The brow ridges gradually decline and disappear. (3) The cranium rises sufficiently above the orbits to house a larger brain. (4) The skull is rounded at the rear. (5) The foramen magnum and occipital condyles are shifted ventrally to join with the upright vertebral column. (6) A mastoid process arises in the ear region. (7) The teeth become smaller in size and are arranged in a U-shaped arc. The canines are moderate in size. (8) The arms with the fingers are proportionately shorter. The feet are nongrasping. The toes are placed in line. The heal bone is elongated to help insertion of muscles in upright posture and walking. (9) The vertebral column shows slight curvature. (10) The ilia are wider than length. Broader ilia help insertion of the big gluteal muscles which is involved in balance.

The great apes can make sounds which indicate some desires and emotions but fail to describe objects. But man can develop sounds into words symbolising things or ideas. The ape-prehuman transition is associated with the descent from trees to the ground which is of great significance in human evolution. This transition freed the hands for making and use of tools to supplement the action of the hands.

History of Mankind

Before the practice of burial of the dead, remains of early man were limited to member of skulls (often partial) and some other bony remains. Remains of complete skeleton became more numerous when the practice of burial of the dead was followed. The work of prehistoric man gave ample materials to draw an inference regarding the activities and manner of life. Limitations in material of early man make the direct line of descent more confusing. The time and place when modern man first originated are controversial. The earliest anthropoids, *Parapithecus*, *Propliopithecus*, etc., (represented by the remains of jaws) were discovered from the Oligocene bed of Egypt. During Miocene period the fossils of anthropoids showed considerable diversity, some possessing prehuman features may have evolved into human line and others leading toward the great apes. An anthropoid fossil, *Dryopithecus* is regarded to stand close to the point of divergence.

Primitive Hominids

Discoveries of remains of prehistoric species and races will give an idea of human evolution. The major forms, as recorded uptil date, are as follows (Fig. 1.28).

Australopithecus, Zinjanthropus, etc., represent the primitive Hominids

The remains of these hominids (*Australopithecus*, *Zinjanthropus*, etc.) were discovered in Mid-Pleistocene or earlier in Transvaal, South Africa in 1925 and in Olduvai Gorge Tanganyika in 1959. Many skulls and some skeletal parts have been discovered. The characteristics are: (1) The skull was smaller in size than that of modern man. (2) The volume of brain ranged from 600–700 c.c. (3) The face was protruding and the forehead was higher than that in apes. (4) The brow ridges were prominent. (5) The occipital condyles were ventrally placed and the rear part of the skull was rounded. (6) The jaws were large with small incisors, large and spatulate canines, large cheek teeth. (7) The ilia of pelvis were wider and the limb bones were slender. (8). The total height was about 5 feet. They used simple chipped pebble tools.

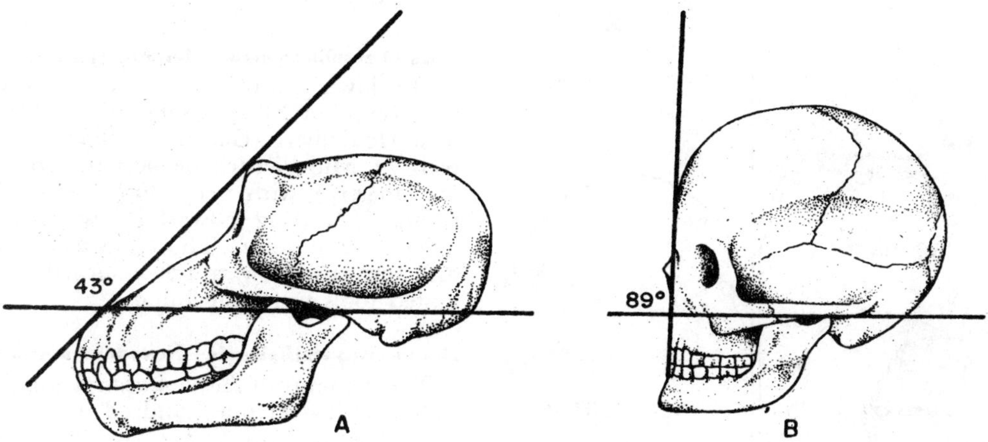

Fig. 1.27. Showing facial angle of chimpanzee (A) and man (B) (after Lull).

Pithecanthropus erectus—Java man

Fragmentary remains of *Pithecanthropus erectus* were discovered in Mid-Pleistocene of Solo River near Trimil, Java since 1891 up to 1945. The characteristic features are: (1) The skull was flattish-topped and projected behind. (2) The brow ridges were solid above the orbits. (3) The brain volume was 775–900 *c.c.* The imprint of brain possibly indicated the ability of speech. (4) The jaws were protruding. (5) The teeth were arranged in even curve but the canines were projecting. (6) The femur reflected its upright posture. (7) The height was about 5 feet. No associated tools were found.

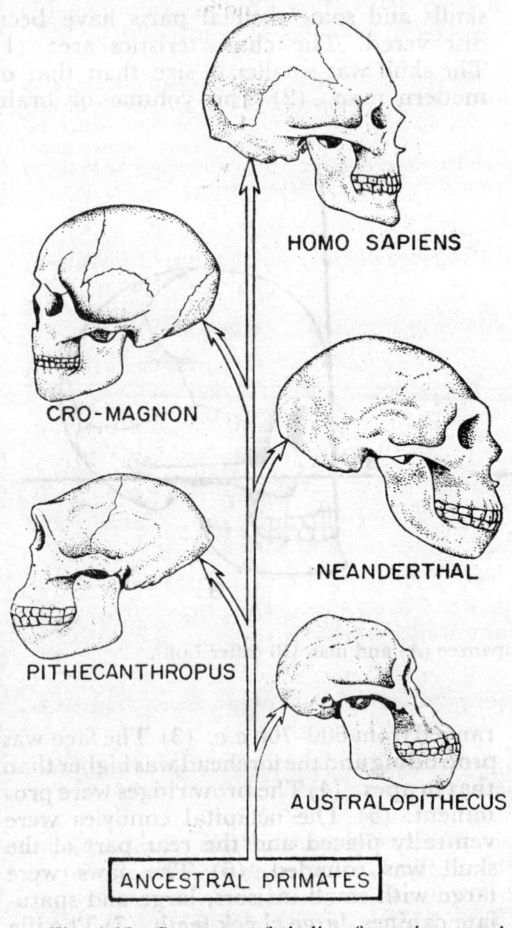

Fig. 1.28. Structure of skull of ancient and modern man. Drawings in some cases are reconstructed (after various sources).

Pithecanthropus (Sinanthropus) pekinensis —Peking man

The remains of skulls and parts, jaws with teeth and some limb bones of *Pithecanthropus* (*Sinanthropus*) *pekinensis* were discovered up to 1943 from the Mid-Pleistocene caves at Choukoutien (South-west of Peking), China. The noted features are: (1) The skull was small and the brain volume was 850–1300 *c.c.* (2) The skull was low-vaulted. (3) The brow ridges were stout. (4) The imprint of brain suggested the ability of speech.

Various implements of quartz and other rocks were discovered. The hearths showed the use of fire.

Homo habilis—Transitional man

The remains of this species were discovered in Pleistocene bed in East Africa. They were the makers of crudely chipped stone tools. They represent an intermediate stage between the *Australopithecus* and *Pithecanthropus erectus*. The mean capacity of brain was 680 *c.c.*

Homo heidelbergensis—Heidelberg man

One lower jaw of *Homo heidelbergensis* was discovered in 1907 in sand pit at Mauer near Heidelberg (Germany). The remains were of Mid-Pleistocene period. The jaw was massive with very broad ascending ramus indicating powerful jaw muscles. There was no chin. The teeth were stout and the canines were not enlarged. Associated tools were not found.

Homo neanderthalensis—Neanderthal man

The remains of *Homo neanderthalensis* totalling well over one hundred individuals were discovered from the late Pleistocene bed (before or during first Ice Age) in Spain and North Africa to Ethiopia, Mesopotamia, Southern Russia, Gilbraltar, Neanderthal Valley near Düsseldorf (Germany) from 1848–1861. The Neanderthal man had (1) massive long and flat-topped skull. (2) The forehead was receding. (3) The brow ridges were heavy. (4) The nose was broad. (5) The orbits were large. (6) The average brain volume was 1450 *c.c.* (7) The jaws were protruding but the chin was receding. (8) The teeth were large. (9) The attachment sites of occipital region of skull and the cervical vertebrae indicated the existence of powerful neck muscles. (10) The limb bones were heavy and slightly curved. (11) The height of males was about 5 feet 3–5 inches. The females were shorter than males.

The Neanderthal man used to live in caves and rock shelters with stone stools and weapons. There was evidence of use of fire. The estimated age was about 100,000 years.

Homo sapiens—Cro-Magnon man

The remains of Cro-Magnon man of estimated age about 30,000–13,000 B.C. were found in late Pleistocene (close of last Ice Age and later) bed of France to Czechoslovakia, East Africa and Eastern Asia. The distinguishing features are: (1) The skull was long and high with no brow ridges. (2) The face resembled the modern man. (3) The occipital region of skull was rounded. (4) The chin was well developed. (5) The average brain volume was about 1590 *c.c.* (6) The height of males was about 5 feet 10 inches.

They were cave-dweller. They had stone implements and they could make wall paintings and sculpture.

Miscellaneous Remains of man

(A) **Ternifine man:** Three jaws and skull fragment of Ternifine man were found in Ternifine and Casablanca, North Africa in 1952. These Mid-Pleistocene remains resembled Heidelberg and Peking materials.

(B) **Mount Carmel man:** Remains of 13 individuals including complete skeleton were discovered in Mount Carmel, Palestine (Israel). These upper Pleistocene remains showed characteristics of both Neanderthal and modern man, but slightly taller.

(C) **Swanscombe man:** The remains (occipital and parietal bones) were found in Swanscombe, Kent, England in 1936–1937. They were of Mid-Pleistocene age. The bones were thick and the brain volume was estimated to be about 1300 *c.c.*

(D) **Solo man (Homo soloensis):** The remains of eleven partial skulls and two femurs of Pleistocene age were discovered from Solo River near Ngandong, Java in 1933. They had low forehead and heavy brow ridges. They exhibited many features which were more modern.

(E) **Rhodesian man (Homo rhodesiensis):** The remains of Rhodesian man of late Pleistocene age were found in 1921 at Broken Hill, Rhodesia (South Africa). A similar skull was also discovered in 1953 in Capetown. The remains consisted of one skull, upper jaw, parts of limb bones, pelvis, sacrum, etc. The brain volume was about 1300 *c.c.* The characteristics of face, brow ridges, orbits, palate and limb bones were much like those of modern man.

(F) **Other remains:** The other fragmentary remains of man include that of: (i) skull fragments of *Pithecanthropus robustus* from Java (1938); (ii) portions of a huge jaw of *Meganthropus palaeojavanicus* from Java (1941) and (iii) three huge molars (five to six times the bulk of those of the present day's man) of *Gigantopithecus blacki* in 1935–1939. These molars were possibly collected from caves in South China.

[*Piltdown Man* (*Eoanthropus dawsoni*) first reported from Sussex, England in 1908 was subsequently proved to be a fake in 1954. The skull with jaw was stained to reflect an antique appearance. The jaw of a modern orang was filed skilfully to remove the simian characteristics.]

Biological trends in Human Evolution

The evolution of man involves the following significant changes:

(1) Switch over from the four gait apes to the bipedal gait of man.

(2) Perfection of hand for tool making.

(3) Increase of intelligence and size of brain.

(4) Change of diet from fruits, hard nuts, hard roots to softer foods.

(5) Increase in their ability to communicate with others and development of community behaviour.

General Consideration on Human Ancestry

Since the discovery of the 'missing link' between apes and men in 1894 by a Dutch anatomist, E. Dubois, a large number of fossils of man have been brought to the limelight. All the newer finds as well as the older ones are being interpreted by different authorities in different ways. The scientists of the past described the fossils in terms of 'individual types' rather than 'populations'. They gave a scientific name of their new find and placed it in a separate species and in a separate genus, whenever applicable. But the modern Anthropologists and Zoologists are trying hard to discard nearly all the names of 'genera' which were coined in the past. They recognise that the ancestors of man have progressed mainly along a single evolutionary line and at times this line became branched to give two or three related species (Fig. 1.29). During the past 600,000 years it consisted of a single species having a common gene pool with a number of races.

The remains of 'Southern apes' (*Australopithecus*) have been claimed to be forerunners of man. These creatures were more like apes than man in respect of their intelligence and way of life. They could walk erect and the architecture of limb and body skeleton was much like those of modern man. An intermediate fossil form, *Homo habilis*, an intermediate form between Australopithecus and the ancient species of man (Java and Peking man), was discovered from the same bed containing East African Australopithecus. This fact gave evidence that the Australopithecus was the direct ancestor of man and they persisted side by side with their derivatives—the earliest men. The transition from apes to man was a gradual process and the series of fossils portrays a gradual but complete transition from apes to modern man.

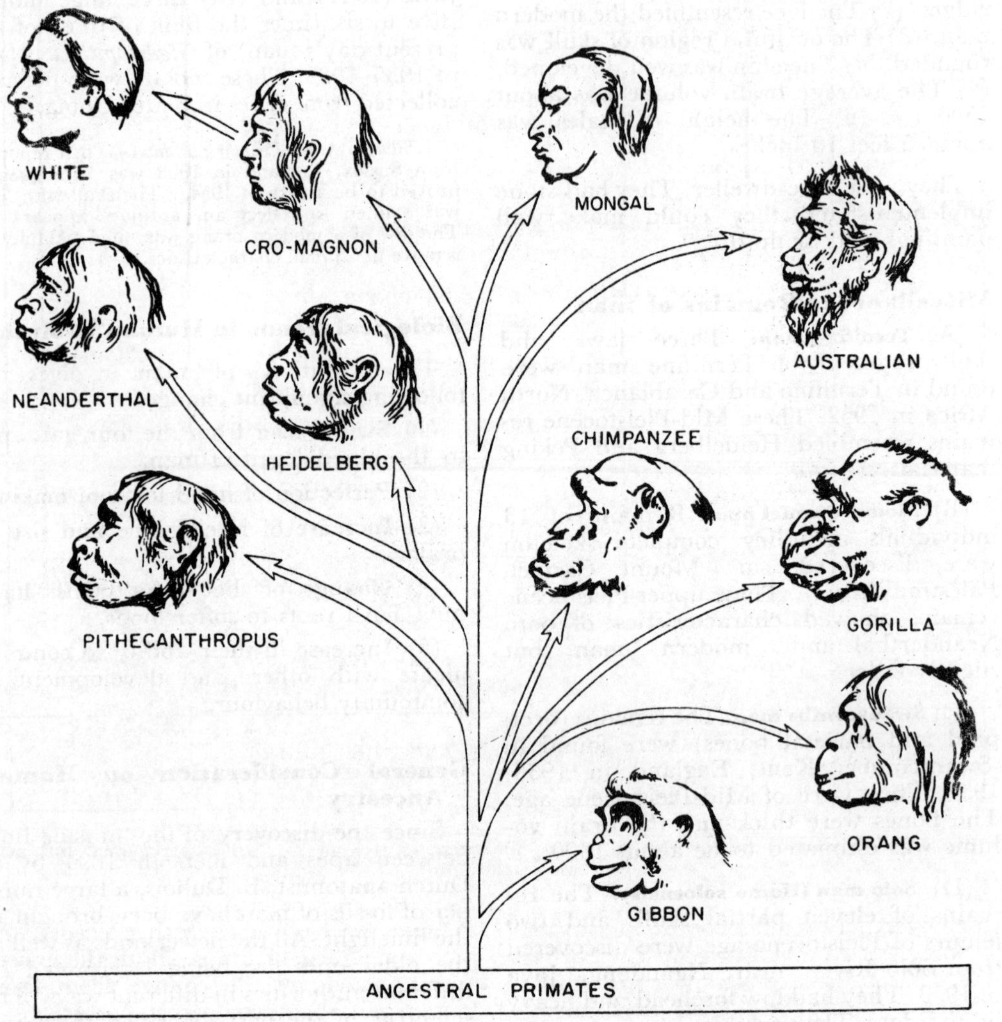

Fig. 1.29. Showing the phylogenetic tree of anthropoid apes, primitive and modern man. Figure shows only the appearance of heads.

Comparative studies on morphology and chemistry of protein have proved that *Homo sapiens*, gorilla and chimpanzee are closely related to each other than other anthropoid apes like orang utan and gibbon. *Homo sapiens*, gorilla and chimpanzee have possibly evolved from a group of apes common in Eurasia and Africa during Miocene. The immediate ancestor of *Homo*, as stated earlier, was the Australopithecus stages without splitting into separate species (Fig. 1.30). The main characteristics which differentiate man from apes evolved at different rates. The use of tools appears

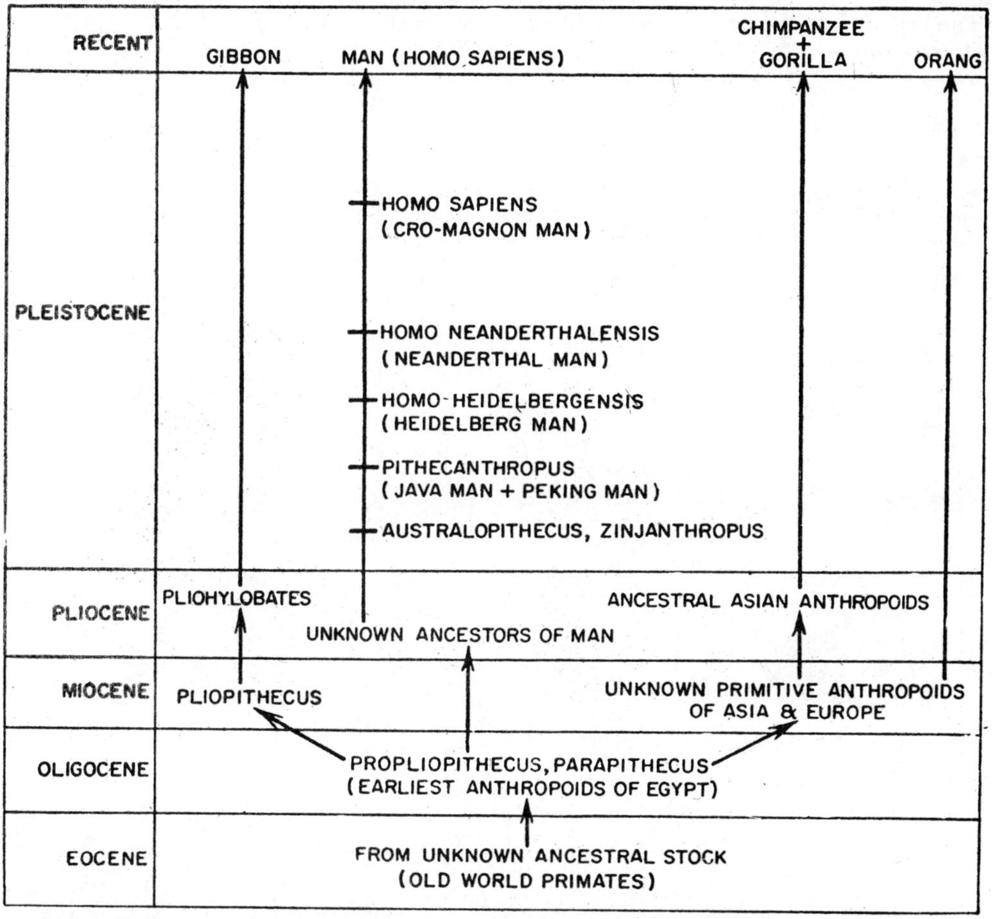

Fig. 1.30. Schematic representation of the phylogenetic tree of the anthroporid apes and man (based on Osborn).

which lived between Pliocene to Pleistocene in North Africa and Eurasia. The earliest man, *Pithecanthropus erectus*, was widespread in Eurasia during Pleistocene possibly evolved into modern man by series of gradual to have evolved first which preceded the increase of size of brain. Both these were accompanied by the change from four-footed gait to bipedal erect posture.

PART TWO

GENETICS

Genetics as a discipline of science was born at the beginning of twentieth century. It is reaching at the helm of Biological Sciences at a stupendous rate by its own right. The genes, physical and chemical basis of heredity, are known. The concerted efforts of Biologists together with Chemists and Physicists are gaining knowledge at a very quick rate. This part deals with fundamental information about the science of Genetics. The knowledge about Genetics may be gathered by the students by further reading.

GENETICS—SCIENCE OF HEREDITY

The twentieth century has dawned with a clarification of the basic principles underlying the production and transmission of characteristics or traits in successive generations of living organisms It is a matter of common and elementary observation that no two individuals, even though they belong to the same species, are exactly similar.

Members of a species show both *similarities* and *differences* amongst themselves. These differences are *variations* which are fundamental characteristics of living beings. Both the similarities and differences are correlated in terms of *biological relationship* amongst the individuals. Biological relationship is only part of the cause of similarities and differences between individuals. The other part of the individuality is formed by the interaction of the hereditary materials, present in individuals and environmental influences. The branch of science which embodies studies of biological make-up, environment and also the analysis of the laws and principles that govern transmission of characters is called *Genetics*.

HISTORICAL BACKGROUND

Early observations on animals and plants advocated the role of natural laws in heredity and variation. But the investigators failed to unravel the basic principles. An adequate body of scientific explanation was, however, available in the beginning of the twentieth century when many investigators furnished evidences to explain the mechanism of heredity and variation by applying effective scientific methods.

Although early animal and plant breeders were not aware of the principles of genetics, they used to practice selection and hybridisation technique to improve the varieties. Chinese farmers, many years prior to the Christian era, recorded how to improve the varieties of rice by applying artificial selection technique. Hippocrates, Aristotle and many other Greek Philosophers speculated genetic principles which are, however, vague and erroneous. Pliny (79–23 B.C.), Gesner (1516-1565) and many Hellenistic and Medieval writers supposed giraffe to be a hybrid between camel and leopard. An ostrich was imagined to appear when a camel mated with a sparrow. Hybridisation between palm trees and acacia resulted in the production of banana trees.

GREGOR MENDEL AND HIS EXPERIMENTS—MENDELISM

Gregor Mendel is regarded as the 'father of genetics'. His experiments with garden peas (*Pisum sativum*) were elegant and the inferences together with his interpretations constitute the foundation of modern genetics.

The present-day knowledge and researches on heredity rest directly on a number of principles and laws advocated by Mendel. Since Mendel, many scientists have added to and extended these principles but it was he who set forth in a clear manner a logical explanation about the ways in which hereditary factors (genes) behave during a cross.

Johann Mendel was born in 1822 of German parentage in Heinzendorf, a village in Silesia. At the age of twenty-three, he joined the Augustanian Monastery of Brunn, Austria (now Brunn, Czechoslovakia) where he spent the remainder of his life (1884). After his entry into the monastery, he chose the name 'Gregor' and was thenceforth known as Gregor Johann Mendel.

Mendel conducted his hybridisation experiments in the garden behind the monastery. He had a scientific turn of mind and with painstaking and methodical researches he built up a tremendous amount of data which forms the very basis of all studies on heredity.

The classic paper of Mendel was written in 1865 and was published in the following year in an obscure journal, "Transactions of the Natural History Society" of Brunn. Here the investigation of Mendel remained buried till 1900 when the great contribution of Mendel was brought to the limelight by three Botanists of different nationalities.

Rediscovery of Mendel's Work. The men who brought to light the work of Mendel were de Vries of Holland, Tschermark of Austria and Correns of Germany. Their rediscovery of Mendel's contribution was published separately only a few months apart and was followed by volumes of papers extending the application of Mendelian principles to both animals and plants.

Mendel's technique of artificial hybridisation. It is essential to know first the technique of Mendel for artificial hybridisation and its difference with normal self-fertilisation. The success of Mendel depends largely on the selection of garden pea as the experimental material. Mendel chose the material because it is an annual plant with well-defined characteristics. Pea plants can be grown and crossed easily. The flowers of this plant have both male and female parts and self-fertilisation occurs ordinarily. Cross-pollination, i.e., transference of pollen from one plant to the stigma of another plant, does not occur in natural state to any great extent. It can be done by an experimenter.

Mendel chose seven pairs of contrasting traits in his experimental studies. They are: (*i*) vines are either tall or dwarf; (*ii*) unripe pods are green or yellow; (*iii*) unripe pods are inflated or constricted between the seeds; (*iv*) flowers are either axial or terminal; (*v*) nutritive portions of the ripe seeds are either green or yellow; (*vi*) seed coats are either smooth or wrinkled and (*vii*) seed coats are white or gray.

Mendel did hybridisation experiments with many plants like beans, maize, snapdragon, etc. Of these, the experiments with garden peas are most fascinating. Peas are self-fertilising plants. Self-fertilisation occurs when two gametes which unite to produce a zygote that develops into a seed and subsequently into the adult plant of the next generation, are derived from the pollen and ovule of the same flower. For artificial crossing of normally self-fertilising flowers of peas, Mendel carefully removed the stamens from one flower while the pollens are still immature and later he transferred to these depollenised flowers, ripe pollen from the flower of another plant of his choice.

The seeds produced by the cross-pollinated flowers were planted and the resultant offsprings (Hybrids) were carefully examined. Mendel then crossed the flowers amongst themselves to obtain another generation. From this second generation of hybrids Mendel formulated the fundamental principles of heredity.

In garden peas Mendel observed seven pairs of contrasting characters. These included smooth seeds and wrinkled seeds, yellow cotyledons and green cotyledons, inflated pods and constricted pods, yellow pods when unripe and green pods when ripe, flowers in the axils of the leaves and flowers at the end of stems, transparent seed coats and brown seed coats, tall plants and dwarf plants.

Mendel's Experiments—Monohybrid Cross

Experiment I. Mendel noted that among the pea-plants some were tall attaining a height of six feet or more while others, though grown in the same soil, were dwarf reaching a height of eighteen inches only. He made a cross between a tall plant from a tall race and a dwarf plant from a dwarf race.

Results. The resulting plants (progeny) of first filial (F_1) generation were all tall.

Experiment II. Mendel crossed two of these F_1 plants together and raised as many plants as possible.

Results. The resulting plants or progeny of second filial (F_2) generation were both *tall* and *dwarf*. There were three times as many tall plants as dwarf ones.

Experiment III. Mendel then self-pollinated the F_2 plants.

Results. (*i*) Dwarf plants when self-pollinated produced only dwarf progeny.

(*ii*) Of the tall plants in the F_2, one-third bred true and two-thirds produced both tall and dwarf in the ratio of 3 : 1 as the F_1 plants did.

The most striking points from the above experiments were (*a*) only one of the two characters concerned in the first experiment showed up in F_1. All plants were tall and not dwarf.

The dwarf character, though disappeared in F_1, was not lost and made its appearance again in F_2.

The logical conclusion arrived at by Mendel from these points was that F_1 plants must have carried a 'hidden factor' for dwarfness. Thus parents: Tall × Dwarf.

F_1 — Tall (Dwarf)

The second conclusion arrived at by Mendel was that if the F_1 plants contain two 'factors' for height (one responsible for tallness which showed up and one for dwarfness which remained hidden) any plant of any generation must possess two 'factors' for height. A tall plant from a tall race contains two factors for tallness and similarly a dwarf plant has two factors for dwarfness. Thus Mendelian experiments could be visualised in the following way:

Pure tall × Pure dwarf
F_1—All Tall
F_1—All Tall × All Tall
F_2—Three tall : One Dwarf

Of the three tall plants, one breeds true that is the plants on being allowed to self-pollinate produce tall plants. Other two on being allowed to self-pollinate or on making a cross between themselves produce tall and dwarf plants in the ratio of 3 : 1. The dwarf plant of the F_2 generation breeds true (Fig.2.1). Fig.2.2 relates the monohybrid cross between black and brown coloured guinea-pigs.

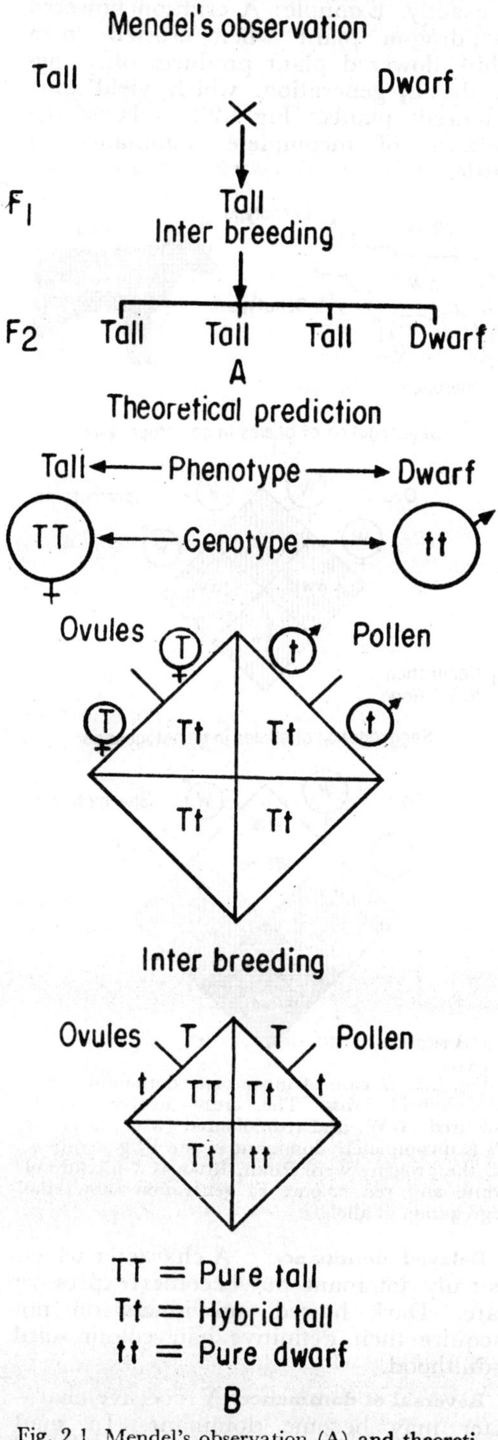

Fig. 2.1. Mendel's observation (A) and theoretical prediction (B) in Monohybrid cross.

Mendel's laws

1. **Law of dominance and recessive.** Mendel designated *dominant*, the character which appears in F_1 masking the effect of the factor for the other character, the character which does not appear is called *recessive*. Tallness is due to a dominant factor (T) and dwarfness is due to a recessive factor (t). In a hybrid, as long as the dominant factor is present, the recessive factor does not produce any obvious effect.

II. **Factors and their segregation.** An organism is the result of the union of the female and male gametes in fertilisation. The gametes contain 'something' which is responsible for hereditary characters—in Mendel's experiment height of pea plants. This something is the hereditary 'factor'. If an organism is true breeding for a character (Tall or Dwarf) the gametes of the parents contain the same hereditary factor. The hybrid receives a (T) factor from one parent and a (t) factor from the other. Among the offsprings of the true hybrid Tt again appears. The Tt hybrid does not form gametes in which T and t are mixed but produces gametes containing either T or t factor; the gametes of the hybrid are pure for one or other gene. In essence, it may be said that (1) there are factors which affect development and these factors retain their individuality from generation to generation without being contaminated when they are present in a hybrid and (2) the factors become sorted out from one another when the gametes are formed.

Types of Dominance. When Mendel crossed a true breeding tall plant with a true breeding dwarf plant, the progeny became all tall. Mendel called such traits for tallness of plants as *dominant* and such traits for dwarfness of plants as *recessive*.

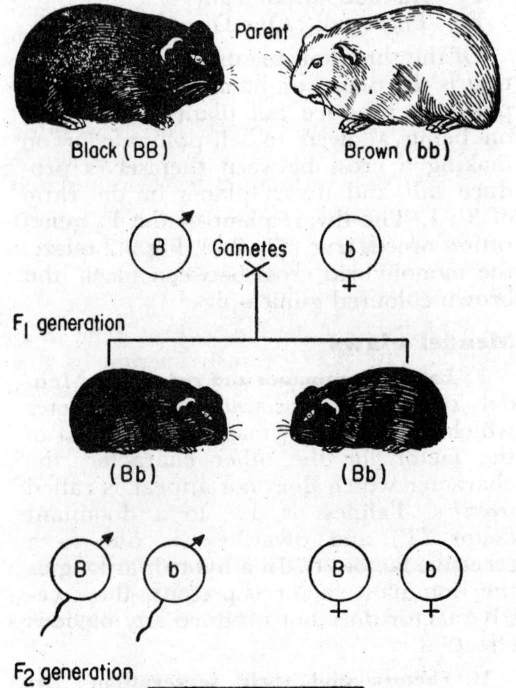

Fig. 2.2. Showing a Monohybrid cross between black and brown coated guinea-pigs.

This is a case of *complete dominance*. Many other characters show such property. But the dominance is never assured. There are different kinds and degrees of dominance. They may be classified as follows:

Incomplete dominance. When the hybrid resembles one parent much closely than the other but may not resemble it exactly. Example: A crimson flowered snapdragon plant when crossed to a white flowered plant produces offsprings in the F_1 generation, which yield pink coloured plants. Fig. 2.3 relates the instance of incomplete dominance in cattle.

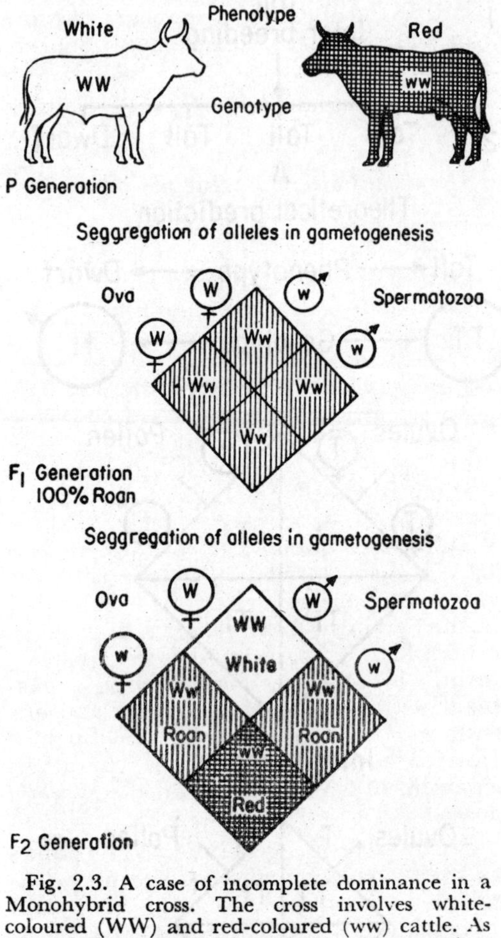

Fig. 2.3. A case of incomplete dominance in a Monohybrid cross. The cross involves white-coloured (WW) and red-coloured (ww) cattle. As W is incompletely dominant in the F_1 generation, all the progeny were Roan. Roan is a mixture of white and red colour. F_2 generation shows the segregation of alleles.

Delayed dominance. A character which is truly dominant but becomes expressive late. Dark haired individuals do not acquire their definitive hair colour until adulthood.

Reversal of dominance. A recessive character may become dominant. In snail (*Helix*) the colour of red shell is dominant over the colour of yellow shell. Sometimes an exception occurs and during crosses snails with yellow shells dominate over red shells.

Conditioned dominance. Morgan described a case in *Drosophila* where an abnormally banded abdomen becomes expressive when the animals are provided with fresh food and moisture. When food becomes dry and there is less moisture the bands in the abdomen disappear. This is a type of dominance conditioned upon environmental factors.

Some dominant and recessive traits in human are:

Dominant traits	*Recessive traits*
Brown eyes	Blue eyes
Long eye lashes	Short eye lashes
Dark hair	Blond hair
Curly hair	Straight hair
Large ears	Small ears
Free ear lobes	Attached ear lobes
Normal vision	Colourblind vision
Six fingers	Five fingers
Normal skin colour	Albino
Freckles	Lack of freckles
Dimpled cheek	Lack of dimple

Dihybrid cross and Mendel's law III. In order to find out how different characters behave in relation to each other in their sojourn from generation to generation, Mendel made another set of experiment. He made a cross between a pea-plant with *round* and *yellow* seeds and one with *wrinkled* and *green* seeds. Such a cross involving two different characters separable in inheritance is called a *Dihybrid cross*.

Results. In F_1 generation, all plants had yellow and round seeds because *yellow* colour is dominant over the *green* and *round* texture is dominant over *wrinkled*.

F_1 plants were crossed among themselves.

Parents: Yellow and Round × Green and Wrinkled.

F_1—All are Yellow and Round.

Parents: All Yellow and Round × All Yellow and Round.

F_2—(1) Yellow and Round—Like grand parents.

(2) Green and Round—New combination.

(3) Yellow and Wrinkled — New combination.

(4) Green and Wrinkled like grand parents. The relative proportion of the types produced are 9 : 3 : 3 : 1 respectively. Figs. 2.4 and 2.5 show the Dihybrid cross in pea plants and guinea-pigs respectively.

Summary of Mendelian Principles. From the various crosses made by Mendel the following principles were formulated.

1. *Principle of unit character*. The inheritance of various traits by an organism is controlled by factors (genes) and these factors occur in pairs.

2. *Principle of dominance*. One gene in an allelic pair may be instrumental in suppressing the expression of the other member of the pair.

3. *Principle of segregation*. In the formation of haploid reproductive cells as gametes or spores, only one gene of each allelic pair is received by the gamete or spore.

4. *Principle of independent assortment*. Hereditary characters are independent units which segregate upon crossing regardless of temporary dominance.

In the F_2 generation, four types of seeds were obtained. They were round yellow seeds, round green seeds, wrinkled yellow seeds and wrinkled green seeds and the approximate ratio of the four types were 9:3:3:1 respectively. Of these four types of individuals two were like the grandparents (yellow round and wrinkled green) while two were entirely new combinations (round green and wrinkled yellow).

Thus, it appears that the seed colour character is independent of seed shape character and they are never tied. The formation of new combinations (round green and wrinkled yellow) shows that the yellow colour can dissociate itself from round shape and *vice versa*. Similarly, green colour can dissociate itself from wrinkled shape and *vice versa*. This property of dissociation leads to the formation of new combination and it is obvious from the above mentioned behaviour of characters that segregation of seed colour is independent of seed shape.

From this experiment (Dihybrid cross), Mendel postulated his third law which states that hereditary characters are usually independent units which segregate

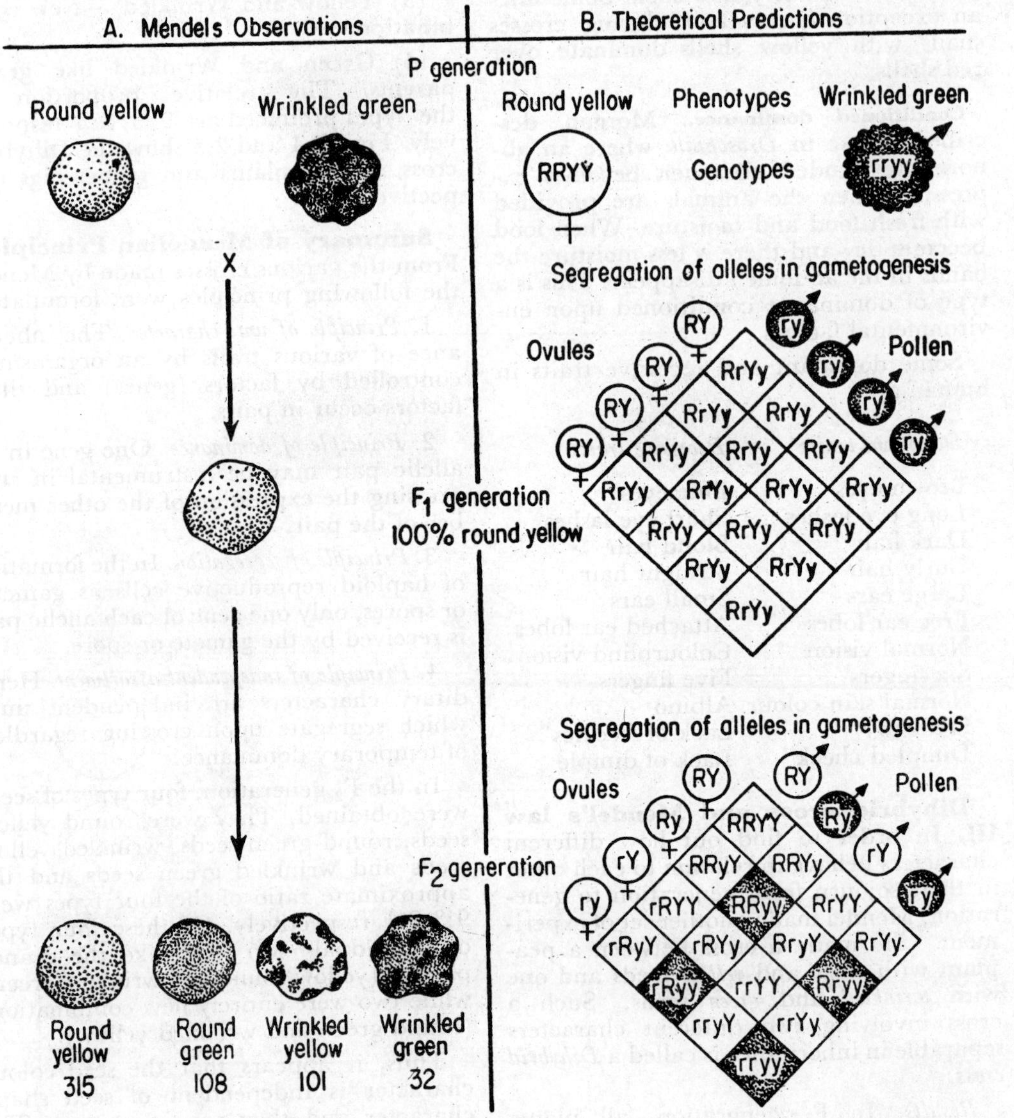

Fig. 2.4. A Dihybrid cross experiment of Mendel. Figures at the bottom indicate the actual number of seeds obtained by Mendel in his experiment. B. A theoretical prediction from the same cross.

upon crossing regardless of temporary dominance. This is known as the law of *Independent Assortment*.

Trihybrid cross

Crosses between parents that differ in three contrasting characters, i.e., combinations of three monohybrid crosses operating together are regarded as the trihybrid crosses. When a cross is made between a seed parent having a tall vine and yellow, round (DDGGWW) with a pollen parent with a dwarf vine and green, wrinkled seed (ddggww), the first generation (F_1) of the cross is represented as follows:

Parents: DDGGWW × ddggww

Gametes: DGW × dgw

F_1 Generation: DdGgWw

(Tall vine and yellow, round seed)

When a F_1 plant is crossed with a full recessive type (DdGgWw × ddggww) eight varieties of gametes (DGW, DGw.

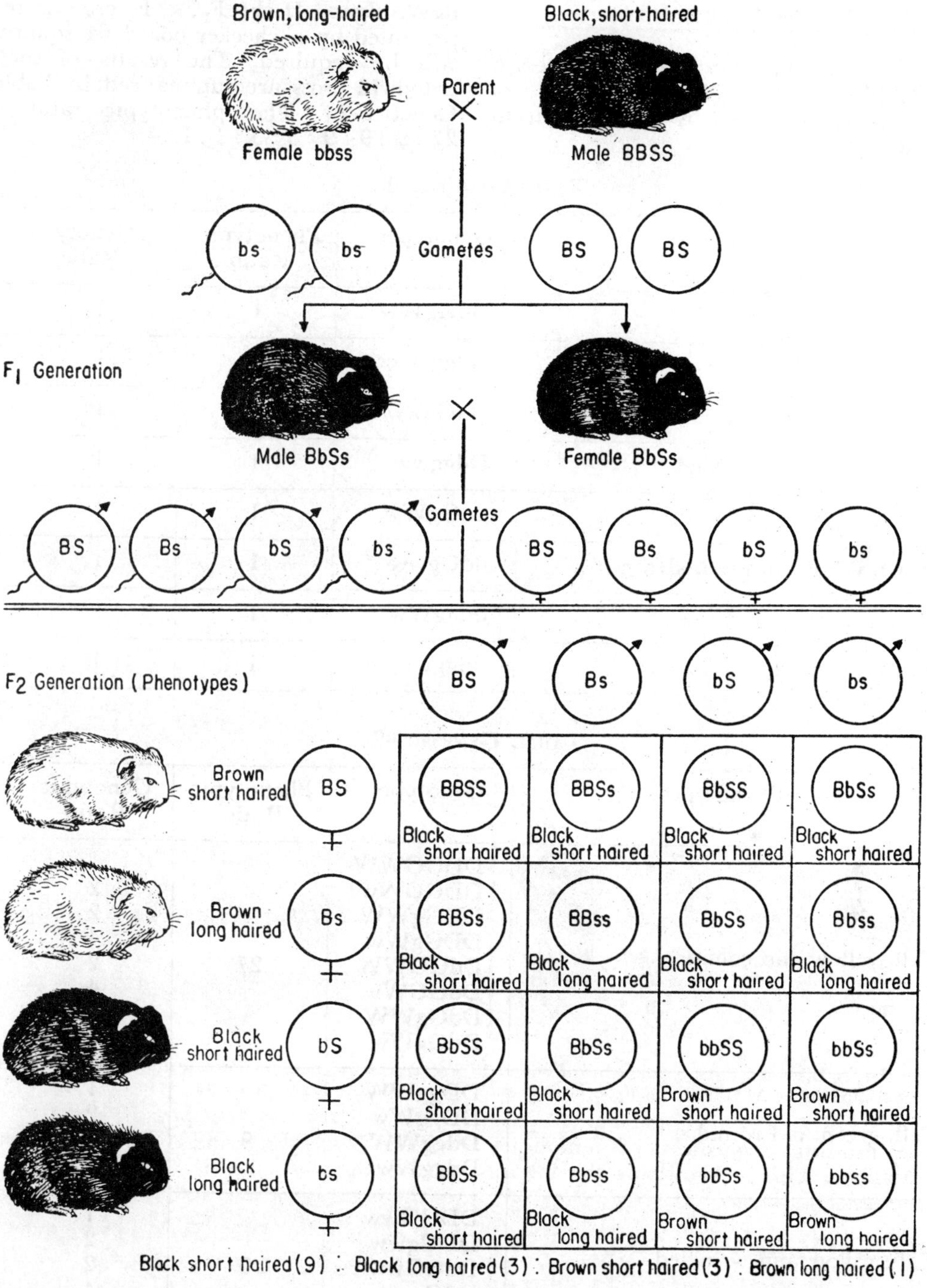

Fig. 2.5. Showing a Dihybrid cross between brown and long-haired guinea-pigs with black and short-haired one. Note the occurrence of Independent assortment.

DgW, Dgw, dGW, dGw, dgW, dgw) will be produced by the F_1 parent while one kind of gamete (dgw) will be produced by the full recessive parent.

The results have been summarised in the Table—Genetics-1.

When the F_1 plants are self crossed (DdGgWw × DdGgWw) eight kinds of gametes are produced by both the parent plants. The gametes are: DGW, DGw, DgW, Dgw, dGW, dGw, dgW, dgw. If the $F_1 \times F_1$ cross is represented by a checker-board 64 squares will be required. The results of such trihybrid cross are summarised in Table Genetics-2. The phenotypic ratio is 27 : 9 : 9 : 9 : 3 : 3 : 3 : 1.

TABLE GENETICS-1

Phenotypes	Genotypes	Phenotypic Ratio	Genotypic Ratio
Tall, yellow and round	DdGgWw	1	1
Tall, green and round	DdggWw	1	1
Tall, yellow and wrinkled	DdGgww	1	1
Tall, green and wrinkled	Ddggww	1	1
Dwarf, yellow and round	ddGgWw	1	1
Dwarf, yellow and wrinkled	ddGgww	1	1
Dwarf, green and round ...	ddggWw	1	1
Dwarf, green and wrinkled	ddggww	1	1

TABLE GENETICS-2

Phenotypes	Genotypes	Phenotypic Ratio	Genotypic Ratio
Tall, yellow and round	DDGGWW DDGGWw DDGgWW DDGgWw DdGGWW DdGGWw DdGgWW DdGgWw	27	1 2 2 4 2 4 4 8
Tall, green and round	DDggWW DDggWw DdggWW DdggWw	9	1 2 2 4
Tall, yellow and wrinkled	DDGGww DDGgww DdGGww DdGgww	9	1 2 2 4
Dwarf, yellow and round	ddGGWW ddGGWw ddGgWW ddGgWw	9	1 2 2 4

TABLE GENETICS–2 (*Contd.*)

Phenotypes	Genotypes	Phenotypic Ratio	Genotypic Ratio
Tall, green and wrinkled	DDggww Ddggww	3	1 2
Dwarf, yellow and wrinkled	ddGGww ddGgww	3	1 2
Dwarf, green and round	ddggWW ddggWw	3	1 2
Dwarf, green and wrinkled	ddggww	1	1

Physical basis of Heredity

Mendel's discovery failed to impress scientists at the time when it was published. For nearly thirty five years his findings remained unknown. The biologists of those ages, as a result of tremendous impact of Darwin's publication of *On the Origin of Species by means of Natural Selection*, were much interested in problems relating to differences between the species and not on differences within the species. Moreover, in those days quantitative methods were unheard of in most areas of Biology. Obviously, a paper on plant hybridisation crowded with numerical ratios and letter symbols were brushed aside as irrelevant.

But re-discovery of the Mendelism became an imperative necessity. Interest on problems of evolution was tremendous. Serious doubts have been cast upon the validity of Lamarckian idea (accepted by Darwin) that variations in an organism are the resultant outcome of impacts between organism and environment. Attention was focussed on variations which are not traceable to environmental differences. Particular interest upon *discontinuous variation* or more precisely on the contrasting unit character studied by Mendel reached its zenith. And once this was undertaken, the rediscovery of Mendelian work became inevitable.

In essence, Mendel's laws are descriptions of what happen when various types of crosses are made. The next question that follows is how or why it happens. The answer to this question is to be found in the behaviour of chromosomes. Chromosomes are thread-like structures found in the nuclei of cells. These are the *threads of life* and form the basis for segregation and other phenomena of heredity. Unfortunately, Mendel knew nothing of the chromosomes and even their existence as the intimate knowledge of cell-structure had to await certain critical technical improvements which were accomplished in the beginning of twentieth century.

The chromosomes. During cell division, chromatin masses present inside the nucleus become hinged together and form definite bodies or structures called chromosomes. Chromosomes are universally present in the nuclei of living organisms. The number of chromosomes present in a given species is constant and they are present in *like pairs* excepting the *sex-chromosomes*. In number, the chromosomes vary in different species. There is a single pair of chromosomes in *Ascaris univalens*. In the fruitfly, *Drosophila*, there are four pairs while in man there are 23 pairs. In many organisms, the number of chromosomes are many and are counted with considerable difficulty. The number of chromosomes of a species is never an index to the complexity or degree of differentiation of the species concerned. Species which are apparently related often differ widely with respect to their number of chromosomes. The chromosomes are capable of maintaining their individuality. This individuality remains unaffected in different physiological conditions such as growth, metabolism and reaction to stimulus. The *genes*, located in the chromosomes, constitute the physical basis upon which heredity hinges. (For Architecture, behaviour of chromosomes during meiosis see any standard book of Cytology).

Salivary gland chromosomes. Most amazing of all chromosomes are the salivary gland chromosomes found in the salivary glands and other larval tissues of two-winged flies (Order Diptera). The chromosomes were first noticed in 1881 and have become an important tool to cytologists and cytogeneticists in the last twenty years.

The salivary gland chromosomes are of great size. They are 70 to 110 times as long as the chromosomes in some oogonial cell. The largest chromosome may reach a length of half of a millimeter. The chromosomes generally lie well separated in the nucleus. The chromosomes are provided with conspicuous transverse bands. These bands can be recognised even in living and unstained nucleus. 2000 such bands may be counted in a long chromosome. The bands form a pattern that is constant for a given chromosome.

The salivary gland chromosomes furnish valuable data regarding the structure and composition of chromosomes. It has formed a sort of biological spectrum indicating the organism's genetical constitution (Fig. 2.6).

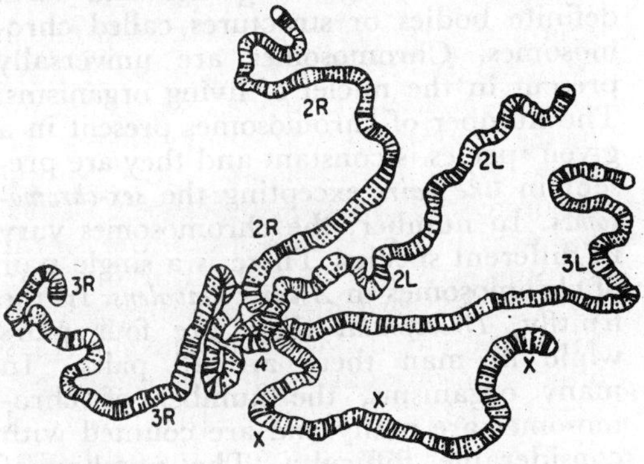

Fig. 2.6. Salivary gland chromosomes of *Drosophila*.

Lampbrush chromosomes. During oogenesis in some animals like shark, amphibia, bird, etc., the diplotene chromosomes become large and are comparable to salivary gland chromosomes of Drosophila. In amphibia, the pachytene stage may prolong for two years and the chromosomes during this long period exhibit peculiar change. The paired chromatids lie side by side and show series of swellings along their length. Sometimes the chromosomes form loops and give off filamentous bodies. The exact nature of this peculiar behaviour of chromosome is still inadequately known.

Chromosome number in some animals and plants are given in Table Genetics–3:

TABLE GENETICS–3

Species	Diploid (2n)	Haploid (n)
Ascaris	2	1
Drosophila	8	4
Trillium	10	5
Pea	14	7
Maize	20	10
Mouse	40	20
Man	46	23
Tobacco	48	24
Tomato	24	12
Gypsy moth	62	31
Rabbit	44	22
Cow	60	30
Horse	64	32
Ass	62	31
Dog	78	39
Monkey	42	21
Gorilla	48	24
Chimpanzee	48	24
Mule	63 (sterile)	?
Protozoa (*Aulocanthe*)	1600 approx.	800 approx.

The gene hypothesis. The ultimate but hypothetical hereditary unit is the *gene*. Castle has defined a gene as the smallest part of chromosome capable of varying by itself. The genes are domiciled on the chromosome in a linear order. They are the beads on the chromosome necklace.

Genes are ultra-microscopic. Still today no optical instrument has been able to bring them within the range of human sense organs. They are made up of complex molecules. The molecules of different genes vary in chemical composition.

Genes are present in duplicate pairs, one member of which comes from the father while the other from the mother. They are present in all the chromosomes of every cell from zygote onwards. Each chromosome houses many genes. Each gene is located in the same particular chromosome and at a definite region of the chromosome. The specific arrangement of genes on chromosome is never thrown at random. Genes are like chemical catalysts and never die in action.

Contrary to atoms and molecules, the genes *grow* by assimilation, *reproduce* their own kind and *mutate* or change by loss, addition and rearrangement. The genes, however, possess a high degree of stability.

A gene can not survive without partner. Genes are not considered today as 'unit determiners' for single trait,in the Mendelian way. They act in conjunction with other genes upon organism as a whole.

True nature of genes is still to be known completely. The genes when remain together in a hybrid do not blend, contaminate or interact with each other. On the other hand, the different genes segregate, separate pure and uncontaminated and transmit to the different gametes formed by the hybrid and go to different individuals in the offsprings of the hybrid. Genes are symbolised in letters. Dominant genes are represented by capital letters (as 'T' in tallness) and recessive genes are represented by small letters (as 't' in dwarfness). This is a convenient method to follow up the transmission and distribution of genes in cross experiments. Thus T and t are allelic genes or *alleles*.

An individual arises from the union of two gametes as in Mendel's experiments. It receives a gene responsible for tallness and a gene responsible for dwarfness. The true breeding tall plant is represented thus by TT and similarly the true dwarf plant is represented by tt. When the two plants are crossed one T bearing gamete, be it male or female, is fertilised by one t bearing gamete. Thus the resultant zygote gets both T and t and its *genetic formula* is written as Tt.

Theory of the genes. T. H. Morgan in his "theory of the genes" has summarised the following:

(1) Characters of an individual are referable to paired genes in the germinal material. They are held together in a definite number of linkage group.

(2) During maturation of gametes each pair of genes separates according to Mendel's first law and as a result each gamete receives only one set.

(3) Genes in different linkage group assort independently according to Mendel's second law.

(4) An orderly exchange of genes occurs between homologous chromosomes by the process of crossing over.

(5) The frequency of crossing over furnishes evidences for the linear arrangement of genes in each linkage group and their relative position to one another.

Ploidy

The chromosomes are basically present in diploid (2n) condition in the cells. The number become reduced to half or haploid (n) during gamete formation. The basic number of chromosomes may be changed in both plants and animals. This alteration in the basic number of chromosome may be found in the following forms:

Polyploid. When the basic number of chromosomes get multiplied, *viz.*, Triploid (3n) Tetraploid (4n), the condition is called polyploidy. When polyploidy arises directly through the increase of lower number—it is called *autopolyploidy*. When the increase in chromosome number is due to crossing with other species it is called *amphidiploidy*.

Monoploid. The haploid chromosomal number (n) which is the lowest chromosome number or the primitive number in a group of animals or plants is called the *monoploid* or *haploid*. This basic number is retained in many plants like corn and insects like bees.

Aneuploid. Instead of showing alteration in the complete set of chromosomes, some time individuals occur with a single chromosome as extra or wanting. Thus a *trisomic* individual possesses the normal chromosomal complements plus one extra. A *monosomic* has one chromosome missing. Such individuals with an incomplete set of chromosomes are called *aneuploids*

Blakeslee and Belling (1924) were the pioneer scientists to record the existence of aneuploidy in a common plant, *Datura stramonium*. This plant has 12 pairs of chromosomes in the somatic cells. They discovered a 'mutant type' having 25 chromosomes rather than 24. At the metaphase of first meiotic division, one of the 12 pairs was observed to have an extra chromosome. In this plant one trisome was present along with 11 disomes. The trisomic plant differs from other plant especially in shape and spine characteristics of the capsule. The chromosome complement is illustrated as 2n+1. Aneuploidy in *Drosophila* with trisomic X chromosome has been illustrated by Bridges. When white-eyed female fruit flies are crossed with red-eyed males, it is expected that only red-eyed daughters and white-eyed sons will be produced. Bridges observed a few red-eyed sons and a few white-eyed daughters. This is because of the failure of the X chromosomes to disjoin or separate in reduction division. This nondisjunction results in the production of some eggs having two X chromosomes and others with only one.

Nomenclature

A set of terms derived from Greek and Latin will be convenient in subsequent studies on genetics.

Gene. The hypothetical 'factor' of Mendel (for details see Chromosomes).

Dominant gene. When a gene, responsible for a certain character, becomes expressive in F_1 (T in Mendel's case).

Recessive gene. A gene responsible for a certain character but remains hidden in F_1 (t in Mendel's case).

Alleles A pair of dissimilar genes responsible for a pair of alternative characters. They hold corresponding positions or *loci* in corresponding chromosomes and are also known as *Allelomorphs.*

Gametes. A mature germ cell.

Zygote. Union of two gametes form a zyygote.

Homozygous. An individual containing a pair of identical alleles, either two dominants or two recessives is called pure or true breeding. Thus TT is homozygous.

Heterozygous. An individual containing a pair of unlike genes, a dominant and its recessive allele. Thus Tt is heterozygous.

Phenotype. The external appearance of an individual.

Genotype. The genetical constitution of an individual, i.e., TT, Tt and tt.

Multiple alleles. Some time more than two alternative factors may affect the same character. These alternative factors are called *multiple alleles*. In domestic rabbit, there are wild coloured individuals, 'Albino' types with white colour and pink eyes and Himalayan albino type with pink eyes and black or dark brown patches on ear, nose and feet. 'Albino' is recessive to wild type, so is the Himalayan albino. When 'Albino' and Himalayan albino are crossed, in the F_1 generation all become Himalayan albino and the F_2 generation shows 3 Himalayan to 1 Albino and there is no reversion to wild type. Obviously, Himalayan and Albino are alleles to one another and they both are alleles to wild colour. In Drosophila 14 alleles for eye colour have been observed.

Mendelian laws in the light of Modern Genetics

Segregation. In the F_1 generation of Mendel's monohybrid cross, the plants are heterozygous and their genetic composition is Tt. The phenotype is tall and shows no visible indication of the presence of the gene t, though it is present in the genotype. The two alleles do not fuse, blend or interact with each other and, according to Mendel, segregate during gamete formation. After gametogenesis half of the gametes become equipped with T and the other half with t. That is the percentage is 50 : 50. If it is considered that there occurs a random combination of gametes in fertilisation and that all the plants grown from the seeds are viable, it will be seen that each combination in TT, Tt, Tt and tt will give rise to approximately one-fourth of the progeny. The homozygous TT plants have only the allele for tallness and will be phenotypically tall. Tt plants have an allele for tallness and an allele for dwarfness. But as T is dominant over t, the phenotypic expression of the plants will be tall. The tt plants likewise have only the alleles for dwarfness and will be phenotypically dwarf. Thus three quarters of the plants will be phenotypically tall and one quarter of the plants will be phenotypically dwarf. Or in other words, tall plants and dwarf plants will appear in the ratio 3 : 1. Thus the phenotypic ratio becomes 3 : 1. But the account of the genotypic ratio is one quarter homozygous tall, two quarters heterozygous tall and one quarter homozygous dwarf. That is the genotypic ratio for TT, Tt and tt will be 25 : 50 : 25.

Independent assortment. If the gene for round seed be represented by R, wrinkled seed by r, yellow seeds by Y and green seeds by y, Mendel's yellow-round seeds parents will have genes RRYY and the green and wrinkled parents will have rryy.

The gametes will obviously be RY and ry respectively. The genotype of the F_1 hybrid zygote will be Rr Yy. Each of the gametes produced by the F_1 zygote will have four kinds of gametes (RY, Ry, rY and ry). Now if the gametes from the F_1 generation are crossed, there will be sixteen possible combinations since there are four kinds of pollen grains and four kinds of egg cells. A record of the combinations will show a ratio of 9:3:3:1. Nine out of these sixteen combinations will be round and yellow in appearance. Three are round and green, three are yellow and wrinkled and only one is wrinkled and green. (*Vide* Figs. 2.4 and 2.5).

CHROMOSOME CHEMISTRY

Chromosomes are made up of two types of chemical substances—*Protein* and *Deoxyribose nucleic acid* (DNA). A second type of nucleic acid, *Ribose nucleic acid* (RNA) is present in the chromosome in small concentration. RNA is present in much larger quantity in the cytoplasm. Synthesis of DNA and chromosome duplications are closely related and it is certain that the DNA in the chromosome is primarily concerned with the transference of hereditary characters from generation to generation.

DNA chemistry

DNA is found to consist of thread-like particles of high molecular weight in isolated condition. A number of units called *nucleotides* become linked together and form the nucleic acid. A nucleotide is composed of a molecule of *Pentose* (5-carbon sugar), a *nitrogen base* (Purine and Pyrimidine) and phosphoric acid. The purine bases are *adenine* and *guanine* and the pyrimidine bases are *cytosine* and *thymine*. The difference in the chemical composition of DNA is due mainly to the difference in arrangement of the nitrogen bases.

In an intact DNA molecule, base remains attached to a 5-carbon sugar (Deoxyribose) to form a *deoxynucleoside*. On analysis of DNA samples it has been found that the bases do not occur at a random fashion and that the sum of the purines is equal to the sum of the pyrimidines. Further, it is seen that the content of adenine equals the content of thymine and the content of guanine equals the content of cytosine.

Model of Watson and Crick

Watson and Crick (1953) have constructed a most satisfying three dimensional model of DNA. According to this model, a DNA molecule consists of two long chains. The chains run in opposite direction and are coiled round each other forming a double helix (Fig. 2.7). Each chain consists of a nitrogen base attahced to pentose and the pentose in its turn is attached to the phosphoric acid. The bases of one chain are joined with those of the other chain

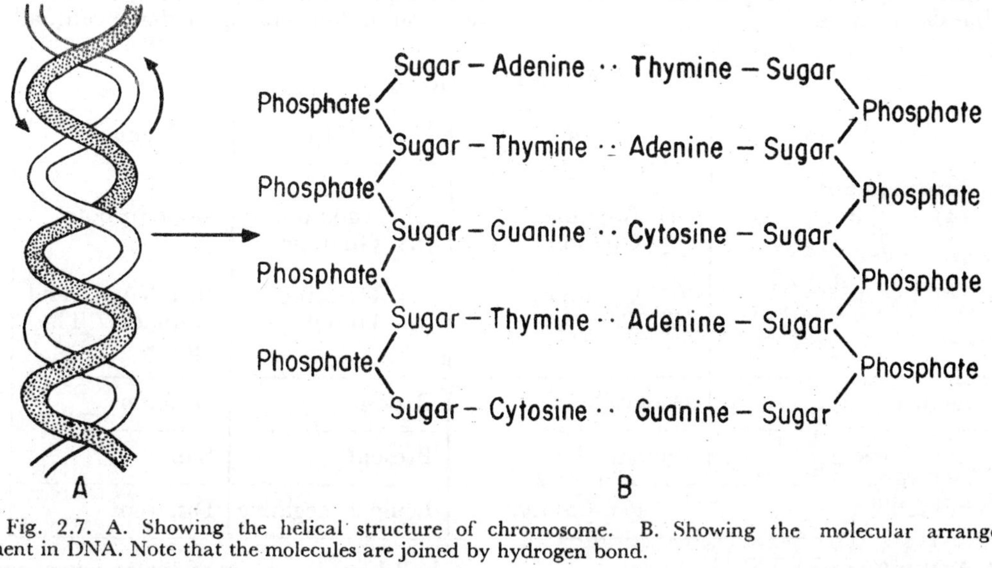

Fig. 2.7. A. Showing the helical structure of chromosome. B. Showing the molecular arrangement in DNA. Note that the molecules are joined by hydrogen bond.

by hydrogen bond. But the bondage between the bases is not at random and their sequences form a regular pattern. Adenine is always linked with Thymine and Cytosine is always linked with Guanine. Thus the sequence of the bases of one strand determines the sequence of the bases of the other strand and this arrangement makes the strands complementary.

It is believed that the base pairs carry all the genetic information in an organism. Thus the four alphabets A, T, C and G (A=adenine, T=thymine, C=cytosine, G=guanine) carry all the codes for the genetical specifications of all organisms. However, it is not known how many base pairs make up a gene.

Differences between DNA and RNA

There are two types of nucleic acids DNA and RNA. DNA is found in chromosomes and RNA is found mostly in the cytoplasm and traces of it are present in the chromosomes. DNA is Feulgen positive and RNA is Feulgen negative. The differecnes between the two types are due to their difference in sugar, pyrimidine bases and the size of the molecules. The differences have been presented in the Table Genetics–4.

TABLE GENETICS–4

Constituents	DNA	RNA	Remarks
Nitrogenous Bases: (*a*) Purines	(*i*) Adenine (*ii*) Guanine	(*i*) Adenine (*ii*) Guanine	Same in both.
(*b*) Pyrimidines	(*i*) Cytosine (*ii*) Thymine	(*i*) Cytosine (*ii*) Uracil	In RNA, Uracil replaces Thymine.
Sugar	Deoxyribose	Ribose	Different
Phosphoric acid	Present	Present	Same
Stainability	Feulgen positive	Feulgen negative	Different

DNA Replication

In 1950, Chargaff showed that in any DNA, the amount of adenine equals to the amount of thymine and the amount of guanine equals to the amount of cytosine. Using X-ray crystallography technique, Watson and Crick in 1953 suggested that the DNA molecule consists of two polynucleotide strands interwined to form a double helix. The two strands are linked together by weak hydrogen bonds between the paired bases. They further advocated that (1) the two polynucleotide strands run in opposite directions to give the observed symmetry and (2) that pairing takes place between one Purine base and one Pyrimidine base. Otherwise the links would become either too small or too large.

This theory has well been substantiated from studies from X-ray diffraction, electron micrograph and enzymatic studies. It is now accepted as a working hypothesis.

The strong point in favour of the hypothesis is that it offers an explanation of of how replication is affected when the chromosomes appear as double structures in the prophase stage of mitosis and meiosis. To quote Watson and Crick "our model of deoxyribose nucleic acid is, in effect a pair of templates, each of which is complementary to the other. We imagine that prior to duplication the hydrogen bonds are broken, and the two chains unwind and separate. Each chain then acts as a template for the formation on itself a new companion chain, so that eventually we shall have two pairs of chains where we had only one before. Moreover, the sequences of the pairs of bases will have been duplicated exactly" (Figs. 2.8 and 2.9).

The experimental support to it

A good experimental support for the correctness of the explanation came from

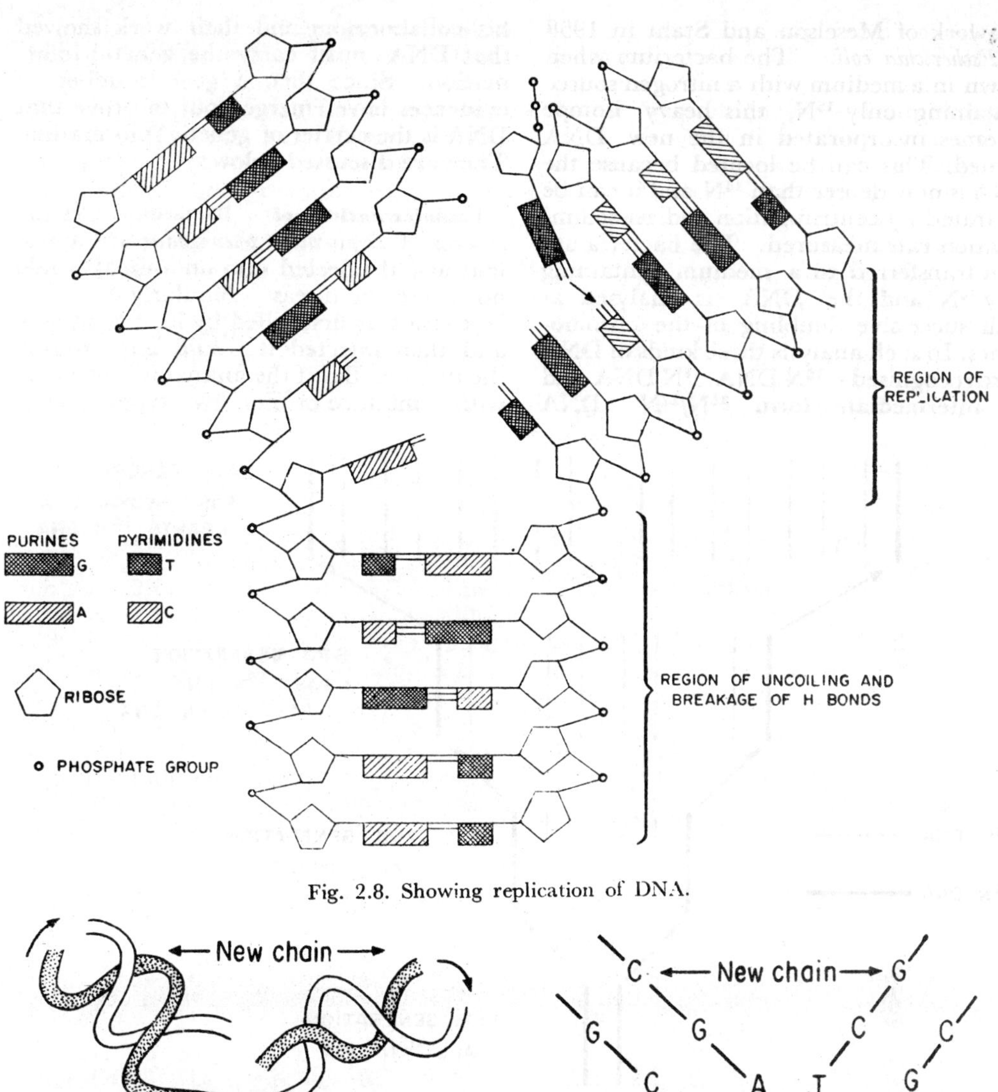

Fig. 2.8. Showing replication of DNA.

Fig. 2.9. A, B. Showing unwind and duplication of DNA by template process. A, T, C, G, are nitrogenous bases.

the work of Meselson and Staht in 1958 on *Escherichia coli*. The bacterium when grown in a medium with a nitrogen source containing only ^{15}N, this heavy isotope becomes incorporated in the new DNA formed. This can be located because the DNA is now denser than ^{14}N and it can be separated by centrifugation and its sedimentation rate measured. The bacteria are then transferred to a medium containing only ^{14}N and the DNA is analysed at each successive doubling of the chromosomes. In such analysis three kinds of DNA were recognised—^{15}N DNA, ^{14}N DNA and an intermediate form $^{15}N/^{14}N$ DNA (Fig. 2.10). The result thus shows complete agreement to the model of Watson and Crick.

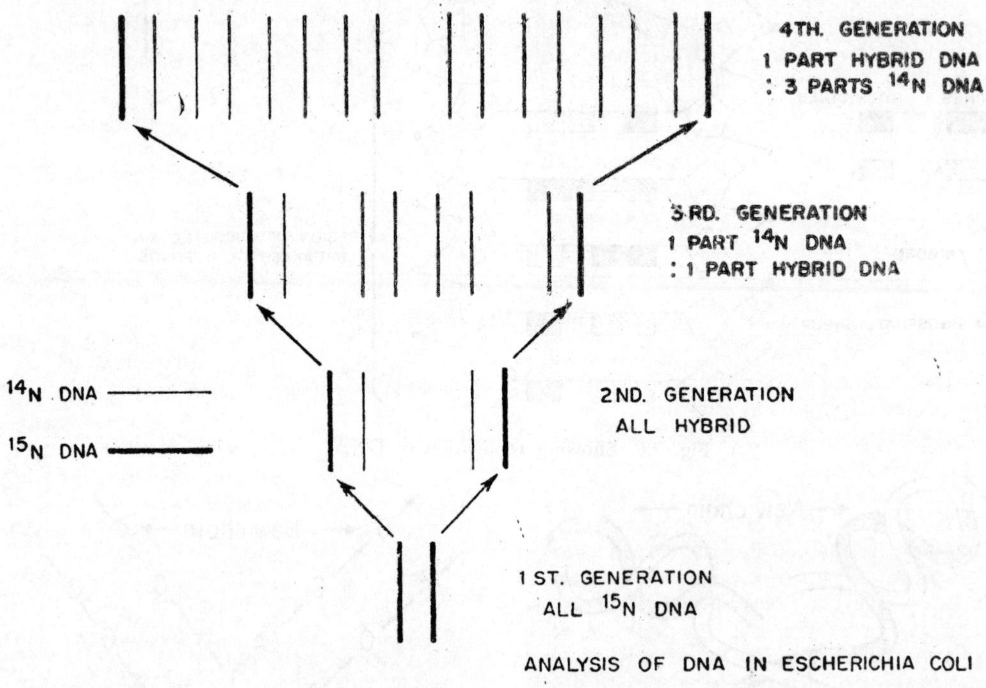

Fig. 2.10. Synthesis of DNA in *Escherichia coli*.

DNA as carrier of Genetic information

The chromosomes consist of proteins, DNA and RNA. In the search for likely genetic material, the proteins were first of all regarded as likely candidates. However, in 1944 work on *Diplococcus pneumoniae*, a bacterium causing pneumonia, was carried out in America by Avery and his collaborators and their work showed that DNA must carry the genetic information. Since then a good number of evidences have emerged out to prove that DNA is the carrier of genetic information. They are discussed below:

Transformation of Bacteria. Certain strains of *Diplococcus pneumoniae* are avirulent and if injected into an animal would not cause the disease. Similarly if a verulent strain is first killed by heat treatment and then injected it would not produce the disease. But if the animal is inoculated with a mixture of these two types (Avirulent and heat killed), the animal shows symptoms of the disease and the avirulent type has now been transformed into virulent type. This genetic change is known as 'Transformation' when carried out in test-tube condition. The other known evidence in favour of DNA's role as hereditary determinant in metabolism comes from the studies of the transformation of *Pneumococci* bacteria. Experiments have established that DNA from one strain of this bacteria can alter the inherited metabolic properties of a second strain of bacteria.

By the application of mutation technique a strain of *Pneumococcus* can be produced which is incapable of utilising Mannitol. These bacteria lack in the enzyme, Mannitol-phosphate-dehydrogenase and are designated as M^- cells. If DNA is isolated from M^+ cells and is introduced in a culture of M^- cells, many M^- cells become transformed into M^+ cells and the progeny of the transformed cells becomes capable of producing the enzyme.

The result proves in a convincing way that the hereditary capacity of the cells can be permanently altered. Further, the experiment makes it clear that at least a part of the genetical material in DNA is taken up by the chromosomes of the host and in certain percentage of the cases this incorporated material is replicated in subsequent generation. However, the mechanism by which the DNA is incorporated into a permanent functional structure is not known.

Certain strains of bacteria are resistant to antibioties such as Penicillin. If purified DNA, extracted from a Penicillin resistant strain of *Pneumococcus* is added to a culture medium in which Penicillin sensitive *Pneumococcus* are being cultured, transformation takes place and the whole culture becomes Penicillin resistant.

Transformation has also been studied in the bacterium *Bacillus subtilis.* A strain of this bacterium is unable to synthesize Tryptophan and as such can not thrive in a culture medium lacking the amino acid Tryptophan. If such a culture is treated with DNA from a normal strain, a number of 'Transformants' appear in the Tryptophan-free medium. From these, many colonies of the new strain arise since the trait is heritable.

Many such experiments involving the transformation of genetic character have been made. These experiments have proved that the transforming agent is DNA and that DNA is the major chemical that controls the synthesis and operation of enzymatic machinery from generation to generation.

Bacteriophage infection. Identification of DNA as the genetic messenger is also proved by experiments with *virus*. Bacteriophage is a virus on bacteria and is popularly called *phage*. The phage is made up of two parts—a *head* and a *tail*. The DNA of the phage lies in the head. Both the head and tail remain capsulated in a protein coat. During infection the phage becomes attached to the bacterial cell wall and injects its DNA into the bacteria. Inside the bacterial cell, the DNA of the phage takes over the metabolic machinery and starts producing phage DNA and phage protein like its own kind.

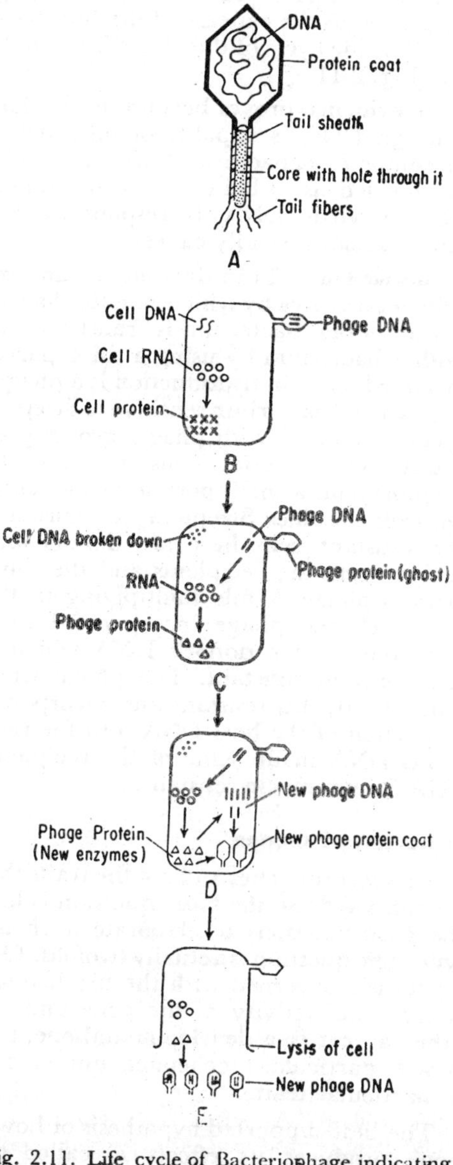

Fig. 2.11. Life cycle of Bacteriophage indicating the role of DNA in heredity. A. Phage morphology. B. Infection by attachment and injection. C–D. Vegetative phage forming new phage DNA and protein. E. Lysis of bacterial cell and emergence of new phages.

Finally, the protein combines with the DNA to form complete phages. After some time the bacteria lyse and break releasing a large number of new phages.

In an experiment the DNA of the phage was lebelled with ^{35}P and these were used to infect the bacteria. The result showed that the host DNA became radio-active. This proves that the DNA of the phage enters the DNA of the bacteria forming new phages at the expense of the host bacteria. The life cycle of the phage is presented in Fig. 2.11.

The evidence proves beyond doubt that a foreign DNA is capable of influencing the synthetic capacities of the host cell. This has led us to believe that an altered DNA or virus DNA is responsible for many diseases specially cancer.

Transduction. Transduction is an experimental device by which genetic characteristics of one bacterium is transferred to another bacterium by using a third party. This third party in transduction is a phage. Thus when a bacterium sensitive to Streptomycin is injected with phages grown previously on bacterial cells resistant to Streptomycin, a small portion of the sensitive cells becomes Streptomycin resistant. The resistant cells here are donors and sensitive cells are recipients and the third party is phage. While multiplying in the donor cell the phage incorporated itself with some of the donor's DNA which is Streptomycin resistant. The phage when transducted, this resistant and incorporated portion of the host DNA into the recipients DNA made some of the recipients DNA Streptomycin resistant.

Function of Genes

These are the genes and not the traits that are inherited. So the basic question is how the gene functions to elaborate a phenotype. The question is actually twofold. One aspect is concerned with the biochemical nature and activity of the gene and the other aspect is a developmental one, i.e., how a particular biochemical unit creates a particular trait.

The best supported hypothesis of how a gene produces its effects is called the '*one gene–one enzyme*' *hypothesis*. According to it the gene controls the production of one enzyme. The enzyme, in turn, regulates a specific metabolic reaction. In the development of a particular phenotype, a number of metabolic reactions occur in a stepwise fashion, each of these steps is regulated by a specific enzyme and each of these enzymes is in turn produced under the direction of a specific gene.

Significant steps in the elaboration of one gene–one enzyme hypothesis

Alcaptonuria. In 1909, A. E. Garrod, an English physician, investigated a condition known as Alcaptonuria in which the urine blackens upon exposure to air and cartilage tissue becomes black and hardened. Family pedigrees showed Alcaptonuria to be inherited by a recessive gene. He found that in Alcaptonuria the substance responsible for the blackening of the urine was *Homogentisic acid* which oxidises upon exposure to air and postulated that Homogentisic acid in normal persons is metabolised (i.e., broken down chemically) but not in persons with Alcaptonuria. He proposed that the normal metabolism of the acid to a simpler chemical substance was under the control of an enzyme. Thus biochemically, according to Garrod, Alcaptonuria results from the absence of an enzyme to metabolise Homogentisic acid.

Garrod also showed that the Alcaptonuria trait behaved genetically as a recessive allele and it occurs only with a homozygous recessive genotype.

Albinism

Garrod also found a consistency with his hypothesis of genes regulating enzymes in his studies in Albinism. Albinism had been recognised as hereditary and is under the control of a recessive gene *a*. The trait appears in genotype bearing *aa*. He postulated that albinism results from a failure on the part of the amino acid, *Tyrosine* to become metabolised to melanin, a black pigment. Absence of melanin pigment results in the condition known as Albinism. Paralleling his conclusions on Alcaptonuria, Garrod advocated that an albino lacks the appropriate gene which produces the enzyme for the conversion of Tyrosine to melanin.

Thus in his studies with Alcaptonuria and Albinism, Garrod arrived at the hypothesis of gene action in which genes regulate or control the presence and working of enzymes.

Garrod's views were not accepted in his own time. In the present-day knowledge, however, the view that genes regulate the presence of enzymes and thereby indirectly control metabolic reactions is accepted.

Work on Neurospora

The 'one gene–one enzyme hypothesis' was restated in 1941 by Beadle and Tatum on the basis of the result of work on the fungus, *Neurospora crassa*. The bread mould, Neurospora can be cultured on a medium containing certain inorganic salts, a carbohydrate like table sugar and vitamin Biotin. As Neurospora can synthesize all its biochemical requirements from a minimal medium, it is reasonable to assume from the gene-enzyme hypothesis that a change in a gene would result in the plant's inability to make one of its particular requirements or in other words, a given metabolic pathway of the plant would be blocked or interrupted by a change in a gene.

Beadle and Tatum were able, by X-ray treatment, to produce mutants which differed from the normal wild type only in their inability to synthesise certain amino acids. One mutant required the addition of arginine to its normal growth medium, another mutant had to be supplied with arginine and citrulline while a third mutant required arginine, citrulline and ornithine for normal growth. Ornithine and citrulline are in fact steps in the biochemical synthesis of arginine.

Glutamic acid ⟶ Ornithine ⟶ Citrulline ⟶ Arginine.

It becomes apparent then that at each stage blockage can be caused by a separate non-allelic mutant gene. This suggests that there is an enzyme for each reaction controlled by its own gene. Thus,

	Gene A	**Gene B**	**Gene C**	
	↓	↓	↓	
	Enzyme I	**Enzyme II**	**Enzyme III**	
	↓	↓	↓	
Glutamic acid	⟶ **Ornithine**	⟶ **Citrulline**	⟶	**Arginine**

Mutation of any one of these genes causes an enzyme deficiency which results in the blockage of the sequences. For their work, Beadle and Tatum were awarded Nobel prize in 1958.

Sickle-cell trait

Linus Pauling and others have done work on 'Sickle-cell trait'—a disease occurring among Africans. The term *Sickle-cell* refers to the peculiar sickle or crescent shape of the red blood cells of affected person. People who are homozygous for sickle-cell gene suffer not only from anaemia but also from such conditions as kidney damage, spleen enlargement and skin lesions and early death. In heterozygous condition, the persons sometimes suffer from anaemia. Thus in the sickle-cell trait blood of the genotype is heterozygous having both Hb^A (normal) and Hb^S (mutant) alleles and the haemoglobin consists of a mixture of both kinds.

Pauling and Itano showed that the Hb^S haemoglobin comes out of the solution when oxygen content is low and also that the normal haemoglobin and sickle-cell haemoglobin respond differently to an electric field—normal haemoglobin moves as a negative ion and sickle-cell haemoglobin as a positive ion.

V. M. Ingram, in correlated studies, has been able to demonstrate the actual differences between the molecules of the two kinds of haemoglobin. He showed that normal haemoglobin had the amino acid *glutamic acid* in certain positions and that a different amino acid *valine* was found in these positions in sickle-cell haemoglobin. Thus normal haemoglobin (and hence normal R.B.C.) is converted by the substitution of certain amino acid (valine) to an abnormal haemoglobin which causes sickling of R.B.C. It is known that amino acids form chains—the Polypeptides—which are collectively referred to as proteins, the change in a gene at the sickle-cell locus results in a very specific change in a protein by the substitution of one amino acid for the other. Thus this piece of research is particularly significant in correlating chemical changes with gene changes and therefore throwing some light on the way gene gives information for protein synthesis.

Another point of interest in sickle-cell trait study is that the heterozygous condition known as sickle-cell trait confers a benefiit on its victims in malarial regions giving immunity against the Plasmodium trophozoites causing malignant malaria. This selection value probably accounts for the high frequency of sickle-cell trait in tropical Africa.

Present position of the hypothesis

Much research is being carried out to find the connection between protein differences and gene differences. Many experiments have been performed with micro-organisms which are most suitable. The fungus Neurospora is normally able to synthesise Adenine but various mutants can only grow if adenine is added to the culture medium. If two of these *non-allelic* adenine requiring mutants (which are of compatible strains) are cultured in the same medium their hyphae fuse and form *Heterokaryons*. In the mycelia thus formed, two kinds of nuclei mingle and the new growth becomes independent of Adenine. This is known as *complementation* since each supplies an enzyme missing in the other.

Most allelic mutants do not show complementation, but in a few rare cases they do and this suggests that the gene can no longer be regarded as an entity. The lozenge trait in Drosophila shows that the gene at the lozenge locus on the chromosome regulates not only the eye-size but also the eye-facets. In addition, the same gene affects the amount of eye pigment and reduces fertility in homozygous recessive females. The several effects of a single gene on different traits are referred to as *Pleiotropism*. The lozenge gene, shows pleiotropism in affecting the three different aspects of the eye (size, facet, pigment) and also fertility. Thus it is more likely that a gene consists of smaller subunits of function. These subunits have been termed *cistrons*. It is likely that a cistron determines one polypeptide chain and a reformulation of the one gene–one enzyme hypothesis has been made. The present reformulated hypothesis is *one cistron–one polypeptide*.

Some unsolved problems of gene action

In a cell there are many genes. But only a few of the genes in a particular cell are active at any one time. The question that arises what calls them into action and how are the other genes held in check? One of the proteins bound to the DNA in the chromosome is histone. It is suspected that histone may play a part in suppressing gene activity until it is required.

The existence of two fundamental types of genes have been postulated by Jacob and Monod. These genes have been termed as *Structural* and *Controlling* genes. The structural genes according to them specify amino acid sequence, while the controlling genes regulate the activity of the structural genes in the DNA. A set of genes is thought to be active only when its controlling gene, the 'Operator' is active. The operator in turn may be controlled by a 'Regulator gene' which is sensitive to chemical changes in the internal environment of the cell. The whole concept is hypothetical in nature but it may be a starting point for further research in the line.

Genetic code

Research in the field of elucidation of genetic code is yet to cross its infancy and it is not proper to give any detailed account of the genetic code. However, a summary of the present state of knowledge and hypothesis is given below.

The linear sequence of bases or base pairs in the segment of the DNA that we call gene, appears to form a *code* for the linear sequences of amino acids in a corresponding polypeptide chain. There are twenty known amino acids and only four bases or base pairs in the DNA. It is thought that each amino acid is coded for by at least a triplet of nucleotide pairs. Such a code would then give $4^3 = 64$ possible symbols or *codons*. The 64 codons are too many for twenty amino acids, so it seems more likely that a doublet code is present which could only give 4^2 or 16 possible codons. Each base pair can only belong to only one codon and there is no overlapping. This is suggested by the fact that a single mutation causes a change in a single amino acid.

Work on artificial polymerisation of amino acids has shown that a triplet sequence of any three bases can bring about specific amino acid formation but in no case it has been found that all four bases are needed for the formation. Some of the possible codons postulated in Nirenberg's work are given in Table Genetics-5.

TABLE GENETICS–5

Postulated code in RNA	Amino acid
CCG	Alanine
GGC, AGA	Arginine
ACA, AUA	Asparagine
GAA	Glutamic acid
UGG AGG	Glycine
ACC	Histidine
UUG UUC	Leucine
GGV	Tryptophan
UGV	Valine

DROSOPHILA MELANOGASTER—

the cinderella of Genetics

After the rediscovery of Mendelism, numerous investigations have been performed with a large array of characters to check up the phenotypic monohybrid ratio of 3:1. Both plants and animals have been used in such experiments.

Plants usually produce large number of offsprings from a single cross. But on the other hand plants produce not more than one generation in a year. In animals the number of progeny produced at a time are ordinarily small although more generations are obtainable in a given time. So the desire for an ideal organism which will have distinct somatic characters, early sexual maturity, free copious breeding in captivity and which will cost less trouble and expense was looming large. This ideal organism is the fruit fly *Drosophila* which satisfied all those requirements.

These flies are very easy to breed in laboratory conditions. Their culture needs little expense and they live happily in half-pint milk bottles. The life cycle of these flies becomes complete in less than two weeks and the fecundity is so high that a pair produces hundreds of offsprings. The larval salivary gland chromosomes of these flies are easy and favourable for observation. This cinderella of genetics was first used by T. H. Morgan in 1909 in his experiments. Uptil now no other materials as suitable as *Drosophila* have been obtained.

LINKAGE

The seven pairs of characters in garden peas with which Mendel worked were located in separate chromosomes. But when genes for different characters stay together in the same chromosome a very confusing phenomenon occurs. Genes in the same chromosome do not abide by the principles of independent assortment. Independent assortment does not hold good in cases where genes are located in the same chromosome. Thus, Mendel's law of independent assortment is not universal but is limited to genes situated in different chromosomes.

The exception to this law was first noticed by Bateson and Punnet in England in 1906.

They bred a dihybrid strain of sweet peas in which one pair of characters was *Purple* (P) and *Red* (p) in the colour of the flowers and the other pair of characters *Long* (L) and *Round* (l) in the shape of pollen grains. The F_1 dihybrids were all purple long when a cross was made between purple long and red round peas because purple and long and dominant characters. But when these F_1 dihybrids having PL, Pl, pL and pl were crossed together to form the F_2 generation the four expected phenotypic kinds in the ratio of 9 : 3–3 : 1 never turned up. What actually turned up was purple long. 4831 purple long, 390 purple round, 393 red long and 1338 red round were obtained out of a total of 6952.

To account for this deviation from the expected assortment, Bateson and Punnet put forward the view that when two dominant characters go in together from the same parent to form the dihybrid offsprings there is a tendency for the same pair of characters to enter the same gamete and to be transmitted together. This phenomenon, they called *coupling* occurs when a dominant and a recessive factor enters the F_1 dihybrid from each parent instead of two dominant factors from one parent and two recessive factors from the other. There occurs always a tendency for two dominant factors entering separately and likewise for two recessive factors to avoid or repel each other. This tendency of unlike pairs (one dominant and one recessive) to stay together avoiding union with their own dominant or recessive sort is termed *repulsion*.

Similar situation encountered by Morgan in *Drosophila* in the year 1910 resulted in his putting forward a satisfactory explanation of coupling and repulsion. Morgan advocated that coupling and repulsion are but two aspects of a single phenomenon called *linkage*. He supposed that this tendency of linked genes to remain in their original combination was due to their residence in the same chromosome.

Thus certain genes located on the same chromosome tend to remain linked together while passing from one generation to another. This tendency of certain genes to stick together in the way in which they entered the cross has been termed *linkage* and such genes are called *linked genes*.

Operation of linkage from Morgan's data

A wild type *Drosophila* having gray body (G) and long wings (L) is crossed with a fly having two recessive mutations of black body (g) and vestigial wing (l). The members of F_1 generation were all like wild types, that is, gray bodied and long winged as gray body colour and long wings are dominant over black body and vestigial wings (Fig. 2.12).

Now a male fly from this dihybrid is back-crossed to a double recessive black vestigial female. Had there been independent assortments, the kinds of flies would have been formed in the following way as shown in Table Genetics—6.

But from actual experiments only two classes of offsprings namely, gray long and black vestigial like two grandparents were obtained (Fig. 2.12). This shows that gray body and long wings entering the dihybrid cross from one parent stay linked together.

Linkage—its characteristics

(1) Linkage is an exception to Mendel's principle of independent assortment.

(2) Linked genes are housed in the same pair of chromosomes.

(3) Lesser the distance between the linked genes stronger is the linkage bond. Genes lying farther apart show less linkage.

(4) Linked genes tend to be transmitted together from one generation to the other.

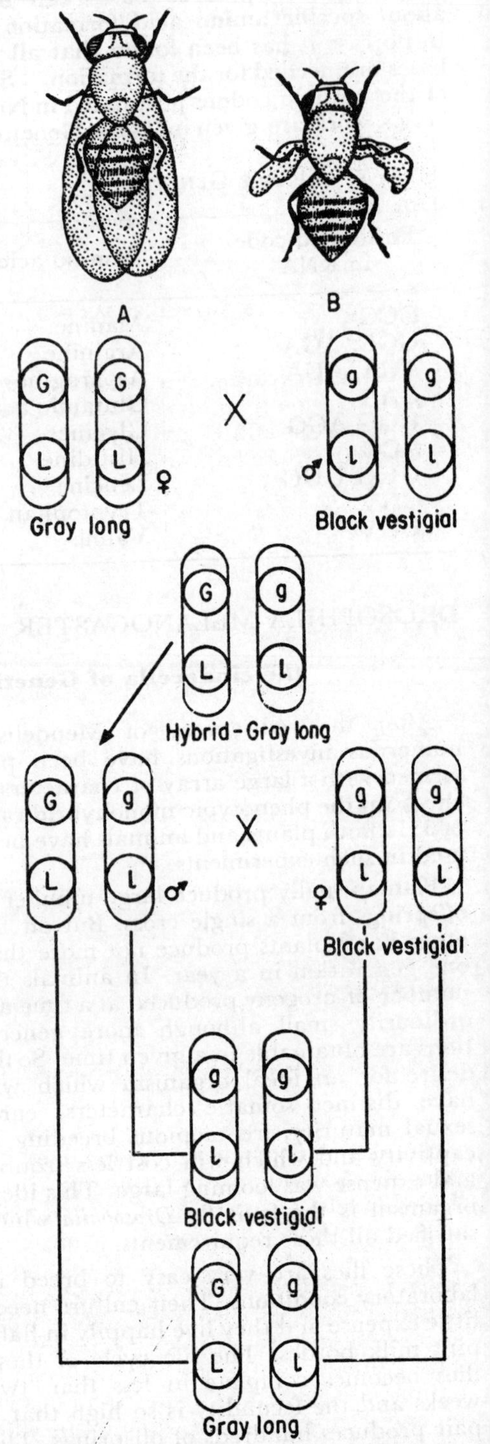

Fig. 2.12. A. *Drosophila* (normal). B. *Drosophila* with vestigial wings. Lower figures showing a case of linkage from Morgan's data.

TABLE GENETICS–6

Male gametes	GL	Gl	gL	gl
Female gametes	gl	gl	gl	gl
F_2 back-cross	GLgl	Glgl	gLgl	glgl
Phenotype	Gray long	Gray vestigial	Black long	Black vestigial

(5) Separation of linked genes occurs very seldom.

Linkage groups. If a certain gene X is linked to two other genes Y and Z, then it is true that Y and Z are linked. In plants or animals where several genes exist, crosses are arranged to ascertain the existence of linkage of pairs or of groups of several pairs of genes. The genes known to exist in a species is thus divided into *linkage groups*.

Crossing over. Linkage is an exception to Mendel's principles of independent assortment and crossing over is in the same way an exception to linkage.

Crossing over means breaks in the linkage of genes within the chromosome and a bodily transhiftment of genes from one chromosome to the corresponding position in its mate (Fig. 2.13). The phenomenon of crossing over closely

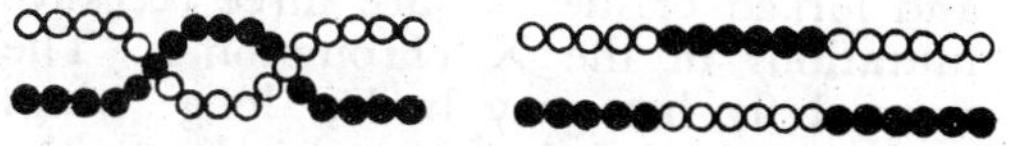

Fig. 2.13. Showing transhiftment of genes from one chromosome to other during crossing over.

resembles independent assortment of Mendel but it is a different thing. Independent assortment is concerned with the whole chromosome while crossing over involves parts of chromosome. It is a sort of shuffling of genes between homologous pairs of chromosomes which always brings forth new combination. The gametes containing the new combinations are known as *cross-over* or accombination gametes. The gametes in which the linked genes remain in their original combinations are called *non-cross-over* gametes.

A case of crossing over in Drosophila

A gray long female obtained by making a cross between gray long and black vestigial fly is back crossed to a black vestigial male. It is expected that in such a cross the two original kinds will be produced in the F_2 generation. But in actual experiment four kinds of offsprings—gray long and black vestigial like the grand parental combinations and two new combinations gray vestigial and black long appeared. The percentage of these four types were: gray long 41·5, black vestigial 41·5, gray vestigial 8·5 and black long 8·5. The percentages show that free and random assortment of all gametes have not occurred because had it been so the ratio would have been 1 : 1 : 1 : 1.

The appearance of new combinations is the resultant outcome of breaks in the linkage of the genes within the chromosomes. This incompleteness of linkage leading to exchange of position of genes from one chromosome to the corresponding position of its partner is due to the phenomenon of crossing over (Fig. 2.14). From the experiment mentioned above it appears that there are 83 per cent (41·5+41·5) of noncrossing over and 17 percent (8·5+8·5) of crossing over.

The percentage of cross-over varies between different genes. But for each pair of genes the percentage remains constant. According to Morgan the cross-over percentage is related to the relative distance on the chromosomes between the two pairs of alleles. Greater the distance, greater will be the amount of crossing over between them. In a simple way it may be stated that breaks occur more frequently in long chromosomes than in short one and between distant points on the same chromosome.

Interference and coincidence. In

crossing over, not only single pair of isolated genes are involved but also the whole blocks of genes which lie close together. Their proximity interferes mechanically with the crossing over of neighbouring genes owing to the limited flexibility of the chromosomes. In other words, crossing over at a particular region of a chromosome tries to prevent another crossing over close to it. This phenomenon is called

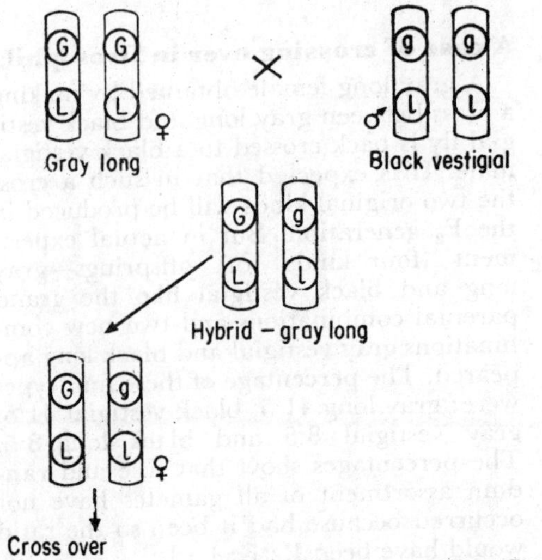

Fig. 2.14. Showing a case of crossing over in *Drosophila*.

interference. It is because of interference that there are no or few double cross-overs within a section of chromosome. 10 units or less in length. The amount of interference becomes less when the distance between two genes increases and there may be no interference when the distance is too great.

The double cross-overs are nothing but coming together or 'coincidence' of two single cross-overs. Thus when double cross-overs occur in expected numbers the coincidence is said to be 100 per cent and in such cases the interference is 0. When there is no double cross-overs the interference is 100 per cent and coincidence is 0. Thus coincidence is inversely proportional to the amount of interference.

Double cross-over

Crossing over just *once* is known as single cross-over and the resultant gametes are called single cross-overs. But sometimes crossing over occurs at two points in the same chromosome pair. This is known as double cross-over and the gametes so formed are called double cross-overs. The amount of double cross-over between two loci increases with the distance apart of the loci. But as a rule double cross-overs are fewer than single cross-overs. Crossing over may also occur at three loci in the same chromosome pair (triple cross-over) but they are still fewer.

A case of double cross-over in Drosophila

A double cross-over involves three linked genes in the same chromosome. In Drosophila yellow body (y), miniature wing (m) and forked bristles (f) are three recessive mutations in the **X** chromosome. The normal fly has gray body, long wings and straight bristles. If we indicate the mutant genes by the symbols and their normal alleles by + signs then yellow, miniature and forked female will be $\frac{y\,m\,f}{y\,m\,f}$, a pure female will be represented as $\frac{+++}{+++}$, and a pure male will be represented as $\overline{+++}$.

A cross between $\frac{y\,m\,f}{y\,m\,f}$ ♀ × $\overline{+++}$ ♂ may give a female of genotype $\frac{y\,m\,f}{+++}$. When reduction division takes place in the female, the following possibilities of gamete formation will be encountered (Fig. 2.15).

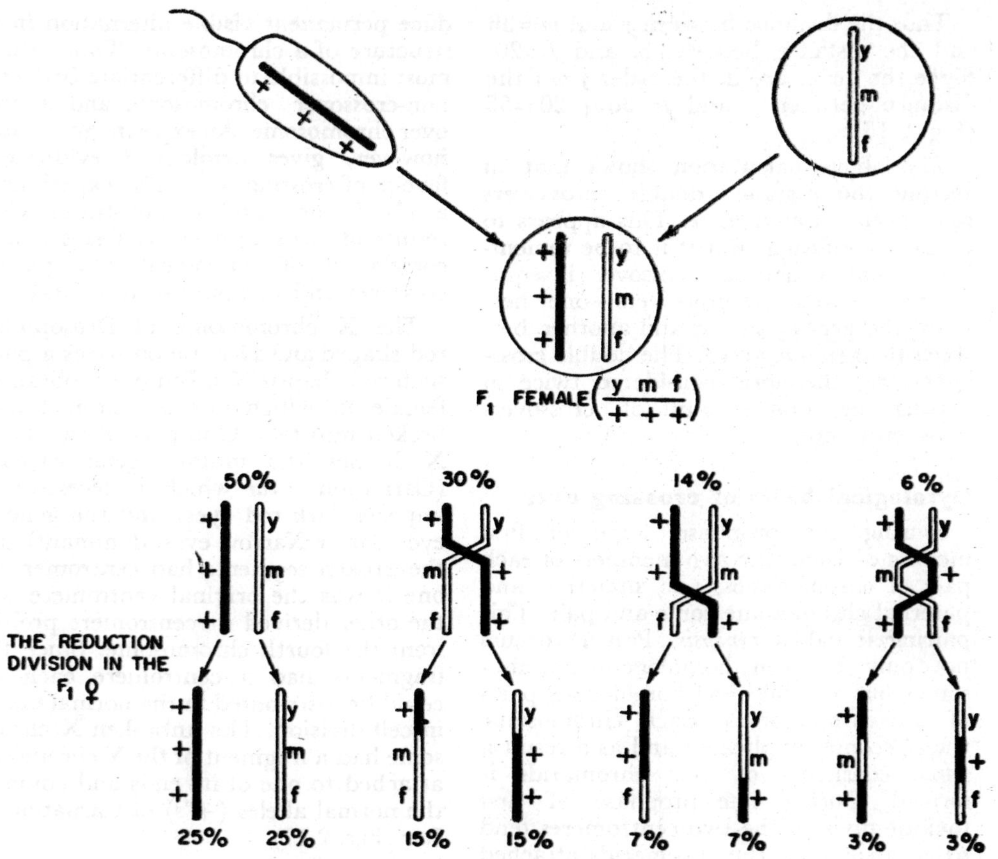

Fig. 2.15. Showing a case of double crossover in *Drosophila*.

1. $\dfrac{y\,m\,f}{+++} = \dfrac{y\,m\,f}{}, \dfrac{}{+++}$ (Non-crossovers 50%)

2. $\dfrac{y\,m\,f}{+++} = \dfrac{+\,m\,f}{}, \dfrac{}{y++}$ (Single crossover between *y* and *m* 30%)

3. $\dfrac{y\,m\,f}{+++} = \dfrac{y\,m\,+}{}, \dfrac{}{++f}$ (Single crossover between *m* and *f* 14%)

4. $\dfrac{y\,m\,f}{+++} = \dfrac{y+f}{}, \dfrac{}{+\,m\,+}$ (Double crossover 6%)

We can now calculate the distances between *y m* and *f*.

Percentage of single crossover between *y* and *m* = 30%.

Percentage of double crossover between *y* and *m* = 6%.

Total percentage of crossover between *y* and *m* = 36%.

Similarly,

Percentage of single crossover between *m* and *f* = 14%.

Percentage of double crossover between *m* and *f* = 6%.

Therefore, total percentage of crossover between *m* and *f* = 20%.

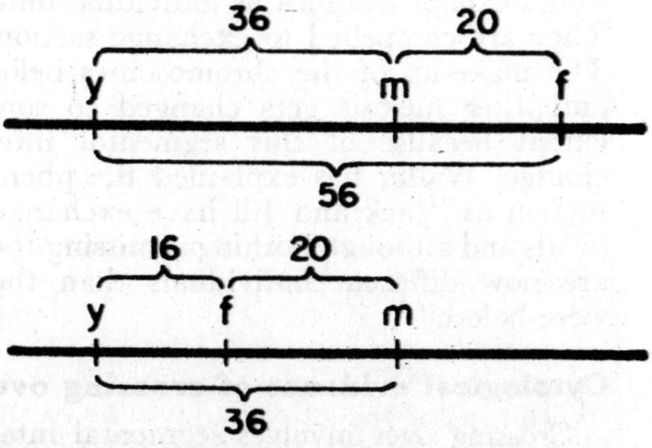

Fig. 2.16. Showing the distance of three genes (*y m f*) in *Drosophila*.

Thus the distance between y and $m=36$ and the distance between m and $f=20$. Since the genes are in the order $y\ m\ f$ the distance between y and $f=36+20=56$ (Fig. 2.16).

The above calculation shows that in getting the distance double crossovers have been counted *twice*. This appears to be little confusing. But it is to be remembered that a double crossover is equivalent to two single crossovers—one between the genes y and m and another between the genes m and f. The double crossovers are, therefore, considered twice in getting the total amount of crossovers between y and f.

Cytological basis of crossing over

During the prophase stage of first meiotic division, the two members of each pair of chromosomes, i.e., maternal and paternal chromosome come and pair. This pairing is called *synapsis*. Pairing occurs not only between homologous chromosomes but also between homologous parts of the chromosomes. Each chromosome then becomes duplicated and as a result a *tetrad* consisting of four chromatids is formed. During late prophase of first meiotic division the two centromeres tend to go apart. But the chromatids attached to the centromeres do not, as a rule, separate uniformly along their entire length. At one or more points along the tetrad, two of the four chromatids appear to lie across each other forming *chiasma*. At each chiasma, two of the four chromatids break and then rejoin, so that newly oriented chromatids are formed out of sections of original ones. Because of this chiasma formation maternal and paternal chromosomes cannot transmit as individual units. They are compelled to exchange sections. The make-up of the chromosomes before and after meiosis gets changed to some extent because of this segmental interchange. Walter has explained the phenomenon as "Jack and Jill have exchanged heads and although nothing is missing they are now different individuals than they were before".

Cytological evidence of crossing over

Crossing over involves segmental interchange between homologous chromosomes. But normally crossing over cannot produce permanent visible alternation in the structure of a chromosome. Thus, it is almost impossible to differentiate between a non-crossover chromosome and a crossover chromosome. An experiment by Stern, however, gives cytological evidence in favour of crossing over. The experiment is a classic one and demonstrates visible results of crossing over. It forms a direct correlation of cytological and genetical crossover and was published in 1931.

The X chromosome of Drosophila is rod-shaped and a female possesses a pair of such rod-shaped X's. But Stern obtained a female in which one X chromosome is broken into two. One part of this broken X houses the mutant gene carnation (Carnation = *car* which is recessive and imparts dark red eyes) and the gene Bar eye (Bar or Narrow eyes, dominant). Both the broken segments had centromeres. In one it was the original centromere while the other derived its centromere probably from the fourth chromosome. Since these fragments had a centromere each they could be distributed in the normal manner in cell division. The unbroken X chromosome had a fragment of the Y chromosome attached to one of its ends and contained the normal alleles (+'s) of Carnation and Bar (Fig. 2.17)

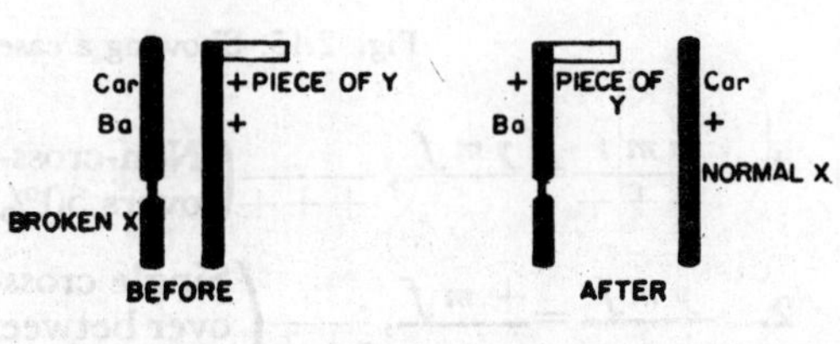

Fig. 2.17. Broken X chromosome as obtained by Stern.

Now if there be no crossover the two X chromosomes will go to the two gametes in their changed (changed from normal) make-up and if there be crossover between Carnation and Bar the broken X chromosome bearing the Bar gene will have the Y chromosome attached to it and the unbroken one will lose the Y chromosome though it will have the carnation gene. The Y chromosome here will act as the marker and thus it will be possible to distinguish the crossovers microscopically.

Now if the X chromosome bearing Car + is added to each of those four classes of eggs produced by the hybrid female only female offsprings will result. When the chromosomes of these offsprings are examined under microscope it is found that (Fig. 2.18)—

(1) The offsprings which appear Carnation Bar are with broken X chromosome.

(2) The offsprings which appear red round (Normal) are with the unbroken X with the Y chromosome attached to it.

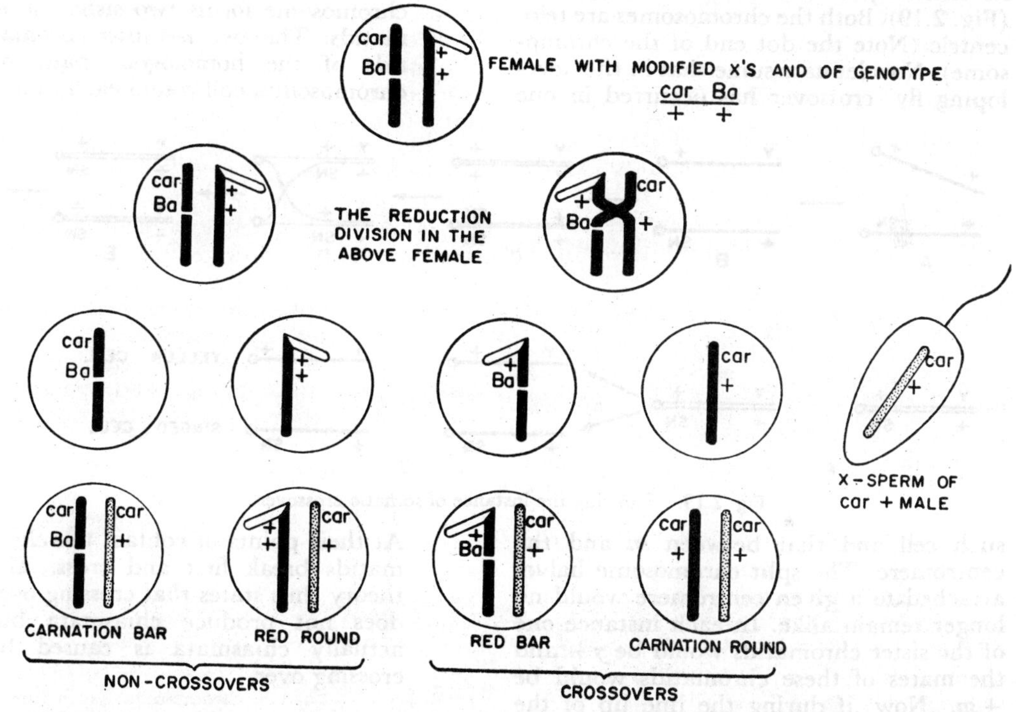

Fig. 2.18. Showing the production of classes of offsprings by hybrid female portraying the cytological evidence of crossing over.

(3) The red Bar offsprings are with broken X with Y chromosome attached to the part bearing the Bar gene.

(4) The offsprings which appear Carnation round are with unbroken X without the attached Y chromosome.

Thus it becomes evident that when two genetic non-crossover classes bear the non-crossover X chromosomes of the mother but the two genetic crossover classes bear the crossover X's of the mother. This is the cytological basis of crossing over.

Somatic crossover

Pairing of chromosomes is restricted to the germ cells and it takes place during the first maturation division. Somatic crossover is a rare phenomenon. In Drosophila such a rare instance of somatic cross-over has been shown by Stern. Such somatic crossovers occur in one or two cells during the course of development of the fly. But these cells give rise to a cluster of cells through the process of division resulting the formation of a patch or spot on the body with the crossover cells. The somatic cells on the other parts of the body will be normal. Thus, the fly will be a *mosaic* of crossover and noncrossover tissues.

A somatic crossover cannot be investigated by taking into consideration the offsprings of the fly since the somatic cells do not give rise to offsprings. So in detecting somatic crossover in any organism it is the organism itself to be examined for any 'crossover spot'. In *Drosophila* it has been possible to detect such spots with the use of certain genes. The suitable genes for the purpose are yellow body colour (y) and 'singed', i.e., short and curly bristles (sn). Both the genes y and sn are recessive mutants and are located on the X chromosomes. Normal alleles for the genes are indicated by + signs.

Stern got a fly of genotype $\frac{y\,+}{+\,sn}$. That is one chromosome of the fly is with $y+$ and its homologous chromosome is with $+sn$ (Fig. 2.19). Both the chromosomes are telocentric (Note the dot end of the chromosome). Now let us assume that in the developing fly crossover has occurred in one

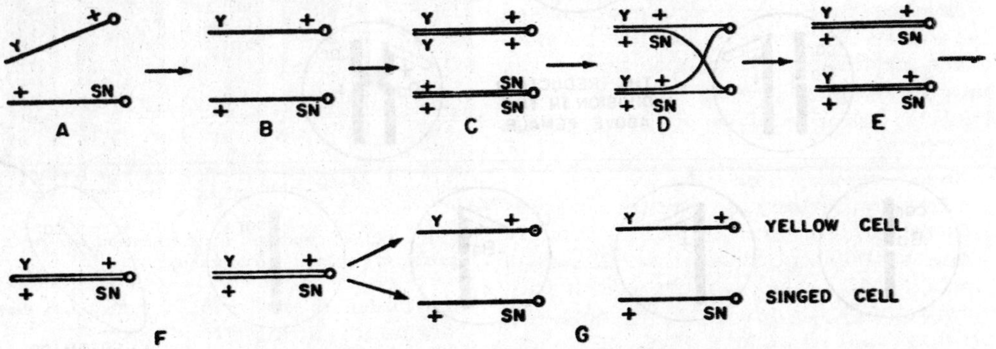

Fig. 2.19. Showing the instance of somatic crossover.

such cell and that between sn and the centromere. The split chromosome halves attached to a given centromere would no longer remain alike. In each instance one of the sister chromatids would be $y+$ and the mates of these chromatids would be $+sn$. Now if during the line up of the chromosomes at metaphase the two chromatids with $y+$ face one pole and the two chromatids with $+sn$ face the opposite pole, at the end of the division we would get two cells—one with genotype $\frac{y+}{y+}$ and the other with genotype $\frac{+sn}{+sn}$. By further division, each of the cells would give rise to cluster of cells. If these two clusters lie close to the surface of the body, $\frac{y+}{y+}$ clusters would form a yellow spot and $\frac{+sn}{+sn}$ would form a spot with singed bristles. The two spots would lie close together since they have been derived from sister cells and they would in this way give rise to twin spots.

The causes for somatic crossover are not known.

Different theories on the mechanism of crossing over

A. **Classical theory**—This theory explains that during the early prophase stage of Meiosis the chromosomes split up longitudinally. Each chromosome forms two sister chromatids. The two *non-sister* chromatids of the homologous pairs of chromosomes coil round each other. At their points of contact the chromatids break first and cross. The theory thus states that crossing over does not produce chiasmata but actually chiasmata is caused by crossing over.

B. **Chiasma type theory**—The theory states that breaking up of chromatids occurs at pachytene stage. After breaking up the chromatids unite again and form a chiasma. Thus according to it the chiasma is the result of crossing over.

C. **Copy choice theory**—This theory is based on the fact that synthetic activity and duplication of chromosomes are intimately associated with recombination. In the mechanism according to the theory the sister chromatids duplicate their genetic parts and non-sister chromatids develop fibres at random. The entire recombinants are formed from newly formed sections. The theory takes for granted that duplication occurs during late meiotic prophase but now it has been established that DNA replication occurs long before synapsis.

Theories that explain the happenings during cross-over

Contact first theory—According to this theory the chromatids destined to undergo cross-over touch each other first and then cross-over to give rise to chiasma. After this, breakage takes place at the point of contact and new attachment of chromatid parts takes place.

Breakage first theory—Muller advocated the theory. According to him the chromosomes destined to cross-over first break into two segments then reunion occurs between non-sister chromatids to give new arrangement.

Strain theory—This theory was advocated by Darlington. The theory states that the chromosomes break as a result of strain at the time of pairing. A sort of strain develops when two chromatids pair, twist round each other and this results in breakage and reunion.

Belling's theory—Belling believes that crossing over occurs between newly duplicated genes and that there is no breakage or reunion during crossing over.

Significance of crossing over

1. Crossing over supports the fact that genes are arranged in a linear fashion on chromosomes.
2. Crossing over provides opportunities for reshuffling of genes and thus brings variations which play major role in the process of evolution.
3. By calculating the cross-over frequency it is possible to plot the genes on the chromosomes.

Chiasma formation—the theories

The process of chiasma formation was first correctly understood by a Belgian Cytologist, Janssens (1909). He suggested that a chiasma represents an exchange of parts between homologous chromosomes. He thought that the exchange involves the whole chromosomes or in other words both chromatids of each homologous chromosome exchange parts with both chromatids of the other. But from the present-day knowledge, we know that the exchange at any point is between single chromatid—one of paternal and one of maternal origin—while the other two chromatids remain unaffected. Fig. 2.20 gives a schematic idea of exchange of genes between chromatids during crossing over.

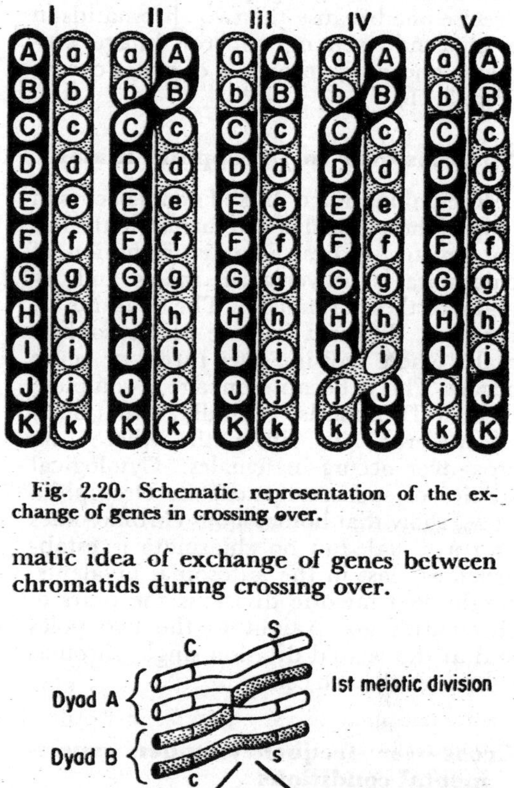

Fig. 2.20. Schematic representation of the exchange of genes in crossing over.

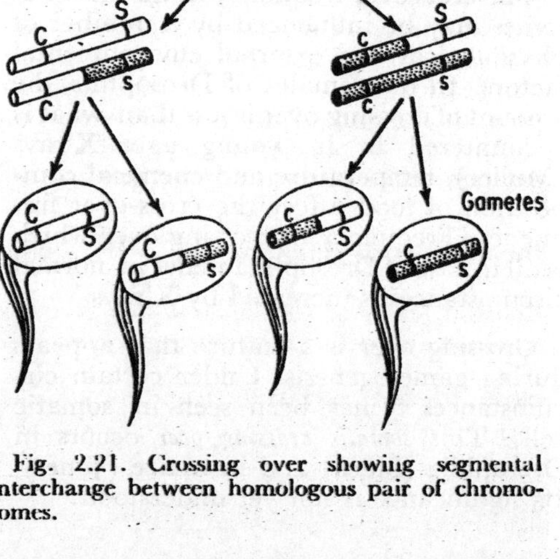

Fig. 2.21. Crossing over showing segmental interchange between homologous pair of chromosomes.

Many speculations have been made to assign causes for breakage of chromatids. The followers of *two-plane theory* advocate that chiasmata causes crossing over. While the followers of *one-plane theory* claim that chiasmata is the consequence not the cause of crossing over. But the real cause behind the breakage and subsequent rejoining of chromatids is still not known. However, the process is a highly precise one because the two chromatids in a chiasma exchange mirror image segments and no gain or loss of genes occurs (Fig. 2.21).

No cross-over in Drosophila males

Recombination of linked genes occurs in most of the organisms that furnish materials for genetic studies. That is formation of chiasmata is universal in males and females of these organisms. The situation is, however, different in case of Drosophila males where crossing over rarely or never occurs. This is because linkage is complete in male Drosophila. A similar situation is encountered in silk-moth where no cross-over occurs in females. Cytological studies of spermatogenesis in Drosophila males show that homologous chromosomes pair as usual. But no chiasmata is established at least in the autosomal bivalents. At the first meiotic divisions the pairs of chromatids go straight to the two poles and at the second division single chromatid passes to each cell.

Cross-over frequency under experimental conditions

The cross over frequency in the chromosomes may be influenced by a number of physiological and external environmental factors. In old females of Drosophila the amount of crossing over is less than what is encountered in its young age. X'rays (Muller), temperature and chemical composition of food affect the cross-over frequency. Frequency of crossing over which is nil in case of Drosophila males in normal circumstances is increased by X'rays.

Crossing over is a feature that appears during gametogenesis. Under certain circumstances it has been seen in somatic cells. This *somatic crossing over* occurs in Drosophila (Stern) and in maize (Jones) Its significance is not yet understood.

Linkage or chromosome map

Constancy of percentages of crossing over and fixed location of genes on chromosomes offer an opportunity to measure the comparative linear distance between any two genes in question. The units for measuring such a distance between two genes is called a 'Morgan'. One unit or Morgan is actually the percentage of crossing over. If two genes show 10 per cent of crossing over it is assumed that the genes for these two characters are 10 Morgans apart along the length of the chromosome on which they are housed. From linkage and crossing over studies it has become possible to determine the specific location of genes on chromosomes with reference to one another. This plotting of specific location of genes on the chromosomes is known as *linkage map*, *genetic map* or *chromosome map*. This has been done thoroughly for hundreds of genes in Drosophila and many other organisms.

Method for construction. Let us assume that three genes K, L and M are located in a chromosome along its length and we want to know the order of the genes on the chromosome and the relative distance between them. If the genes K and L upon breeding back to the recessive show 10 per cent of crossing over with k and l, we can say that they are 10 Morgans apart. If the genes L and M upon breeding back to the recessive show 30 per cent of crossing over with l and m, then it is clear that L and M are 30 Morgans apart. Thus it becomes obvious that the relative distance between K and M will be 40 units (10+30) in case K, L and M are serially arranged. In case K, L, M are not serially arranged along the length of the chromosome the relative distance between K and M will be (30−10) or 20 units and the order of the genes will be M, K and L.

The work of gene localization is an amazing work. It ranks equal to the accomplishment of the mathematicians and astronomers who have been able to measure the vast distances that separate the stars. The wise application of genetic map stands in a good way to discover a good many facts about the orderly mechanism of heredity (Fig. 2.22).

SEX DETERMINATION

Sex is an emblem of life. It is a fundamental quality recognised in living organisms

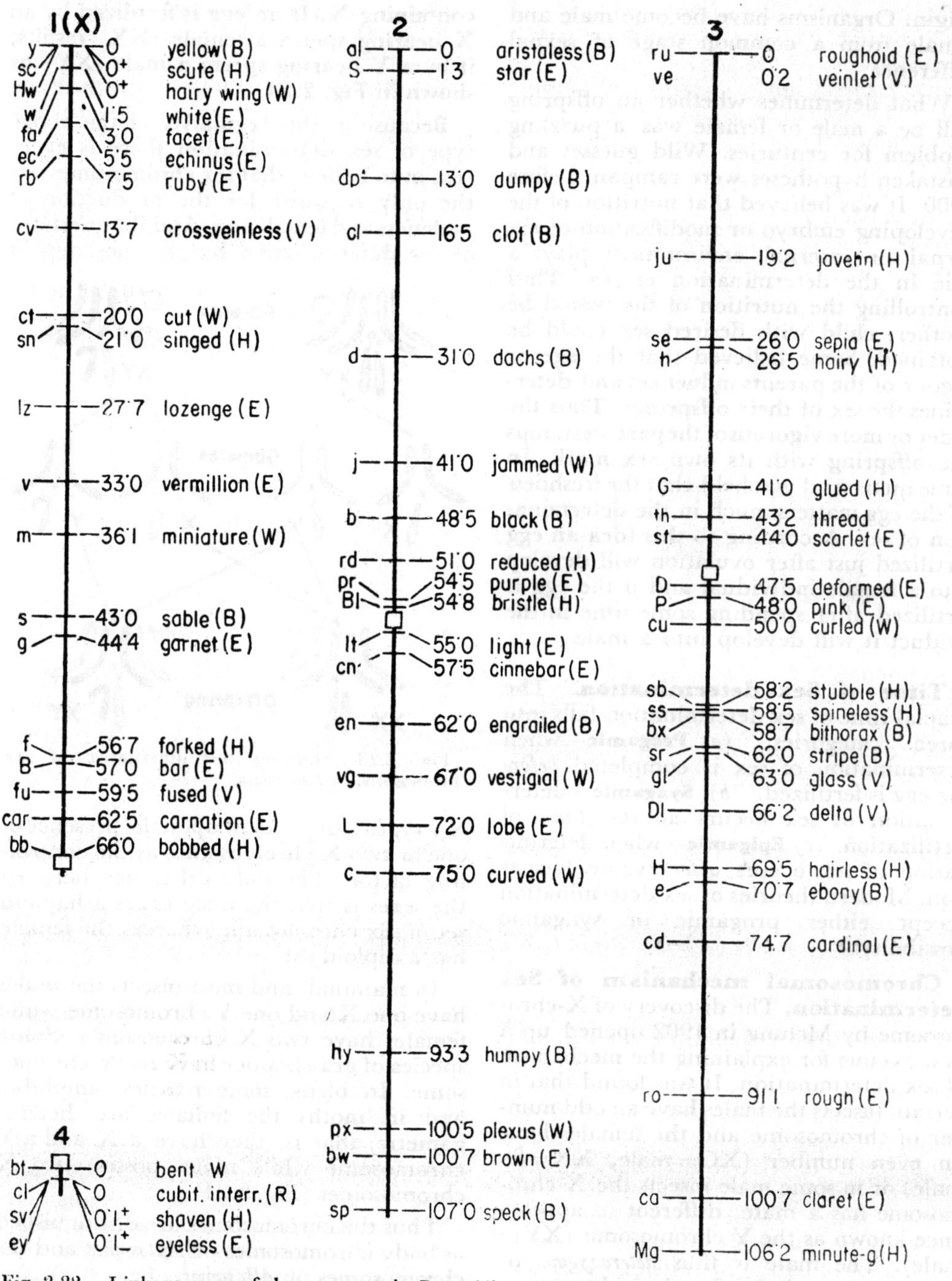

Fig. 2.22. Linkage maps of chromosomes in *Drosophila*. 1 (X) indicates the X chromosome and 2, 3, 4 indicate the autosomes. The location of some of the numerous genes have been shown.

specially in higher forms of life. All higher animals are present in two distinct sexes—male and female. Individuals of same species show remarkable hereditary differcences amongst its male and female members. Existence of higher form of life in two distinct forms—male and female, is an evolutionary outcome. There are evidences which prove that dimorphism of sexes has developed from a single origin. Organisms have become male and female from a common stage of sexual difference.

What determines whether an offspring will be a male or female was a puzzling problem for centuries. Wild guesses and mistaken hypotheses were rampant before 1900. It was believed that nutrition of the developing embryo or modification of external or internal environment plays a role in the determination of sex. Thus controlling the nutrition of the would be mother, child with desired sex could be obtained. Some believed that the age or vigour of the parents influences and determines the sex of their offsprings. Thus the older or more vigorousof theparentsstamps the offspring with its own sex mark. In some quarters it was held that the freshness of the egg matters much in the determination of sex. According to this idea an egg fertilized just after ovulation will develop into a female individual and if the egg is fertilized after spending some time in the oviduct it will develop into a male.

Time of Sex determination. The crucial time of sex determination falls into three categories. (*a*) **Progamic**—when determination of sex is completed *before* the egg is fertilized. (*b*) **Syngamic**—determination of sex occurs *at the time* of fertilization. (*c*) **Epigamic**—when determination of sex occurs *after* zygote formation. Modern theories of sex determination accept either progamic or syngamic possibility.

Chromosomal mechanism of Sex determination. The discovery of X-chromosome by Mclung in 1902 opened up a new avenue for explaining the mechanism of sex determination. It was found that in certain insects the males have an odd number of chromosome and the females have an even number (XO=male, XX=female) or in some male insects the X chromosome has a mate, different in appearance known as the Y chromosome (XY=male). The male is thus *heterozygous* or *heterogametic* and the female is *homozygous* or *homogametic* as regards sex chromosome.

In Drosophila the males have XY, while the females are XX. After maturation two kinds of sperms are formed. One is X bearing and the other is Y bearing. They are formed in equal number. On the other hand, only one kind of egg is formed containing X. If an egg is fertilized by an X bearing sperm a female (XX) results, if by a Y bearing sperm a male (XY) as shown in Fig. 2.23.

Because of this regularity of the above type of sex determination there is every reason to believe that sex chromosomes are the only requisite for the production of maleness and femaleness. And the questions of sex determination by chromosomes in

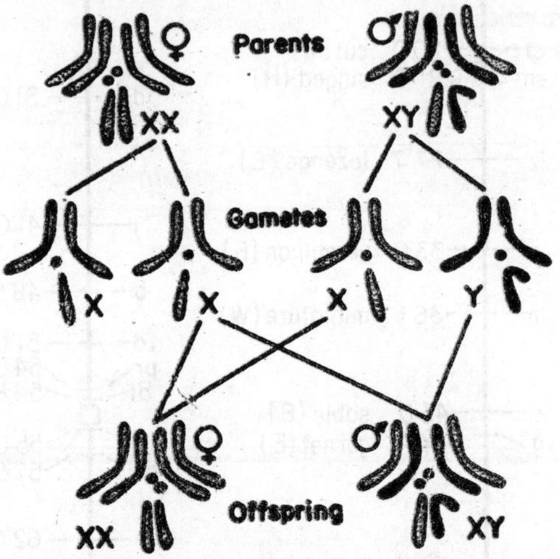

Fig. 2.23. Showing the inheritance of sex chromosome in *Drosophila*.

XY types is dependent upon the presence of one or two X chromosomes as the controlling factor. The only difference between the sexes is that the male bears a haploid set of sex chromosome whereas the female has a diploid set.

In mammals and most insects the males have one X and one Y chromosome while females have two X chromosomes. Some species of grasshopper have no Y chromosome. In birds, some reptiles, amphibia and in moths the females are heterogametic, that is, they have a X and a Y chromosome while males possess two X chromosomes (Fig. 2.24).

Thus the chromosomes are distinguished as body chromosomes or *Autosomes* and sex chromosomes or *Allosomes*.

Sex chromosomes in different organisms is shown in Table Genetics–7.

Theories of Sex determination

Castle's theory of Alternate dominance. Castle advanced a tentative hypothesis regarding sex determination in the year

TABLE GENETICS–7

Organisms	Heterozygous Sex	Gametes		Zygote	
		sperm	ovum	male	female
Drosophila	Male	X and Y	all X	XY	XX
Man	Male	X and Y	all X	XY	XX
Protenor (grass-hopper	Male	X and O	all X	X	XX
Birds	Female	XX	X and Y	XX	XY
Moth	Female	XX	X and Y	XX	XY

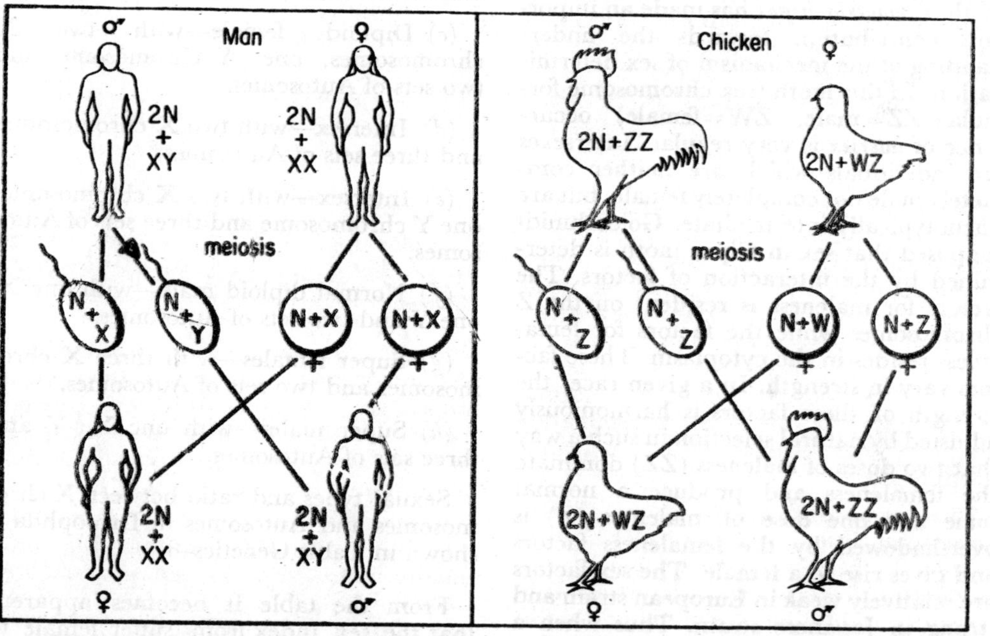

Fig. 2.24. Mechanism of sex determination in XY type male and ZZ(XX) type male. N stands for autosomes.

1903. He assumed that germ cells are heterozygous for sex and after maturation male and female eggs as well as male and female sperms are formed. Secondly, he assumed that during fertilization the gametes always prefer to unite with their sexual opposites and never with their like. Thirdly, the sex characters follow the law of alternate dominance, i.e., in males the male determiner dominates, M(F), while in the female, the female determiner dominates F(M). When two determiners of same kind unite (in 50% cases) lethality results.

Castle's theory cannot explain the cause of alternate dominance neither it accords with discoveries in regard to the chromosomal structure. It has been abandoned.

Corren's theory of Heterogamesis. To avoid the difficulties in explaining the theory of alternate dominance, Corren in 1906 supposed that only one parent is heterozygous with respect to sex while the other is homozygous. Corren's theory is a sort of modification of Castle's view and it is accepted today as heterogamesis has stood the testimony of cytological researches.

Riddle's theory of metabolic differentiation. Riddle (1915) basing his conclusions upon experiments on pigeons and doves, put forward the view that sex determination is depdendent upon the degree of metabolism involved. According to him "The chromosomal constitution is not the cause of sex, it is but a sign or index." Riddle's theory is more a physiological one rather than a cytological explanation for the occurrence and determination of sex.

Goldschmidt's quantitative theory. The classical work of Goldschmidt on Gipsy moth, *Lymantria dispar* has made an important contribution towards the understanding of the mechanism of sex determination. In this moth (sex chromosome formula=ZZ=male, ZW=female) occurrence of *intersex* is very regular. Intersexes are individuals which are neither completely male nor completely female, but are phenotypically intermediate. Goldschmidt supposed that sex in Gipsy moth is determined by the interaction of factors. The factors for maleness is resident on the Z chromosomes while the factors for femaleness resides in the cytoplasm. These factors vary in strength. In a given race, the strength of these factors is harmoniously adjusted by natural selection in such a way that two doses of maleness (ZZ) dominate the femaleness and produce a normal male and one dose of maleness (Z) is overshadowed by the femaleness factors and gives rise to a female. The sex factors are relatively weak in European strain and strong in Japanese strain. Thus when a female of European strain is crossed to a male of Japanese strain—the females (ZY) become intersexes as the strong Z chromosome for maleness overpowers the relatively weak femaleness W.

European Weak female, when crossed with Japanese strong male — the sons become normal while the daughters become intersex.

He postulated the existence of certain theoretical enzymes *Andrase* for maleness and *Gynase* for femaleness. If in the near future such enzymes could be demonstrated, practically it will be possible to prove the quantitative dominance of one over the other or their mutual balance. This isolation will certainly explain the occurrence of sexes and intersexes in a better way.

Bridges's Genic balance theory of sex. In 1922 Bridges found some females of *Drosophila melanogaster* in which every chromosome including the X chromosomes is present in triplicate. These Triploid females resemble the normal diploid females and are fertile. When he made a cross between these types of female to normal diploid males—eight sexually distinct kinds were formed. They are:

(*a*) Triploid females—with three X chromosomes and three sets of Autosomes.

(*b*) Normal diploid female—with two X chromosomes and two sets of Autosomes.

(*c*) Diploid female—with two X chromosomes, one Y chromosome and two sets of Autosomes.

(*d*) Intersex—with two X chromosomes and three sets of Autosomes.

(*e*) Intersex—with two X chromosomes one Y chromosome and three sets of Autosomes.

(*f*) Normal diploid male—with one X, one Y and two sets of Autosomes.

(*g*) Super females—with three X chromosomes and two sets of Autosomes.

(*h*) Super males—with one X, Y and three sets of Autosomes.

Sexual types and ratio between X chromosomes and Autosomes in Drosophila is shown in Table Genetics–8.

From the table it becomes apparent that the sex index from super female to super male runs in a descending scale.

Considering this ratio between allosomes (sex chromosomes) and autosomes, Bridges put forward the view that both autosomes and sex chromosomes are responsible for the determination of sex. According to him it is the genic balance or ratio between autosomes and sex chromosomes that determine the sex. He thought that femaleness is resident on the genes of X chromosomes and maleness is resident on the genes of autosomes. XY is a male because one female tendency (X) is not strong enough to overcome the male

TABLE GENETICS–8

Sex	Chromosome	Sets of Autosomes	Sex index ratio X/A
Super female	3	2	1·5
Normal diploid female	2	2	1
Diploid XXY female	2	2	1
Triploid female	3	3	1
Intersex	2	3	0·67
Intersex with Y chromosomes	2	3	0·67
Normal male	1	2	0·50
Super male	1	3	0·33

tendency in the autosomes (2X). Similarly XX is a female because the double dose of X is strong enough to overcome the male tendency in the autosomes.

Moreover, the ratio of sex chromosomes and autosomes furnished in a satisfactory way the mechanisms to account for the presence of super sexes and intersexes.

Further evidences in favour of genic balance

Dobzhansky and Schultz have shown that when pieces of X chromosomes are added to intersex (2X, 3A) a shift of characters occurs towards female direction. The larger the piece the greater will be the shift. This fact clearly indicates that sex-determining genes are present on various parts of the chromosome.

Role of Y chromosome

In Drosophila XY chromosomes give rise to males. But Drosophila males without Y chromosome have been discovered. Even females with Y chromosomes (XXY females) are obtainable. Similar occurrence has been recorded in human. Experiments with sex-linked inheritance have shown that the Y chromosome is empty of genes. Considering all these it is natural to ask what then be the role of Y chromosome in Drosophila? In general the Y chromosome is regarded merely as a degenerated X chromosome that has lost its sex determining genes and other genes. It plays a dummy hand in the game of heredity. It is sexually neutral and in all probabilities its presence renders mobility to sperms.

Gynandromorphs

Gynandromorphs are usually female on one side and male on the other. Gynandromorph occurs in many organisms. In Drosophila gynandromorphs arise from zygotes having two X chromosomes. Due to misdivision one X chromosome in some cells gets lost during cleavage. The parts with one X chromosome develop as male side while the part with two X chromosomes develops as female side of the

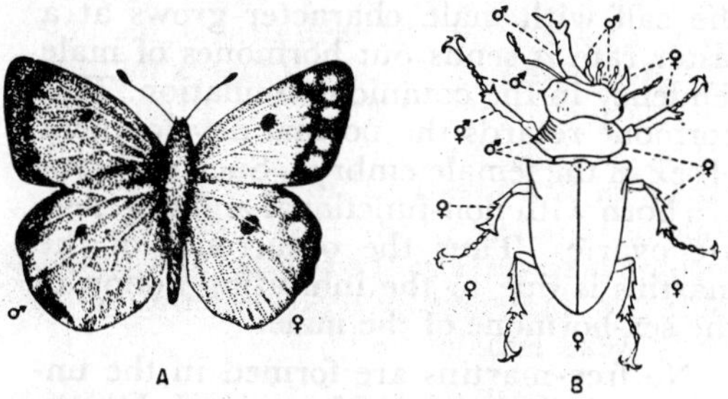

Fig. 2.25. A. Gynandromorph butterfly, *Colias* sp. B. A gynandromorph beetle, *Lucanus* sp. In both male features are on the left and female features on the right. (A—after Dobzhansky and B—after Wigglesworth).

gynandromorph. Gynandromorphs are not to be confused with intersexes which are neither male nor female (Fig. 2.25).

Role of Environment on Sex-determination

External. In *Daphnia* culture if the container becomes over crowded more males are formed (Banta). By manipulation of food and/or oxygen, modification of sex ratio is possible in rotifers and aphids (Whitney).

A remarkable case of sex-determination in case of *Bonelia* has been described by Baltzar. Female Bonelia is about the size and shape of a small plum with a long proboscis that projects above mud in which the body remains embedded. In contrast with the females, the males are very minute and live as parasites within the body of the female near the oviduct. From fertilized eggs free-swimming larvae hatch. The larvae which settle in mud develop into females. If any larva takes refuge in the proboscis and then into the body cavity near oviduct it develops into a male. This is because it finds itself in an environment which is chemically different from sea-water and mud.

Intersexes of Bonelia are formed when the larvae are detached from the proboscis of female Bonelia after temporary attachment and allowed to grow in mud.

Internal. Another conspicuous instance of modification of sex by hormones without changing the chromosome make up is encountered in 'free-martin formation'. Free-martin is a sterile female with a tendency towards maleness. In cattle twin calves of opposite sexes are often found with a common embryonic circulation.As the calf with male character grows at a faster rate it sends out hormones of male tendency in the common circulation. This hormone retards the normal development of sex in the female embryo. Such embryos are born with non-functional and degenerate ovaries. Thus the occurrence of the martins is due to the inhibiting action of the sex-hormone of the male.

No free-martins are formed in the unlike twins of goats, sheep and human beings as there is circulatory union between the embryos.

Sex-linked inheritance

Genes are responsible for the expression of characters and they are located on the chromosomes. Sex-chromosomes are responsible for the sexual demorphism in higher plants and animals. There are certain characters which appear to be associated in inheritance with sex chromosomes. This association is due to the genes lying in the X or the Y chromosome. *Sex-linked inheritance* is thus a type of inheritance in which characters other than sex are inherited along with X or Y chromosome because these characters or genes for these characters are located on the sex-chromosomes. Inheritance of sex-linked characters is therefore related to sex determination.

Sex-linked inheritance in Drosophila

Sex-linked genes in Drosophila was first discovered by Morgan in 1910. He found in a culture of red-eyed Drosophila the presence of a single white-eyed individual. It was a male. When this white-eyed male was bred to red-eyed females, the progeny of F_1 became all red-eyed. The F_2 generation consisted of red-eyed and white-eyed individuals in the ratio of 2 : 1. This result resembles Mendelian monohybrid ratio with the red-eyed allele as dominant. But the most interesting outcome in this experiment was that in the F_2 generation all the white-eyed individuals were male but not all the males were white-eyed. Half of the males were red-eyed.

Thus the cross presents a striking exception to the expected sort of behaviour because in all crosses involving contrasting characters each variety of offsprings appears in equal number of males and females. But in this case the white-eyed character occurred only in males.

The genetic peculiarity of the white-eyed trait can only be explained on the basis of chromosome theory of heredity. The odd behaviour of the white-eyed trait can easily be accounted for if it is assumed that the alleles for white and red eyes are located on the X-chromosome.

Explanation. Let us assume that in the white-eyed male the gene 'W' for white eye is located on the X chromosome. Since the sex chromosome composition in male Drosophila is XY and Y chromosome carries no gene, the normal allele for white eye in the Y chromosome is absent. The white-eyed male thus carries a W gene on its single X chromosome. The red-eyed females on the other hand carry two genes W one in each of the two X chromosomes since the sex-chromosome composition in female Drosophila is XX. The progeny of the F_1 generation are all red-eyed. But all the female members are heterozygous and the males as they receive their X chromosomes from the female are all red-eyed. When these are bred together, white appears in one-fourth of the F_2 offsprings indicating that red- and white-eye colours are due to an allelic pair of genes of which red acts as dominant. Such a result can be obtained if it is assumed that the Ww alleles are carried on the X chromosome and X chromosome alone. The theoretical ratio of three red flies to one white fly and that the white fly should be male is all that was found in the actual cross. Production of white-eyed females—another evidence for sex-linked character.

The white-eyed male observed in Morgan's culture is a product of mutation. From white-eyed males it is possible to have white-eyed females through cross experiments. The F_1 females in the cross between red-eyed females and white-eyed males receive one X chromosome from the mother and one X chromosome from the father. When these F_1 females are crossed to white-eyed males half of the female members will be white-eyed. Thus the production of white-eyed females lend further support to the consideration that the character for white eyes is located on the sex chromosomes.

Thus the peculiar pattern of genetic behaviour shown by white-eye character in Drosophila indicates that the gene for this white character is carried on the X chromosome. Moreover, in Drosophila the sex-linked traits such as eye colour follow a *criss-cross* inheritance. The males transmit their sex-linked traits to their grandsons through their daughters and never to or through their sons. The traits cross from one sex to the other during their sojourn from generation to generation.

The sex-linked inheritance in Drosophila gives additional support to chromosome theory of heredity. Uptil now 140 other genes in *Drosophila melanogaster* are known which follow the same path of inheritance as to the white-eye colour and it is obvious that these genes are housed in the X chromosomes.

Sex-linked genes in Man. In human beings about 20 genes are known which are inherited in the same way as the white or its allele in Drosophila. These are borne in the human X chromosome. They furnish the most interesting expressions of heredity.

Haemophilia. Haemophilia is considered as a disease in which the clotting of blood, in case of injury, does not occur. It is mere due to a genetic defect. Men rather than women are the victims of haemophilia.

In human beings the sex chromosomal constitution is like that of Drosophilia, i.e., males are XY and females XX.

The X chromosome may carry a recessive gene for haemophilia. This gene does not show up if there is a dominant gene for the clotting of blood in the other chromosome. Thus in women haemophilia will not be expressive in case the gene for haemophilia is present in one of the X chromosomes. But such women are carriers of haemophilia.

In males there is a single X chromosome. If the bad gene for haemophilia is present on that chromosome he becomes a bleeder.

If a haemophilic man marries a normal woman, then all the daughters born out of the marriage will carry in one of their X

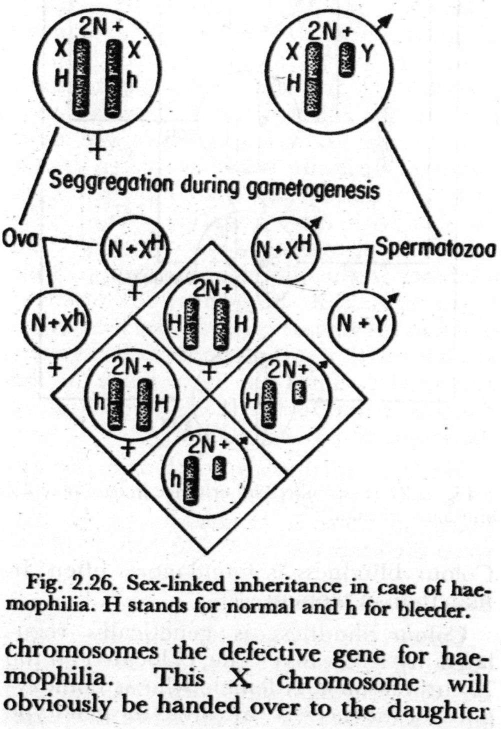

Fig. 2.26. Sex-linked inheritance in case of haemophilia. H stands for normal and h for bleeder.

chromosomes the defective gene for haemophilia. This X chromosome will obviously be handed over to the daughter by the father. And such daughters are the carriers of haemophilia.

The carrier daughter when married to a normal man will produce 50% sons who will be haemophilic like their maternal grandfather. A woman will become haemophilic only if her father is haemophilic and her mother is a carrier. This is again a case of criss-cross inheritance (Fig. 2.26).

Colour blindness. Colour blindness is a common sex-linked human trait. In effect the trait produces inability to differentiate between red and green colour.

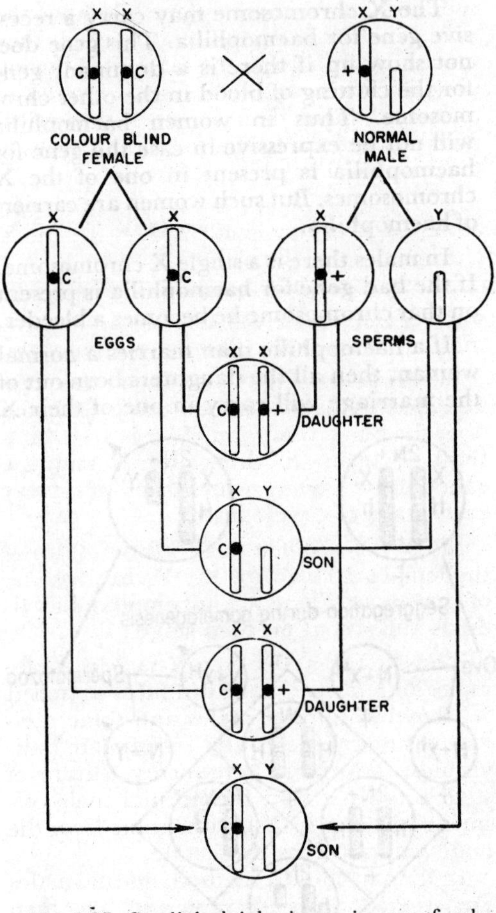

Fig. 2.27. Sex-linked inheritance in case of colour blindness in man.

Colour blindness is found more often in man than in woman.

Colour blindness is genetically regulated by a recessive gene, c, located on the X chromosome. A female who is homozygous for the recessive allele has the genotype cc. A colour blind woman transmits X chromosome to each of her offsprings irrespective of their sex. Contrary to that the male always transmits his X chromosome to all his daughters only.

With those concepts in view we can now analyse the results of marriage between a colour-blind woman and a normal man. In such a marriage all the daughters will be normal while all the sons will be colour blind. Now if these *carrier* daughters are married to colour-blind man 50% of the sons and daughters will be normal and 50% of the sons and daughters will be colour blind. Fig. 2.27 shows the inheritance of colour blindness in man.

MUTATION

Heredity results accurate reproduction of genes. During cell-division the chromosomes divide to give rise to daughter chromosomes. The chromosomes are beset with genes and consequently chromosome duplication means in a way the duplication of genes. Genes always arise from existing genes although the material for the synthesis comes from nutritional sources. The newly formed gene is an exact replica of the parent gene.

The process of gene reproduction though exact, sometimes presents an error in copying. The copy of the gene differs from the original and the modified gene reproduces its changed structure. This error in copying is *Mutation*. Thus mutation is a change in gene, potentially capable of being transmitted in the changed form. Mutation occurs in genes (gene mutation) as well as in chromosomes (chromosomal mutation). Chromosomes usually reproduce faithfully and accurately. Sometimes a chromosome may alter by loss of some parts of genes, by reduplication, by translocation or by inversion.

History. Instances of sudden appearance of new hereditary types in plants and animals were in record by breeders. These new types were referred to as *sports*. The appearance of Ancon lamb (lamb with short and bowed legs), hornless cattle, double toed cats, albino rats are all examples of spontaneous mutations.

In Darwin's time these *sports* were regarded as insignificant. They were considered as monstrosities rather than the origin of new viable types. Appearance of a number of sports in *Oenothera lamarckiana* led de Vries to put forward his mutation theory. The striking mutants in *Oenothera* are forms like giants which had large size, nanella which were dwarfs and many other forms which differ in colour, size or shape of various parts. De Vries introduced the term '*Mutation*' or '*Saltation*' to ascribe for these sports.

Aniridia in man as a case of mutation

Genes are responsible for characters. In man the existence of gene is known which regulates the formation of normal iris of the eye. Sometimes this gene becomes so altered that the iris fails to develop. Failure of the iris to develop is known as *aniridia*. Mutation of this gene occurs in the germ cell of a man and is transferred to the zygote which finally develops into a new man.

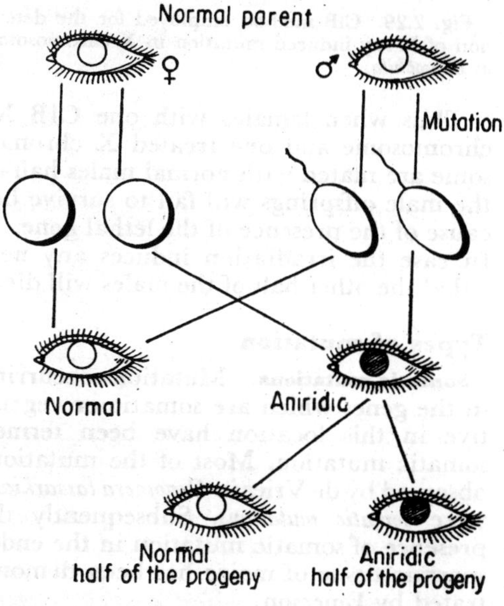

Fig. 2.28. Aniridic condition in man due to mutation.

The mutated gene is a dominant gene and as a result the new individual is born with no iris in the eye. The mutated gene persists in the germinal cells of this aniridic man who will eventually produce half of the germ cells with this mutant gene (Fig. 2.28).

Detection of mutation

Mutations produce both *visible* and *lethal* effects and it is possible to detect both these effects.

Detection of visible mutation

Mutations may be produced artificially by applying different methods. If a visible nonlethal mutation is produced it will be detected readily provided the gene that mutates is a dominant one.

In one peculiar strain of *Drosophila melanogaster* known as 'attached-X' strain a very favourable condition is present and because of this condition a visible but recessive mutation can be detected very easily. In the 'attached-X' strain the females have two X chromosomes attached to one another at the ends nearest the centromere and a Y chromosome. Because the two X chromosomes are attached they always tend to go to the same pole during meiosis. When such a female is crossed to a normal male then comes a theoretical possibility of the formation of four types of individuals. They are:

(*a*) Attached X+Y—when the two attached X of the female unite with the Y chromosome of the male.

(*b*) X+Y male—When the Y chromosome of the female unites with the X chromosome of the male. This is a normal male.

(*c*) Attached X+X—When the two attached X of the female unite with the X of the male. Such individuals are with a total of three X chromosomes and are called *super females* which dies at a very early stage of development.

(*d*) Y+Y—When the Y chromosome of the female unites with the Y chromosome of the male. These are incomplete individuals and die at an early stage.

Thus a cross between attached X females and normal males produces 'attached X' females, normal males and some nonviable super females and incomplete individuals. The most noteworthy feature of such a cross is that the normal male offspring receives its X chromosome from the male and not from the female.

In the attached X method, normal males are treated with X'-rays and are then mated with attached-X females. As the sons in such a cross receive their X chromosome from the father, it becomes very easy to detect any mutation that may occur to the treated X chromosome of the male.

Detection of induced lethals

In order to estimate the frequency of the induced lethal mutations that are produced in X chromosomes, male Drosophila flies are exposed to radiation of known kind and intensity. They are then mated with non-radiated or normal females. If a lethal is produced in the X chromosome of the male it will not be detected in the first generation of the females because they will be 'masked' by the 'non lethal' of the X chromosome of the female parent. But if the resultant female offsprings are again mated to normal males the mutation will reappear in the next generation and will be manifested as a deficiency in the male population. It is obvious that the treated X chromosome of the male passes into a female and subsequently again into the male. These grandsons have no X chromosome other than the one received from the grandfather and the presence of a lethal in that X chromosome results in the early death of all males.

A very simple and standard technique has been devised by Muller for the detection of new lethal in the X chromosome of *Drosophila melanogaster*. The technique is known as ClB method.

The method employs a special stock of female Drosophila fly. One X chromosome of such a female fly bears a recessive lethal gene (l) a dominant gene (B) for bar eye and a dominant gene (C) which is a cross-over suppressor besides other genes. Such females with one X chromosome as marker are mated to irradiated males. One fourth of the offsprings produced out of such a cross will receive a ClB X chromosome and an irradiated X chromosome and will be females. One fourth of the offsprings, will be male with a ClB chromosome and will eventually die because of the presence of the lethal character (l). The remaining half of the progeny will not have the ClB chromosome and as such they will be normal in their eye structure. Such flies will be discarded. The ClB females with treated X chromosome will then be mated with normal males. It is to be remembered that in a normal female crossing-over takes place between the two Xchromosomes. But in ClB chromosome such a cross-over is not possible because of the presence of the gene, C, which is a cross-over suppressor. Thus the chance of transference of the induced lethal from the irradiated chromosome to the non irradiated chromosome becomes nil. The use of B gene is to identify the flies that have the cross-over suppressor (Fig. 2.29).

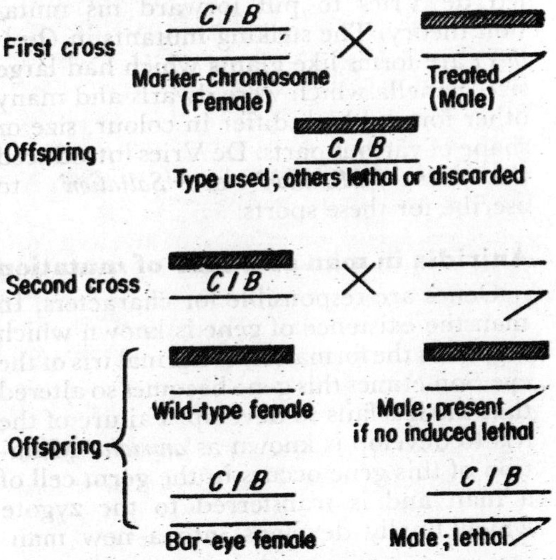

Fig. 2.29. ClB method employed for the detection of X-ray induced mutation in X-chromosome in *Drosophila.*

Thus when females with one ClB X-chromosome and one treated X chromosome are mated with normal males half of the male offsprings will fail to survive because of the presence of the lethal gene, l. In case the irradiation induces any new lethal the other half of the males will die.

Types of mutation

Somatic mutations. Mutations occurring in the genes which are somatic or vegetative in this location have been termed somatic mutation. Most of the mutations observed by de Vries in *Oenothera lamarkiana* were *somatic mutations*. Subsequently the presence of somatic mutation in the endospermic tissue of maize has been demonstrated by Emerson.

Germinal mutation. Mutation occurring in the germ cells (sperm or ovum) has been termed *germinal mutation*. Such mutations are again divided into *gametic* when occur in the gametes and *zygotic* when occur in the zygote.

Spontaneous mutation. The naturally occurring mutations are the *spontaneous mutations*. Natural agencies are responsible for the production of such mutations. The appearance of a white-eyed male in a culture of wild type Drosophila of Morgan is a classical case of spontaneous mutation. Presence of such mutation speaks in favour of evolution.

Induced mutation. It is possible to induce mutation by the application of artificial means. Any agent that can produce mutation is called a *mutagenic agent*. Many of such agents are known. The important ones are X-rays, ultra-violet rays, radium, heat and temperature, mustard gas, coal tar, formaldehyde, caffine, etc.

Anomozygous mutation. Mutations due to structural changes of a chromosome or a set of chromosomes have been termed *anomozygous mutation*. Structural changes in the chromosomes are caused by translocation, inversion, duplication and deficiency.

Biochemical mutation. Mutations that affect specific biochemical processes have been called *biochemical mutation*. Such mutation occurring in an individual inhibits its ability to synthesise essential materials like vitamin or amino-acid. The inhibition may be total or partial. In partial inhibition there occurs a partial blockage in the steps of a synthetic process. Biochemical mutations of *Neurospora* is well studied. The alkaptonuria and phenylketonuria of man are examples of *biochemical mutations*.

Lethal mutation. Most mutations are lethal. Such mutations cause loss or alteration of a function during embryogenesis. Biochemical mutations are mostly lethal. But in some cases, organisms with biochemical mutations can thrive well if the metabolite which they can not synthesise is supplied exogenously and in such cases biochemical mutations are not regarded as lethal.

CYTOPLASMIC INHERITANCE

Main role in the transmission of hereditary tracts is taken up by the genes on the chromosomes. But cases are known which indicate that cytoplasmic or extranuclear elements may also take part in inheritance. In some cases, presence of cytoplasmic DNA has been proved and this DNA takes active role in carrying genetic information. Some cases of cytoplasmic inheritance are cited below.

Inheritance of plastids. The cytoplasm of plants bears minute pigments called plastids. Plastids arise from smaller DNA particles of the ovum and multiply by division. During the formation of gamete a plastid enters the egg and the pollen receives none. Thus plastid is entirely a maternal contribution.

Cytoplasmic inheritance of plastids has been studied in Four-o'clock plant, *Mirabilis*. *Albomaculatus race* of this plant have variegated leaves with white and green patches. Sometimes a branch is formed which bears leaves which are either green or white or variegated. A seed from a uniformly green branch produces a plant which is totally green and a seed from a white branch grows into a white seedling which dies before maturity. Similarly the variegated branch gives rise to variegated offsprings. The pollen is devoid of plastid and cannot influence the inheritance of plastid. Thus the colour depends entirely upon the egg and what the egg produces is dependent upon the plastid of the cytoplasm. Thus it is an evidence of cytoplasmic inheritance.

Kappa particles of Paramoecium. Presence of certain particles in the cytoplasm and their role in inheritance speaks

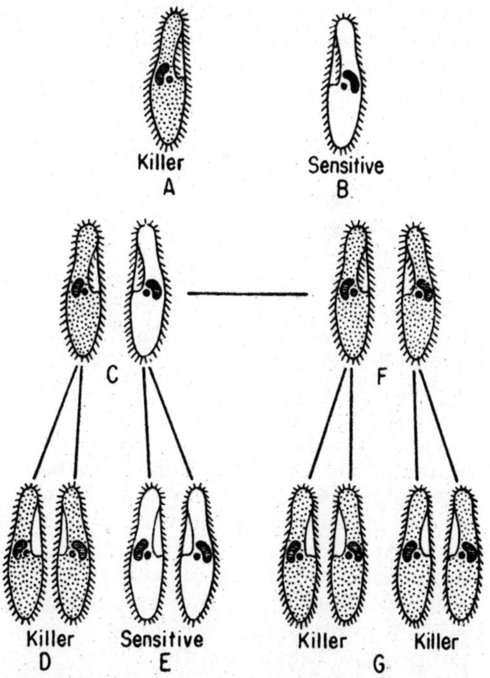

Fig. 2.30. Cytoplasmic inheritance of Kappa particles in *Paramoecium*. A—Killer, B—Sensitive. C. Conjugation without cytoplasmic exchange. D—Killer, E—Sensitive. F. Conjugation with cytoplasmic exchange. G—All killers.

in favour of cytoplasmic inheritance. These particles are called "*Plasmagenes*". Some races of *Paramoecium aurelia* possess a substance in their cytoplasm. The substance is called "Paramecin or Kappa particles". Races of Paramoecium with Kappa particles in the cytoplasm are designated as killer strain. Other members of the race which do not have Kappa particles are sensitive to the Kappa particles. Recent studies have shown that presence of Kappa particles is dependent upon the presence of a dominant gene K in the killer strains. Further, the Kappa particles are made up of nucleic acid and protein.

Fission is the usual mode of reproduction in Paramoecia. Killer individuals in such cases produce killer offsprings and sensitive individuals produce sensitive offsprings. If killer strain is mated with the sensitive strain normally the products that follow such a mating do not show any blending that is killer parent produce killers and sensitive parent produce sensitives. This shows that the exchange of nuclear genes during mating has nothing to do with the inheritance of Kappa particles. In case the conjugation between the killer and sensitive strains become prolonged and there occurs considerable cytoplasmic exchange along with nuclear exchange, all the exconjugants become killers. This shows that Kappa particles if enter into the cytoplasm of sensitive strains make them killers. When killers KK conjugate with non-killer kk for a prolonged period and exchange cytoplasmic Kappa particles, the exconjugants become all Kk heterozygous and become killers as K is a dominant gene (Fig. 2.30).

PART THREE

ECOLOGY

Ecology is a Greek word which means the study of the habitation of living organisms (*oikos*=habitation, *logos*=discourse). The word ecology has been defined variously by different authors. Some prefer to define it as "scientific natural history" or "the science of community population" or "the study of biotic communities". The most comprehensive definition of ecology will be "a study of animals and plants in their relation to each other and to their environment." This part of the book deals with the different aspects of ecology.

PART THREE

ECOLOGY

Ecology is a Greek word which means the study of the habitation of living organisms (*oikos*=habitation, *logos*=discourse). The word ecology has been defined variously by different authors. Some prefer to define it as "scientific natural history" or "the science of community population" or "the study of biotic communities". The most comprehensive definition of ecology will be "a study of animals and plants in their relation to each other and to their environment." This part of the book deals with the different aspects of ecology.

STUDY OF ECOLOGY

Ecology is studied with particular reference to plants or to animals, hence the topics *Plant ecology* and *Animal ecology*. Since plants and animals are intimately interrelated, study of plant ecology or animal ecology alone is bound to be imperfect and inadequate. So plant and animal ecology are to be given equal emphasis and it is better to study them under the term *Bioecology*. The term *Synecology* denotes ecological studies at the community level while the term *Autecology* denotes ecological studies at the species level.

CHANGEFULNESS OF ANIMALS AND ENVIRONMENTS

A continuous change is encountered in animals and environments and this change makes the ecological studies a difficult issue. Every animal digests, respires, works, fights against diseases and excretes and in so doing it maintains its metabolic processes. Failure on the part of the animal to maintain these metabolic activities will bring forth death. The nature of metabolism varies with age, sex and other conditions. Environments, on the other hand, vary with seasonal and diurnal cycles, climatic changes and other fluctuations. The uniqueness of living organisms lies in the fact that they can withstand such fluctuations.

HISTORY

In a sense ecology is the new name of 'Natural history'. Man's interest in natural history dates back to prehistoric times. The carving and pictures discovered in France and Spain speak on the observation of the cave-dwellers about the fauna and flora around them. The writings of Romans and Greeks bear evidences of their interest in natural history. 'The Histories of Animals' of Aristotle (384–322 B.C.) is a famous contribution in this line.

The first naturalist to give a systemised knowledge about the relation existing between living organisms and environment was Buffon. In a series of work in 1749, he stressed on habits and adaptations. After this, outstanding advances were made in the study of natural history in the eighteenth and nineteenth centuries. Darwin's Naturalistic voyage round the world, Wallace's Island of life and many other work stimulated the knowledge of Biology to a great extent. However, the term ecology was first coined by German Biologist, Haeckel in 1878.

The science of ecology after undergoing a several hundreds of years gestation period has emerged today as a matured, honoured and scholarly discipline in biological science.

SCOPE OF ECOLOGY

The solution of a particular ecological problem requires several lines of approach. None of this constitute an end in itself but each one of these makes important contribution in making the picture complete.

These various lines of approach towards the ecological problem can be translated as:

(*a*) **Biotic**
(*b*) **Quantitative**
(*c*) **Climatic (both physical and chemical)**
(*d*) **Taxonomic**
(*e*) **Genetic and evolutionary.**

Biotic factors are the direct outcome of the various types of activities amongst the animals. A competition for food and shelter always exists amongst the members of a community. This competition demands various types of activity amongst the animals.

Quantitative study includes an assessment of the population density in a given area and also an estimation of the number of members present in different communities. Information of this kind is of immense value in solving many problems like food availability and movement within a particular colony.

Climatic factors include both physical and chemical conditions present in a habitat. These factors are ever changing in nature. Physical factors include mainly temperature, light and humidity. Chemical factors include acidity or salinity that are specially present in aquatic habitat. Some animals are so sensitive that a minute climatic change becomes fatal to them. Climatic factors play an important role in the distribution of animals.

Taxonomy means classification, naming and description of organisms. A mere naming of a large number of animals of a given area, as was done earlier in ecological surveys, is meaningless without a consideration of the circumstances that enable them to live there. Thus a complementary observation of the various ecological factors together with taxonomy is emphasized in ecology.

The **genetic** and **evolutionary** aspects have taken a rightful place in ecological problems. In recent years the knowledge of heredity and the mechanism of the operation of Natural Selection have increased to a considerable extent. Evolution is no longer regarded as a thing of the past and it has been proved that evolution is a dynamic process though the progress is very slow. In certain circumstances it has become possible to detect and to measure the rate of evolution in wild population.

The above subdivisions form the backbone of the study of ecology. The inter-relationship existing between these subdivisions can be best understood with the help of an example. Let us assume that we want to study the ecology of a given species of edible fish inhabiting a large lake, with an object of establishing a new colony of these fishes to be started elsewhere. In so doing, the first information that we need is that whether the food available in the new place is to be taken by these fishes. Our second enquiry would be to find out whether predators are present in the locality. These two are included within the biotic factors. We will have to determine the number of fishes that are to be let loose in the new locality and the number is to be determined in such a way that they can live there without being overcrowded. Herein lies the involvement of the quantitative aspect. We will have to study the water itself and to find out the extent of the fluctuations in its constitution such as salt content, acidity or alkalinity in order to determine the tolerance of the fishes in the changing factors. If the first lake is a very old one and the fish in question had been isolated there for a great period of time, it is possible that a subspecies or local race might evolve there. In such cases the taxonomist might come forward and help identifying the species. Such a situation opens up a case for the Geneticists and Evolutionists to find out how and at what rate the new forms have evolved.

ECOLOGY IN THE ATOMIC AGE

Ecological studies have gained a new momentum with the coming of the atomic age. The promise and threat of atomic energy as well as the promise and threat of human population explosion in the present century have pressurised the study of Environmental Biology to a great extent. Rapid industrialisation has made the environment polluted. Population explosion has led man to encroach upon otherwise unsuitable habitats. Ponds, lakes and even marshy lands are being filled in or drained out to provide house sites. Safe disposal of radio-active waste materials is and always remains a problem. All these involve the principles of *Energy flow* and *Biogeochemical cycles* which are the most redeeming features of the study of Ecology. A full knowledge of these aspects is necessary for the survival of the human race. And for this, Environmental Biology has become a most fascinating subject by its own right.

ANIMAL HABITAT

Animals have colonised all the available places on the surface of the earth. Some species have adapted themselves to the most unlikely conditions such as the nematode worms which inhabit vinegar vats. The available space on the surface of the earth is collectively called **biosphere.** There are three major divisions of the biosphere. They are: **atmosphere, hydrosphere** and **lithosphere.** There are no permanent and characteristic inhabitants in the atmosphere though the air is traversed by many flying animals and plant spores. The hydrosphere is divided into three subdivisions. They are: **marine**: **fresh water** and **estuarine.** Lithosphere includes only land.

A **habitat** is a specific set of physical conditions that surround a single species, a group of species or a large community. The ultimate division of the biosphere is the **microhabitat** which is nothing but the most intimately local and immediate set of conditions surrounding a single individual. Thus the burrows of a mouse and the decaying log are the microhabitat of a mouse and insect respectively. Other individuals or species are considered as part of the community but not part of its habitat.

CHARACTERISTICS OF DIFFERENT HABITATS

On a world-wide basis three major types of habitats or ecological environments are recognisable—the marine, fresh water and land. Estuarine or brackish water is intermediate between marine and fresh water and usually occurs in such places where the river flows into the ocean. *Swamps*, on the other hand, represent a blending of aquatic and terrestrial habitat. The characteristics of the different habitats are given below.

Marine habitat

The marine habitat is the largest of all habitats. The seas, oceans and bays have occupied about 70% of the earth's surface. The physical features of the marine habitat are relatively stable.

The depth varies from *intertidal zone* (a zone covered by water only part of the time) to depth as great as 35,400 feet or 6·7 miles. The average depth is about 12,500 feet. In tropical seas, the usual temperature is about 32°C. and that in the arctic region is about −2·2°C. In a given area this temperature rarely varies more than 5°C. during the year (Fig. 3.1).

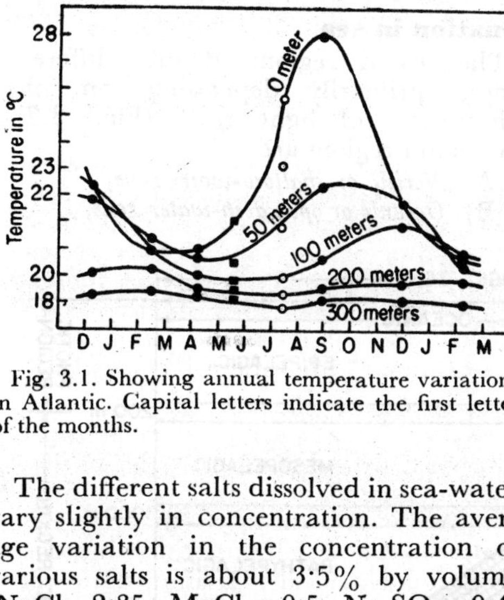

Fig. 3.1. Showing annual temperature variations in Atlantic. Capital letters indicate the first letter of the months.

The different salts dissolved in sea-water vary slightly in concentration. The average variation in the concentration of various salts is about 3·5% by volume ($NaCl=2{\cdot}35$, $MgCl_2=0{\cdot}5$, $Na_2SO_4=0{\cdot}4$, $CaCl_2=0{\cdot}11$). Various gases, namely, O_2, CO_2 and N_2 remain dissolved in sea-water. The amount of gas dissolved is dependent upon depth and temperature. So far as the concentration of oxygen is concerned it is more in surface water as also in cold water at a certain depth. Penetration of light into the sea-water is dependent upon factors like turbidity and surface motion. On an average light can penetrate a depth up to 6000 feet and below it there is permanent darkness (Fig. 3.2). Pressure at different depths of the sea is different. It is known that pressure increases at the rate of one atmosphere, i.e., 14·7 lbs per square inch per 10 m of water. This means that organisms in the depth of the sea are exposed to tremendous pressure. It has been estimated that an organism residing at a depth of 3500 feet experiences a pressure of 257 tons per square inch on the body. Living organisms exhibit much stratification in

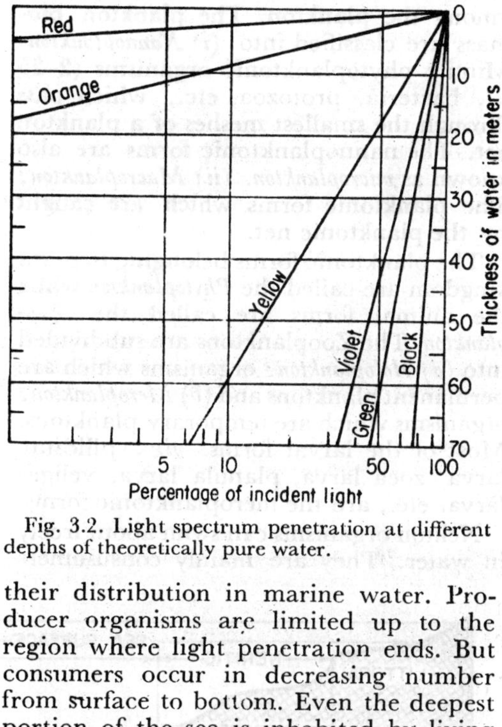

Fig. 3.2. Light spectrum penetration at different depths of theoretically pure water.

their distribution in marine water. Producer organisms are limited up to the region where light penetration ends. But consumers occur in decreasing number from surface to bottom. Even the deepest portion of the sea is inhabited by living organisms. All the phyla (excepting amphibia) have representatives present in the sea.

Ecological classification of marine organisms

The marine biota is varied. The animal communities are grouped into two vertical components—*Pelagic* and *Benthic* forms. The pelagic forms include the communities inhabiting the open sea, while the benthic forms are the bottom dwellers. All the communities are grouped into two heads—*Producers* and *Consumers*.

Pelagic organisms. Pelagic organisms are found in the open sea. Pelagic organisms are classified ecologically on the basis of their means of locomotion and the depth at which they live. The recognised groups are listed below. They are divided into two groups: **Plankton** and **Nekton.** Plankton includes organisms that lack the power of locomotion by their own. They float in water and move passively from one place to another with the help of wind or water current. Most of them are microscopic in size and are represented by protozoa and larval forms of crustacea, helminths, coelenterates and molluscs. Both producers and consumers are present among the plankton. The plankton biomass are classified into (*i*) *Nannoplankton*: Minute phytoplanktonic organisms (2–25 μ), bacteria, protozoa, etc., which pass through the smallest meshes of a plankton net. The nannoplanktonic forms are also known as *microplankton*. (*ii*) *Macroplankton:* The planktonic forms which are caught by the planktonic net.

The planktonic forms belonging to plant kingdom are called the *Phytoplankton* while the animal forms are called the *Zooplankton.* The Zooplanktons are subdivided into (*a*) *Holoplankton:* organisms which are permanent planktons and (*b*) *Meroplankton:* organisms which are temporary planktons. Most of the larval forms, *viz.*, pilidium larva, zoea larva, planula larva, veliger larva, etc., are the meroplanktonic forms.

Nekton organisms can swim about freely in water. They are mainly consusumers and are represented by squids, whales, seals, fishes, turtles and many sea birds.

Benthic organisms. Marine benthos or bottom dwellers are characterised by numerous sessile or relatively inactive animals. The benthonic forms exhibit marked zonation and they are quite distinct from each other in three primary regions—supratidal, intertidal and subtidal zones. The benthonic animals are divided into two classes: (*i*) *Epifaunna:* inhabiting the bottom-surface either attached or moving freely on the surface. (*ii*) *Infaunna:* dig into the substrate or construct tubes or burrows. The epifaunna reaches maximum development in the intertidal zone. The infaunna is more fully developed in the subtidal zone and below. The distribution of benthonic subcommunities depends largely on the type of bottom, whether sandy/rocky/muddy. A sandy beach is better populated. Most of the large animals are specialised burrowers. The diatoms, amphipods and other infaunna live amongst the sand grains. The rocky coasts show three distinct zones—(*a*) periwinkle zone or high-tide area; (*b*) barnacle or mussel zone and (*c*) sea-weed zone or low-tide area.

Zonation in sea

The sea is regionated into different zones, primarily depending on the penetration of light rays (Fig. 3.3) The main regions are:

(A) *Neritic or shallow-water zone,*
(B) *Oceanic or open deep-water zone.*

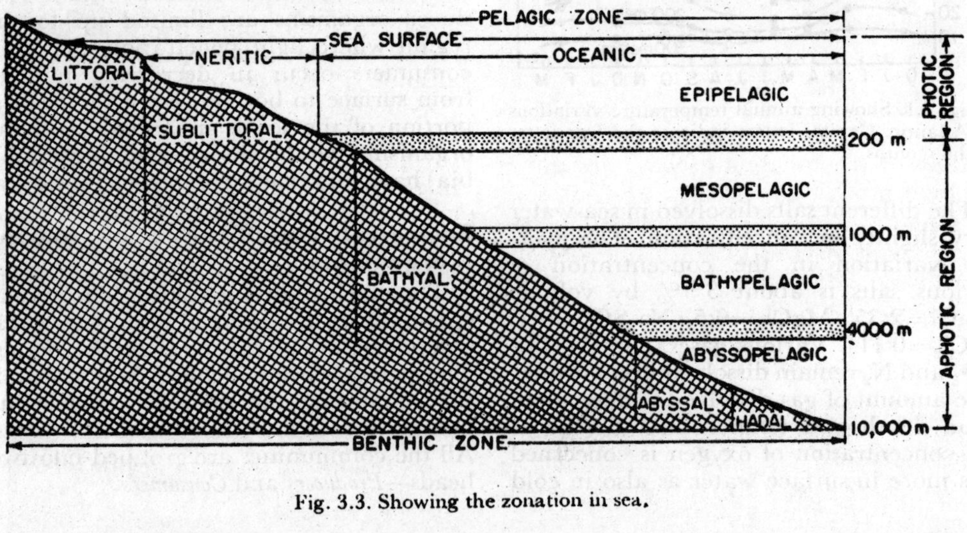

Fig. 3.3. Showing the zonation in sea.

Neritic zone. It is the region of shallow water above the continental shelf. This region receives enough light and as a result producer organisms are present in profuse number. This zone has been divided into three subdivisions.

(i) **Intertidal zone.** Intertidal or Littoral zone is alternately exposed to air and covered by water usually twice daily. This region is most favourite of all habitats with much light, oxygen, carbon dioxide and minerals. All the major phyla from Protozoa to Chordata are found here. Some of the inhabitants are motile and some are sessile.

(ii) **Subtidal zone.** This zone extends between low tide mark and down up to a depth of 150 feet. This region is populated by both producers and consumers.

(iii) **Lower neritic zone.** This zone extends on to the edge of the continental shelf and is characterised by the presence of some light but lesser organisms.

Oceanic zone. The region of the open sea beyond the continental shelf embraces the oceanic division. It is further subdivided into: *bathyal, abyssal* and *hadal* on the basis of depth.

Bathyal zone. This zone extends from the edge of the continental shelf (600 feet) to a depth of 6,000 feet. The water of this region is quiet and becomes progressively colder with depth. There is little or no light and fewer number of organisms are present there.

Abyssal zone. This zone extends from 6000 feet level to the deeps. The water is cold, quiet and has little oxygen concentration. The zone is permanently dark. Very few animals live there. The organisms living there usually have dark body colours. Eyes, when present, are large. Many of the inhabitants have the power of light production.

Hadal zone. The hadal zone includes the deep areas of the ocean lying bottomwards.

Besides the zonation, the sea can be divided into two regions depending on the penetration of light. The zones are:

(*i*) *Euphotic zone:* Light zone extending from the surface of sea up to the depth of about 150–200 m.

(*ii*) *Photic zone:* It is the lightless zone of open sea including the bathyal, abyssal and hadal subdivisions.

Fresh-water habitat

Fresh-water habitat is the smallest of the three primary environments. The habitat is extremely variable. In depth such habitats vary from shallow rain puddles to Lake Baikal of U.S.S.R. that reaches a depth of 5712 feet. Temperature of these habitats varies from frozen to boiling condition. Dissolved salts vary widely in concentration. Dissolved gases vary from almost none to about 7 c.c. per litre. Though the freshwater habitat is small in area and less varied than terrestrial habitat it contains many more kinds of animals than does the terrestrial habitat.

Estuarine habitat

Estuarine or Brackish zone is formed in those regions where a river meets a sea. The composition of water in this zone undergoes constant change. The concentration of dissolved substances in such habitats is unstable. At high tides such habitats experience maximum salinity. Salinity decreases during low tide and periods of heavy rain. The rate of flow of the water or current varies greatly in different estuaries, also in various regions of any one. The constant turnover in the water of estuaries brings about considerable changes in temperature of comparatively short duration. Thus the conditions prevailing in different estuaries vary greatly depending upon topography and other factors.

Terrestrial habitat

Of the three major types of habitats, terrestrial habitat is the most variable. In altitude it ranges from below sea-level (Death-valley) to mountain peaks more than 28,000 feet in height. A considerable variation of temperature is encountered in terrestrial habitat. The lowest recorded temperature is −60°C. and the highest recorded temperature is 60°C. in certain deserts. The chemical and physical nature of the soil, sand and surface rocks varies greatly. The amount of moisture or relative humidity and rate of precipitation are extremely variable in different regions of the globe. The major subdivisions of the terrestrial habitat are given below:

(i) **Tundra.** The arctic region of North America, Europe and Asia is known as

Tundra region. This treeless region is characterised by long-cold winter with no or little sunlight. The summer is cool, short and is with long daylight hours. Frosting is very common and may appear at any time in this region. Soil is never thawed beyond a depth of few inches from the surface. Bogs, marshes and ponds are the common features of this region. Mosses, Lichens and low herbs are the common vegetation. Migrant birds, like water fowls make their summer nesting grounds in the region. Resident birds of North America are Snowy owls and Ptarmigan. Amongst the mammals, Lemmings, Arctic hares, Arctic foxes, Caribou, Musk-oxen and Weasels live in this region.

The *alpine* community on the top of many high mountains of both temperate and tropical regions resemble the tundra biomes in many respects. The plant and animal species of tundra and alpine regions are almost identical.

(ii) Taiga. The broad band just south of the tundra region of Eurasia and North America is known as taiga region. The taiga region in the southern hemisphere is poorly developed because of the absence of land mass. The region bears evergreen coniferous forests. Climatic condition of this region is represented by bleak winter and cool summer. Precipitation is of moderate type. The summer lasts from three to six months. Red foxes, Lynx, Caribou, few reptiles and birds are found here. Spruce, Firs, Pines and Cedars are the common vegetation.

(iii) Temperate deciduous forest. This region is characterised by warm summer and cold winter. There is about 40 inches of rainfall throughout the year and the rainfall is uniform throughout the year. The dominant trees of this region bear broad leaves. The leaves are shed in summer and new leaves develop in the following spring. Beech, Maples, Oaks, Walnuts are the common plants of this region. Deer, Gray foxes, Racoons, flying squirrels, many snakes and amphibia are the major vertebrates present in this region.

(iv) Grass lands. Large areas in both temperate and tropical regions do not support trees but remain covered with heavy growth of grass. This is because of lack of sufficient moisture. All the grass lands are ecologically similar in appearance. These grass lands are also known as Steppes, Prairies, Savanas etc. The rainfall in these regions is usually 12 to 40 inches per year. The rainfall is limited to few weeks in the year. Dominant plants of this region are Blue stem and Grama grasses. Characteristic grass land vertebrates are Bison, Pronghorn antelope, Coyotes, Prairie dogs, Rabbits, Larks, several snakes and lizards.

(v) Deserts. Deserts occur both in temperate and tropical regions. Precipitation of moisture is the controlling factor in all deserts. The average rainfall in a desert is less than 10 inches per year and that too is very erratic with half or more than half of the annual rainfall occurring in one or two heavy downpour. Summer temperature is very high. The rate of evaporation too is very high. The plants of the deserts are highly modified for the purpose of conservation of water. The leaves are reduced or absent or modified into thorns. The roots are long and go deep inside the soil. Many plants possess internally spongy tissue which collect and store water obtained during rainy season. Cacti, Yucca are the dominant plants. Kangaroo, rats, Foxes, Coyotes, many lizards, several snakes and toads are the common vertebrates present in the desert.

Deserts may be snowy too. The sage brush area of North America and the Sierra-Cascade mountain system are deserts of this type. The climate in these deserts is very dry. The deserts are hot in summer and cool in winter. Pronghorn antelope, Coyote, ground squirrels and many reptiles are found in these snowy deserts.

(vi) Rain forests. Abundant rainfall in the tropics, subtropics and few regions in the temperate zone results in the formation of luxuriant vegetation of broad-leaved evergreen trees. Vines, Orchids and Epiphytes are plentiful in these regions. Characteristic vertebrates of these regions are: Monkeys, Sloths, Ant-eaters, Bats, many colourful birds, frogs and salamanders, turtles and snakes.

A close similarity between terrestrial zonation and the zonations in mountains of high altitude is found. The zonation in a high mountain consists of a number of vegetational belts located at various altitudes along a mountain slope. Temperature and rainfall play a role on the distribution of these vegetational belts. In a high mountain temperature decreases correspondingly at higher altitudes and thus produces communities of both animals and plants located along the slopes similar to more spacious biomes that occupy large areas within certain latitudes from the equator to the poles (Fig. 3.4).

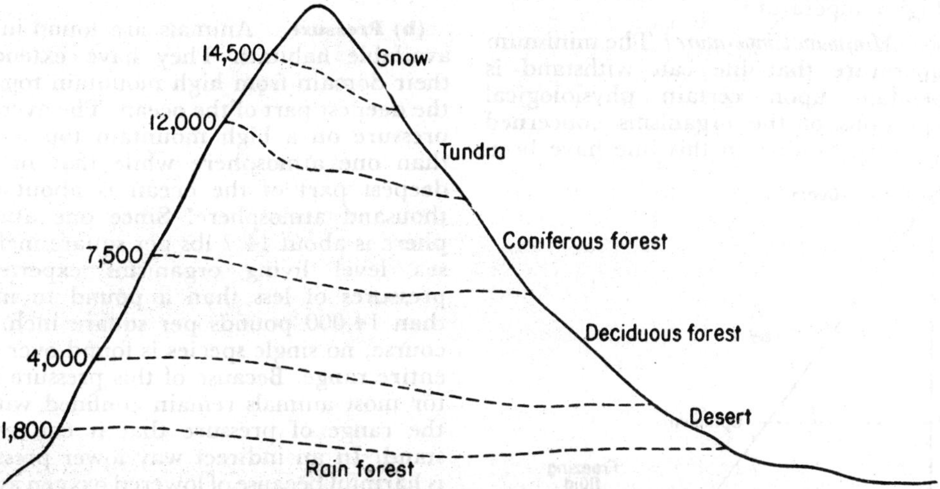

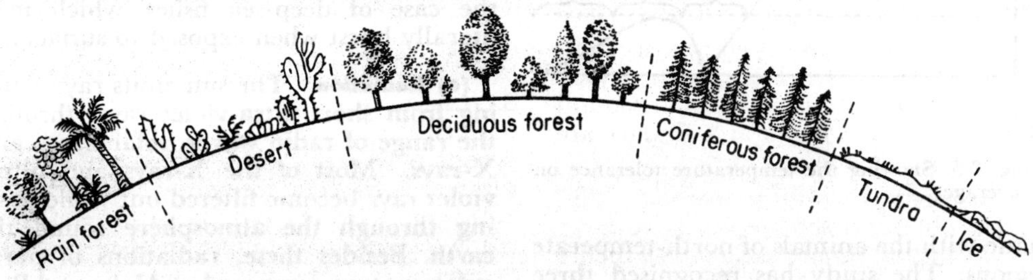

Fig. 3.4. Showing the similarity in the vegetation along latitude and altitude. Elevation in feet.

FACTORS INFLUENCING HABITATS

Animals and plants are not isolated organisms living independent of each other. Both of them are surrounded by a host of *Abiotic* or non-living factors. These factors are measured in terms of chemical composition, texture, pressure, temperature, humidity, etc. Moreover, under natural conditions no organism is divorced from other living object and as such in a community the different members develop various sorts of interrelationship concerning food, shelter and reproduction. These interrelationships are collectively known as *Biotic* *factors.* These two factors are discussed below.

Abiotic factors

(a) Temperature. Life can exist within a narrow range of temperature. Even within this narrow range, temperature tolerance of various species of animals varies. The temperature tolerance of a given species at different stages of its life history may vary. The narrow range of tolerable temperature for animals can be distinguished into three levels. These levels are:

(i) *Maximum temperature:* The maximum temperature at which some organisms are known to survive is about 88°C. Hot springs often bear a temperature of this kind. Living bacteria have been found in such hot springs. Cysts of protozoans and seeds can withstand this temperature. Multicellular blue algae live successfully in waters of Yellowstone National Park in Wyoming at a temperature of 85°C. The maximum temperature that can be tolerated by most of the animals ranges between 40°C.–45°C. One species of snow algae dies when the temperature rises above 4°C.

The death of animals at a temperature above the lethal maximum is due to a number of factors. The protein of the cells in most cases becomes denatured or coagulated at high temperatures. The chemical composition and reaction of the chemicals in the protoplasm also change at high temperature.

(ii) *Minimum temperature:* The minimum temperature that life can withstand is dependant upon certain physiological adaptations of the organisms concerned (Fig. 3.5). Studies in this line have been

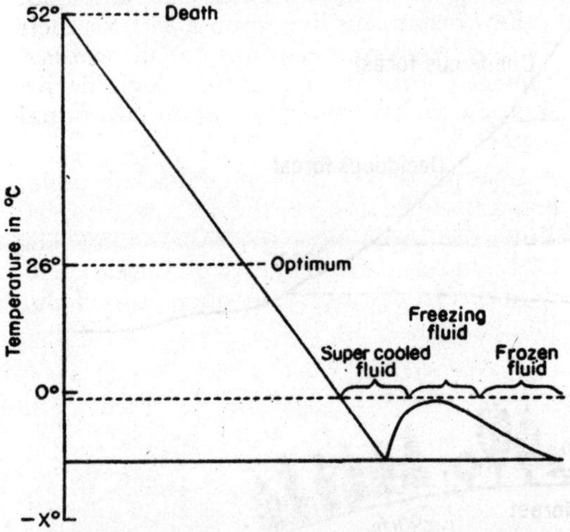

Fig. 3.5. Showing the temperature tolerance on an average.

made with the animals of north-temperate regions. The study has recognised three types of animals: (*i*) those which can survive in temperature approaching absolute zero (—273°C.), (*ii*) those that die at or near the freezing point (—4°C.) and (*iii*) lastly those that die at some temperature level above the freezing point. It has been observed that death is due to the freezing of body fluid.

Animals included in the first group are cysts of Protozoa, Rotifers, Nematodes. The second group includes the *ectothermic* animals. Ectothermic animals are those which have little or no ability to maintain a body temperature above that of surrounding environment. The third group includes mammals and birds where body temperature is independent of environmental temperature.

(iii) *Optimum temperature:* Optimum temperature lies between the two lethal limits of temperature and at this temperature life processes function smoothly. It is to be remembered that optimum temperature is not midway between the maximum and minimum levels for a given animal but it lies nearer the maximum end of the range.

(b) Pressure. Animals are found in all available habitats. They have extended their domain from high mountain tops to the deepest part of the ocean. The average pressure on a high mountain top is less than one atmosphere while that in the deepest part of the ocean is about one thousand atmosphere. Since one atmosphere is about 14·7 lbs per square inch at sea level living organisms experience pressures of less than a pound to more than 14,000 pounds per square inch. Of course, no single species is found over this entire range. Because of this pressure factor most animals remain confined within the range of pressure that it can withstand. In an indirect way lower pressure is harmful because of lowered oxygen availability. In a direct way, it is harmful in the case of deep-sea fishes which may literally burst when exposed to surface.

(c) Radiation. The sun emits rays ranging from short ultra-violet rays through the range of radio waves, solar flares and X-rays. Most of the X-rays and ultra-violet rays become filtered out while passing through the atmosphere round the earth. Besides these, radiations of particulate nature composed of Alpha and Beta particles, protons, neutrons and mesons are also emitted by the sun. Several of these together with Gamma rays constitute the cosmic rays which reach us. Moreover, disintegration of radio-active elements emits Gamma rays and X-rays. Particulate radiations, X-rays and Gamma rays have damaging effects on the protoplasm. X-rays, however, can be tolerated in small doses. Under careful medical supervision, X-rays in small doses is used in the cure of localised tissue growths, tumours and warts. Many animals require ultra-violet rays for the activation of vitamin D. Too much ultra-violet rays cause sun-burns and too much of it is lethal to unicellular animals.

The infra-red light is important for the life, because in this particular band of light 60% of the solar energy remains concentrated. Our earth is bombarded with radio waves of different kinds. Most of the waves have no effect on the biological system. Most of them irritate us by interfering with our own radio communication.

(d) Light. Light is essential in most animal habitats because the process of photosynthesis is directly dependent upon light and without photosynthesis food resources cannot grow. It is a popular saying that cows are assimilating grass and grass is assimilating sunlight, carbon dioxide, water and salts. Light brings advantages as well as disadvantages in every habitat. Light undergoes diurnal and seasonal variation and influences the behaviour of animals in many ways.

Biotic factors

Food relationship amongst organism. The activity of all living organisms is centered round food. Amongst the living organisms, plants and only a few animals are capable of producing their own food by photosynthesis or chemosynthesis. The organisms which derive raw materials from the environment and produce their own food are called *producers* or *autotrophs*. Most animals and few plants are directly or indirectly dependent upon the producers. These organisms are called *consumers* or *heterotrophs*. The heterotrophs are again divided in *herbivorous* and *carnivorous* animals. Herbivorous animals feed upon plants as such are directly dependent upon the producers. Carnivores are dependent upon the herbivorous animals for their food and as such they are indirectly dependent upon the plants. The carnivorous animals that kill other animals are called *predators* and the animals that are killed called *prey*. *Omnivorous* animals are those which are adapted to both animal and plant food.

Few organisms live upon a *host* for their food. These organisms are called *parasites*. *Saprozoic* animals are those which derive their nutrients from dead plant or animal matter.

Food chain. Any habitat, if it is habitable, contains both autotrophs and heterotrophs. The autotrophs depend on raw materials of the habitat and heterotrophs are dependent directly or indirectly upon autotrophs.

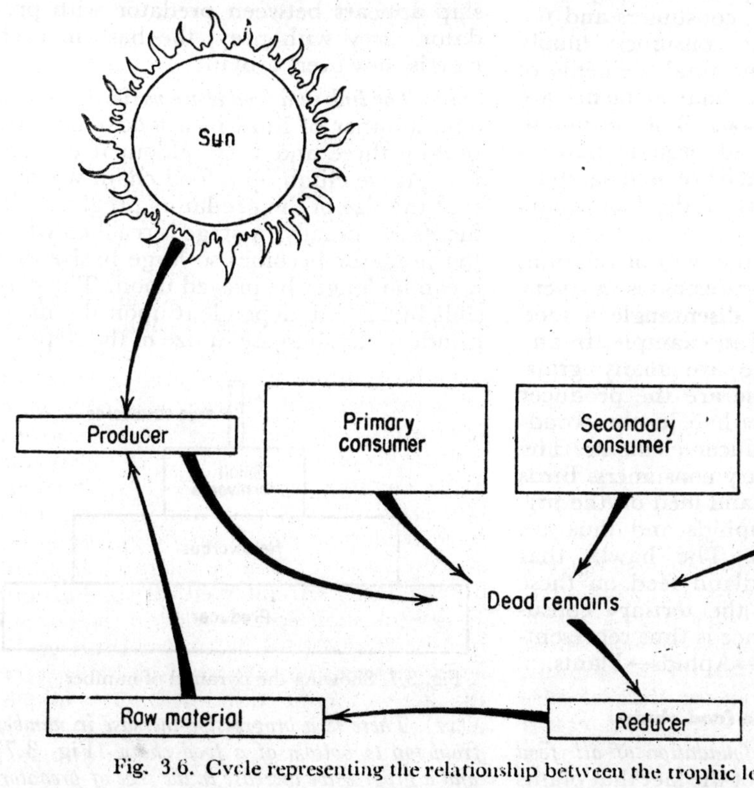

Fig. 3.6. Cycle representing the relationship between the trophic levels in nature.

These relationships pertaining to the nutritional factor form a *food series* or *food chain.* The food chain is never a closed circuit but bears many ramifications.

Several trophic levels form the knots in a food chain. The producer elements or autotrophic biota form the very foundations of a food chain. These photosynthetic organisms must exist if other levels are to be maintained.

Other trophic levels of the food chain consist of *consumer elements*. The consumer elements include a group of organisms having diverse food habit and preference. Herbivores, carnivores, omnivores, parasites, saprozoics are all included in the consumer elements. Any consumer that is dependent upon the producer elements for its food is called a *primary consumer*. The next step in the food chain is occupied by *secondary consumers*. Secondary consumers are dependent upon the primary consumers. A step above this remains occupied by *tertiary consumers*. Thus a survey of a food chain reveals the position of an organism in a food chain. The final consumer in a food chain will eventually die due to old age, injury, disease or any other factor. The producers, consumers and the waste products of the consumers finally become exposed to the final elements of the food chain. These final elements are the *decomposers* or *reducers*. The decomposers transform the dead organic matters into raw materials which are used again by the producer elements of the food chain (Fig. 3.6).

The food chain or the web of relations between different organisms is a very complex one. Let us disentangle a food chain with the help of an example. In any open woodland there are many grass, herbs and trees. These are the producer elements. Aphids that live in the woodland feed on the producer elements. The aphids thus are primary consumers. Birds that live in this woodland feed on the primary consumers or aphids and thus are secondary consumers. The hawks that may live in the woodland feed on these birds and thus form the tertiary consumers. The food sequence is thus represented by: Hawks→Birds→Aphids→Plants.

Common features of the food chain

(i) *Plants lie at the foundation of all food chains*. This is because of the fact that plants are the only organisms that can manufacture their own food.

(ii) *Food chains are of complicated nature*. A variety of food is used by the animals. Thus each chain has many side branches. Each branch divides and subdivides giving the food chain a very complex appearance. Hardy has established the minute details of the complexity of the food chain from the study of the life cycle of herring. When $\frac{1}{2}$ to $\frac{1}{4}$ inch in length, the young herrings feed on diatoms, larvae of molluscs and crustacean copepods. At the same time they are preyed upon by the jellyfishes. When the herrings are $\frac{1}{2}$ to $1\frac{3}{4}$ inches in length, they feed mainly on copepods and at this stage they are not preyed upon by the jellyfishes. In the next stage of its life (length about 5 inches) its food habit changes slightly but food preference becomes wider. Invertebrate predators cannot take them as their prey at this stage. From this stage onward and uptil the attainment of adulthood the fishes switch over from copepods to young sand eels, larger crustaceans and tunicates (Oikapleura). Thus at the different stages of the life history of herring a complex relationship appears between predator with predator, prey with prey, the base in each case is, however, a plant.

(iii) *The link in a food chain varies from* 3–5. The number of links in a food chain are usually three and very seldom it exceeds five. As we climb up a food chain we find that the size of the predators progressively increases. Finally, a stage is reached when the predator becomes so large in size that it can no longer be preyed upon. The possible link is thus dependent upon the magnitude of the increase in size in the steps.

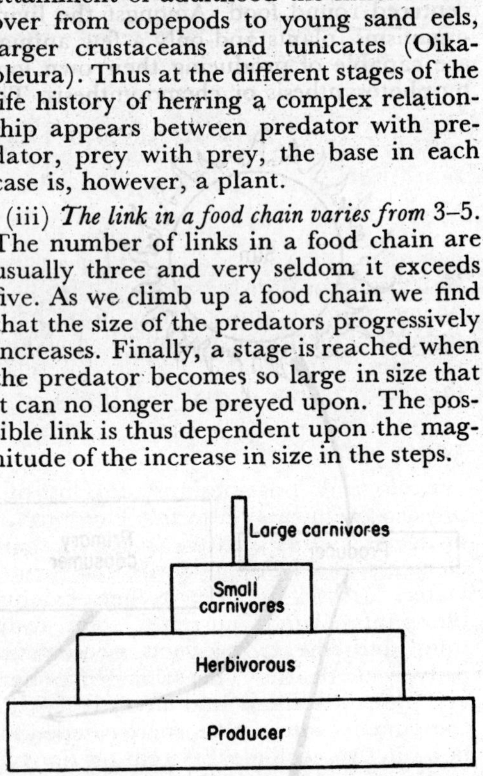

Fig. 3.7. Showing the pyramid of number.

(iv) *There is a progressive increase in number from top to bottom of a food chain* (Fig. 3.7) *and a progressive increase in the size of predators from bottom to top.* At each step of a food chain the size of the predator becomes large and the number become small. The mortality rate among smaller animals is usually higher. Such a progression or *pyramid of number* as it is sometimes called is a characteristic feature of all animal communities (Fig. 3.8).

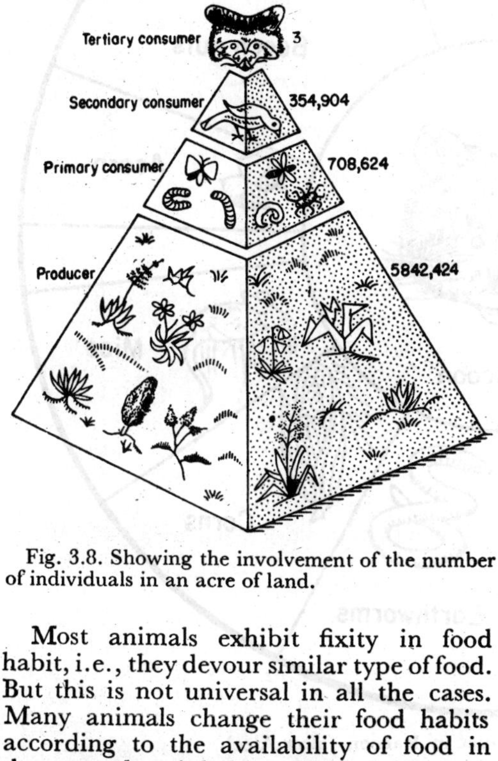

Fig. 3.8. Showing the involvement of the number of individuals in an acre of land.

Most animals exhibit fixity in food habit, i.e., they devour similar type of food. But this is not universal in all the cases. Many animals change their food habits according to the availability of food in the area they inhabit. Fig. 3.9 relates the seasonal shift of diet in Racoon.

ECOSYSTEM

Ecological relationships develop on a physico-chemical platform. The physical or *abiotic* factors include basic inorganic elements and compounds in the form of water, carbon dioxide, oxygen, calcium, phosphates, wind, currents, solar radiations and the by-products of organism activity or death. The *biotic* components are plants, animals and microbes. These components interact in energy-dependent fashion. An ecological system or *Ecosystem* is thus comprised of the abiotic physico-chemical environment and the biotic assemblage of plants, animals and microbes. Ecosystems are real but at the same time abstract being made up of conceptual schemes developed from a knowledge of real system.

The structure and dynamics of ecosystem

In the final analysis all ecological relationships are energy oriented. *Radiant energy* in the form of sunlight is the prime source of energy for any ecosystem. This energy is used in the process of photosynthesis by which carbon dioxide is assimilated to give rise to energy rich and complex carbon compounds. The chlorophyll-bearing plants are the only organisms that are capable of performing this vital function. These plants have been termed *producers*. The energy incorporated within the producer by the process of photosynthesis is utilized in the synthesis of other molecules that meet the nutritional requirements of the producer itself. So the producer may be referred to as *autotroph*, i.e., one who is capable of self-feeding. Any organism whose nutritional requirements are met by feeding on other organism is called a *heterotroph*. A primary consumer then is a heterotroph that derives its nutrition directly from the plants. A secondary consumer is in the same token a heterotroph which derives its energy from the primary consumer. The autotroph and heterotroph relation may be present in an extended form.

Besides these there is another group of heterotrophs in an ecosystem. This group is represented by micro-organisms like bacteria and fungi and has been termed as *decomposers*. The decomposers do not take food in the ingesture manner as is the practice of the heterotrophs. The enzymes secreted by the decomposers act upon the dead plants and animals and some of the products of decomposition are then absorbed. The decomposers thus render an invaluable service to ecosystem. The exo-enzymatic digestive activity of the decomposers releases the basic elements bound in protoplasm to the environment and makes the elements available to the plants for reuse. The producer-consumer arrangement in an ecosystem is called *Trophic Structure* (Trophic=Food). Each food level is known as trophic level. The amount of living material present in different trophic levels or in a component population is called standing crop. The standing crop is expressed in terms of *Biomass* or number

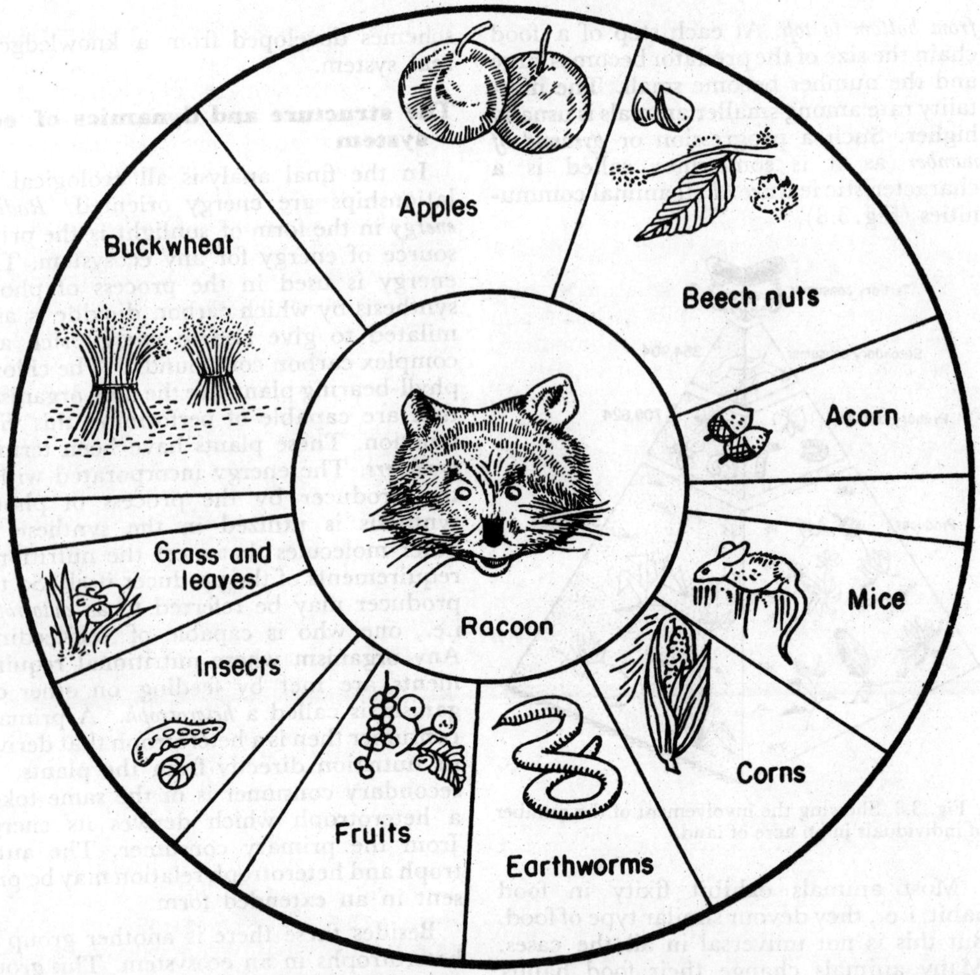

Fig. 3.9. Seasonal shift of diet in Racoon (after Alee).

per unit area. Biomass is measured as living weight, dry weight, ash-free dry weight, carbon weight, calories, etc.

The abiotic materials that remain present at any given time is considered as standing quantity or *Standing state.* In studies of ecosystem it is important to distinguish between quantities of material and organisms present at any one time or the average over a period of time and also the change in the standing states and standing crops per unit of time.

It is now evident that there are two processes operating in an ecosystem. One is energy transfer from autotroph to heterotroph and the other one is movement of nutrient elements. The energy transfer is unidirectional and non-cyclic while movement of nutrition elements is cyclic.

Energy flow in ecosystems

Living organisms utilize solar energy. In photosynthesis, this radiant energy is transformed into chemical energy and in cellular metabolism the chemical energy is again transformed into mechanical or heat energy.

Calculation on the basis of meteorological data has revealed that only about one fifty millionth of the tremendous solar energy reaches the outer atmosphere of the earth at a constant rate. This constant is called *Solar flux*. The solar flux has been defined as the amount of radiant energy (of all wave-lengths) that cross a unit area or surface per unit of time. The value has been estimated to be 2 calories per square centimeter per minute (2 cal/cm^2/min.).

Energy fixation by autotrophs

The basic equation of Photosynthesis is adequate for the measurement of the amount and rate of energy fixation by the autotrophs.

$$6CO_2+12H_2O \xrightarrow[\text{chlorophyll}]{\text{673 kilo cal.}} C_6H_{12}O_6+6CO_2+6H_2O.$$

The uniqueness of the equation is such that if the amount of one of the components is known, the other components of the equation can be worked out. If the amount of the carbon dioxide utilized in a given period of time is measured it will not be too difficult to know the amount of carbohydrate produced because it is known that for each six units of CO_2 assimilated one unit of carbohydrate is produced.

Measurement of primary production

A good number of techniques have been developed by the ecologists to measure primary production, i.e., the amount and rate of energy fixation by autotrophs.

A. Harvest method

The common and at the same time oldest method of measuring primary production is the *Harvest method.* In this method the plants grown on a particular field are clipped at ground level and their weight is taken. This is done at periodic intervals. After drying to constant weight the harvest is then expressed in terms of *Biomass* or mass per unit area per unit time (the unit of mass= gram, unit of area=meter, and unit of time=year). And now if the exact caloric content of the material is known the biomass can easily be converted and expressed in terms of energy.

Though widely used, the harvest method has a number of limitations. In this method the roots of the plant are not taken into consideration and as a result considerable error is introduced. The method fails to account for the amount of the material consumed by herbivores and also the energy utilized by the autotroph in its own metabolic activity. However it cannot be denied that harvest method can throw some light on the *gross estimate* or *gross production.*

B. Carbon dioxide assimilation method

The incorporation of CO_2 in photosynthesis or its loss in respiration under natural condition can be determined by using *infra-red gas analyzer.* With the help of the analyzer it is possible to measure the amount of CO_2 entering or leaving an air-tight chamber of known area or volume. In this method it is assumed that any CO_2 taken from the incoming air has been incorporated into organic matter during the period of observation thereby indicating both the amount and rate of photosynthesis. As there had been respiratory activity during the period of observation the use of infra-red gas analyzer alone cannot be full proof. So a comparable study is established using a light-tight container (where no photosynthesis but respiration alone will take place). The amount of CO_2 released from this chamber in a given period will be an index of the amount and the rate of respiratory activity. This value when added to that obtained in the light exposed chamber will give the total or gross production of the system.

The use of infra-red gas analyzer is not feasible in an aquatic ecosystem.

C. Oxygen production method

In aquatic ecosystem the usual procedure is the measurement of oxygen evolution by the aquatic plants. In this method samples of water containing the autotrophs are collected from a given depth of pond, lake or ocean. The sample is then distributed into smaller and paired sample bottles—the capacity of each bottle being 125–300 ml. One of the paired bottles is made up of light-transmitting material, permitting photosynthesis to take place while the other is made up of light-tight material to preclude photosynthesis but not respiration. The bottles are then suspended usually at the same depth from which samples were collected and are allowed to incubate for a given period of time. At the end of this time the oxygen content of the light-transmitting bottle is determined either by standard chemical titration or by electromagnetic devices. The value thus obtained is then plugged into the photosynthetic equation and the amount and rate of photosynthesis is determined. The normal consumption of oxygen in respiration is calculated from the light-tight bottle. By combining this amount with the value obtained in the light bottle, the total or gross production is approximated.

D. Radio-isotope method

Another method to study the primary production of aquatic autotrophs is the *radio-isotope method*. The method is similar to the light-dark paired bottle technique for measuring oxygen. But here radio-active form of carbon known as carbon-14 (^{14}C) is the extra material used. Water samples containing autotrophs are collected from a desired level of the pond, lake or ocean and are then distributed into light-dark paired sample bottles. A known quantity of carbon-14 (^{14}C) in the form of a bicarbonate ($NaH^{14}CO_3$) is then introduced into the bottles and the bottles are resuspended at the same level of water from which the samples were taken. The bottles are kept there for six hours. During the period of exposure the stable carbon present as CO_2 or HCO_3 as well as the unstable ^{14}C are assimilated into carbohydrate and incorporated into the protoplasm of the autotrophs. At the end of the incubation period the samples are filtered to capture the stable and radio-active carbon that has been incorporated biologically. After drying the filters are placed in a counting chamber to determine their level of radio-activity or in other words, the amount of radio-active carbohydrate produced. Calculations are then made from photosynthetic formula depending upon the assumption that the uptake of ^{14}C is proportional to stable carbon.

Estimation of energy accumulation by the primary producer

Edgar Transeau (1926) made the pioneer venture to estimate the energy accumulation by the primary producers. He based his calculations on an estimated harvest of 10,000 plants (corn) per 10,000 sq. meter of land in one growing season (of about 100 days). The plants weighed 6000 kg and their chemical compositions were known. He calculated that the corn plants gave 2675 kg of carbon. It is obvious that the carbon entered the plants through photosynthesis and it would be equivalent to 6687 kg of glucose. This he termed as *net production*(NP). He further calculated the amount of glucose that had been spent in the metabolism of the plants. This loss due to cellular respiration (R) came to be 2045 kg. Thus according to his calculations the gross production was (NP+R= 6687 kg + 2045 kg) 8732 kg. Now it is known that in the production of 1 kg of glucose 3760 kilo calories are needed. So in the gross production of 8732 kg of glucose 33 million kilo calories will be required. He further calculated that 408,000 gallons of water were transpired by the plants and this can be done at the expense of 910 million kilo calories of energy.

The total solar energy available to the corn field was 2043 million kilo calories (Fig. 3.10).

	GLUCOSE (kg)	KILOCALORIES (millions)	SOLAR ENERGY (%)
INCIDENT SOLAR RADIATION		2043	100·0
BIOLOGICAL UTILIZATION			
ENERGY INCORPORATED			
NET PRODUCTION (NP)	6687	25·3	1·2
RESPIRATION (R)	2045	7·7	0·4
GROSS PRODUCTION (GP) (=NP + R)	8732	33	1·6
ENERGY USED IN TRANSPIRATION	—	910	44·4
ENERGY NOT UTILIZED	—	1100	54·0

Fig. 3.10. Energy budget for 100 days in an acre of corn field.

From the above data Transeau calculated the efficiency of energy utilization and that came to be only 1·6 per cent of the total energy available. He was further able to calculate the utilization of energy that has been incorporated.

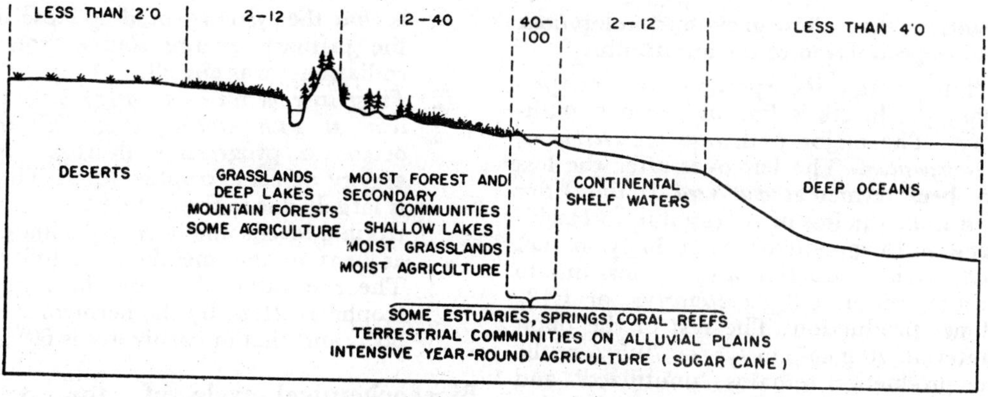

Fig. 3.11. Gross primary production of the world (in kilo calories per square meter per day).

Calculations of energy utilization by Transeau:

$$\frac{\text{gross production}}{\text{solar radiation}} \times 100 =$$

$$\frac{33 \text{ million k. cal.}}{2043 \text{ million k. cal.}} \times 100 = 1{\cdot}6 \text{ per eent.}$$

Calculation of utilization of incorporated energy by Transeau:

$$\frac{\text{energy of respiration}}{\text{energy of gross production}} \times 100 =$$

$$\frac{7{\cdot}7 \text{ million k. cal.}}{33 \text{ million k. cal.}} \times 100 = 23{\cdot}4 \text{ per cent.}$$

The data calculated by Transeau was well ahead of his time and by the application of the data it has been possible to calculate the gross primary production in kilo calories of the world per square meter per day (Fig. 3.11).

Flow of energy beyond the producers

The autotrophs incorporate and utilize energy. What happens to this incorporated energy (or in other words the fate of the energy beyond the initial trophic level) is a pertinent question. To provide the answer to this question an analysis of the following data will be of much help. The data was prepared by Raymond Lindeman (1942).

	g cal/cm²/yr.
Incident solar radiation	118,872·0
Plant utilization	
Net production (NP)	87·9
Respiration (R)	23·4
Gross production (GP)	111·3

Annual energy budged of Ceder Bog Lake of Minesota estimated by R. Lindeman.

The table shows that in the Ceder Bog Lake of Minesota (Fig. 3.12) there is a gross production (GP=NP+R) is 111 g cal/cm²/yr. with an efficiency of energy capture of 0·10 per cent. Of this energy 23·4 g cal/cm²/yr. or 21% is consumed in

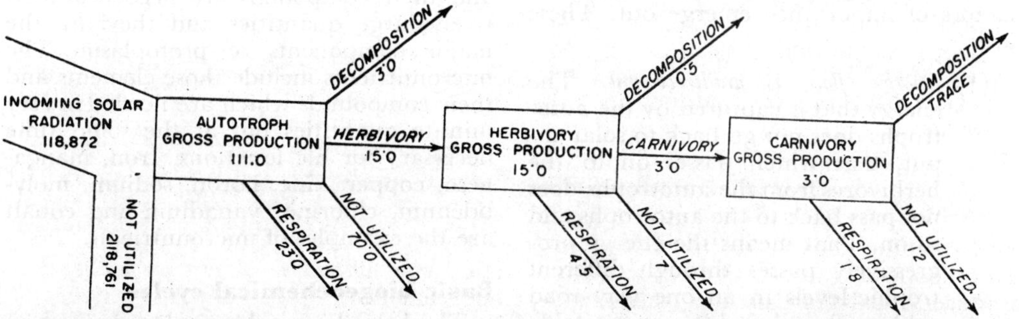

Fig. 3.12. The sequences of the fate of energy flow in gram calories per square centimeter per year in Ceder Bog Lake, Minesota.

reactions that allow growth, development and reproduction of the autotrophs.

The energy incorporated by the primary producers is lost in various pathways. The major pathways are *Herbivory*, *Decomposition*. The left over after the loss has been termed as *unutilized*. In the Ceder Bog Lake the loss of energy due to Herbivory is 15 g cal/cm²/yr. or 17%. of net autotroph production and the loss due to decomposition is 3 g cal/cm²/yr. or 3·4% of net production. The rest of the plant material, 70 g cal/cm²/yr. or 79·5% of the net production remains 'unutilized' and finally it becomes part of the accumulating, sediments.

Fate of energy beyond the primary consumers.

The herbivores are the primary consumers. In the Ceder Bog Lake the herbivores incorporate as much as 15 g cal/cm²/yr. of energy. Of this the herbivores lose 4·5 g cal/cm²/yr. or 30% of energy due to metabolic activity. This leaves 10·5 g cal/cm²/yr. of energy available to the carnivores. But this energy is not utilized. It has been calculated that only 3 g cal/cm²/yr. of the net production (NP) passes to the carnivores.

Again at the level of the carnivore about 60 per cent of the carnivores energy intake is consumed in metabolic activity and the rest 40 per cent becomes the part of the non-utilized sediments. Yearly loss due to decomposition is insignificant.

Nature of energy flow in an autotroph based ecosystem

From the calculations made above in the Ceder Bog Lake energy flow several factors of importance emerge out. These are:

(1) *Energy flow is unidirectional.* The energy that is captured by the autotrophs does not go back to solar input, that which passes on to the herbivores from the autotrophs does not pass back to the autotrophs and so on. That means the energy progressively passes through different trophic levels in an one way road as the energy is in no case available to the previous trophic level. The danger that lies with any ecosystem is that the system would collapse if the primary energy source (solar radiations) was cut off.

(2) *Progressive decrease in energy utilization at each trophic level.* There occurs a progressive decrease in energy at each trophic level. This is largely due to the fact that at each trophic level some energy dissipates as heat in the metabolic activity. The respiratory loss by the autotrophs is 21%, by the herbivores is 30% and that in carnivores is 60%.

Biogeochemical cycle of the ecosystem

Three major types of cycles are encountered in the ecosystem. One of these is the *Hydrologic cycle* involving the movement of compounds and the *Chemical cycle* involving movement of elements.

The chemical elements that constitute the essential components of protoplasm tend to circulate in the biosphere in characteristic paths from environment to organisms and back to the environment. In so doing the elements follow more or less a circular route and form what is known as *Biogeochemical cycle*. The word 'Bio' here refers to living organisms and the word 'Geo' refers to the rocks, air and water of the earth. The movement of those elements and compounds that are essential to life has also been termed as *Nutrient cycling*. The elements and dissolved salts essential to life have been termed *Biogenic salts* or *Nutrients*. The nutrients have been divided into two groups—*Macronutrients* and *Micronutrients*. The macronutrients are exemplified by carbon, hydrogen, oxygen, nitrogen, potassium, calcium, magnesium, sulphur and phosphorus. These elements and their compounds are needed in relatively large quantities and they are the major components of protoplasm. The micronutrients include those elements and their compounds which are needed in very minute quantities but at the some time necessary for life functions. Iron, manganese, copper, zinc, boron, sodium, molybdenum, chlorine, vanadium and cobalt are the examples of micronutrients.

Basic biogeochemical cycles

The basic biogeochemical cycles can be resolved into two compartments or pools—the *Reservoir pool* and the *Cyclical pool*. The reservoir pool is large, slow moving and is formed of non-biological components. The cyclical pool, on the other hand, is small but more active portion that is exchanging (i.e., moving back and forth) quickly between the organisms and its environment. From the standpoint of the biosphere, in its entirety the reservoir pool can further be divided into *Gaseous types* and *Sedimentary types*. The gaseous type of reservoir is represented by the atmosphere or hydrosphere. The gaseous type of reservoir shows little or no permanent change in its distribution and abundance of elements. Carbon and nitrogen are the chief representatives of the gaseous type of reservoir. The sedimentary type of reservoir is the lithosphere or the earth's crust. The sedimentary types are exemplified by phosphorus, sulphur, iodine and others show a tendency to stagnate.

Hydrologic cycle

In the hydrologic cycle there occurs an interchange of compounds between the earth's surface and the atmosphere via *precipitation* and *evaporation*. The biota of the ecosystem plays an accessory role in the cycle and the presence or absence of the biota does not affect the movement of the cycle. However, it is an established fact that a significant amount of water is incorporated by the biota of the ecosystem in protoplasmic synthesis and also there is a substantial return to the atmosphere by way of transpiration.

Hutchi on (see Hutchinson G. E.—*A Treatise on Limnology, Geography, Physics and Chemistry, Vol. I*, John Wiley and Sons, 1957) has calculated the world precipitation per year and water contents of the various parts of the earth (Fig. 3.13). According

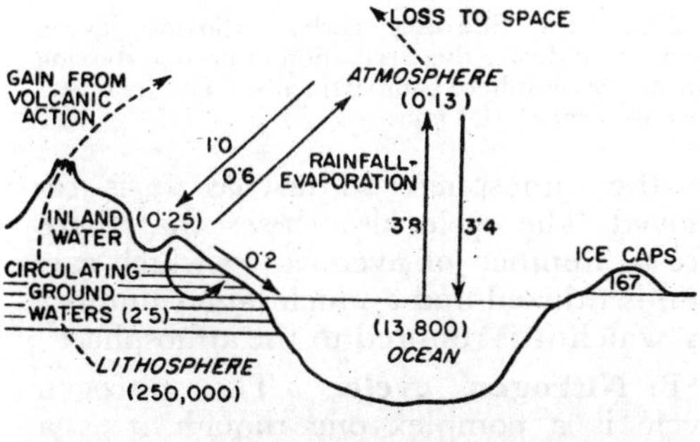

Fig. 3.13. Showing hydrological cycle. Arrows indicate the circulation of water between the components of biosphere units in geogram (10^{20}).

to him the world precipitation per year amounts to 4·46 geograms (1 geogram= 10^{20} gm). Of this amount 0·99 geogram falls on land and 3·47 geogram falls on ocean surface. The water content of the earth's surface is 266,069·88 geogram. The water content of the various parts of the earth is given below.

	Parts	contents (geograms)
1.	Primary lithosphere	250,000
2.	Ocean	13,800
3.	Sedimentary rocks	2,100
4.	Polar caps	167
5.	Circulating groundwater	2·50
6.	Inland water	0·25
7.	Atmospheric water vapour	0·13
	Total	266,069·88

Characteristics of hydrological cycle

1. The general world precipitation pattern is dependent upon the interaction of several forces. Primary of these forces is the interaction between atmospheric circulation and the topography. The distribution of the major ecosystems is dependent upon the interactions. The rate and amount of precipitation being as critical as those of evaporation, the ratio of these two factors forms the crucial factor in determining the distribution of particular types of ecosystems.

2. More water evaporates from the ocean than that returns to it via rainfall and conversely less water evaporates from land and more returns to it via rainfall. This indicates that a part of the rainfall which supports land ecosystem comes from the water evaporated from the ocean. It has been estimated that in Mississippi valley as much as 90% of the rainfall comes from the ocean.

3. There is about 0·25 geogram of water in fresh-water ponds, lakes and rivers. The run-off is 0·2 geogram per year and the annual rainfall is 1·0 geogram. Thus the annual recharge rate of ground water is 1·0−0·2 geogram or 0·8 geogram.

4. The circulation pattern of atmosphere determines the pattern of precipitation distribution. The trade wind may be cited to explain this. The trade winds move

from cooler latitudes towards the equator carrying moisture and depositing the moisture in the equatorial region. This makes the equatorial region very wet whereas the coast line to the north and south of the equator remains relatively dry. The other example that may be cited is that of *Rain shadow*. When moist air moves over a mountain range it rises and cools to supper saturation. Precipitation occurs in such cases on the windward side of the range. As the moisture depleted air continues to move it comes down, warms and picks up moisture by evaporation from ground and water surfaces. This region thus receives less moisture than the windward region forming the rain shadow. Examples of rain shadow region are Rocky mountains and south side of the Himalayas.

5. In the hydrological cycle the H_2O atmospheric compartment is small and it has a more rapid turnover rate and shorter time of residence in the atmosphere than CO_2 cycle.

Gaseous nutrient cycle

A. **Carbon cycle.** The carbon cycle is the simplest of all nutrient cycles. The major reservoir of carbon is the ocean where it remains stored as bicarbonate. The oceans contain more than 50 times as much carbon as that of air and the oceanic reservoir regulates the amount of carbon in the atmosphere. Atmosphere is the minor reservoir though the basic movement of carbon is from atmosphere to producers, from producers to consumers, from consumers to decomposers and then back from the decomposers to the atmosphere. In the atmosphere carbon dioxide concentration is from 0·03 to 0·04 per cent. Annual fixation of carbon dioxide by green plants is about 4 to 9×10^{13} kg.

A considerable amount of biologically fixed carbon as gaseous CO_2 is returned to the atmosphere by the respiratory activity on the part of the producers and consumers. But the most substantial return is accomplished by the respiratory activity of the decomposers and the way of their processing the waste materials and other trophic levels.

It has been estimated that 8 billion tons of CO_2 is injected annually into the atmosphere. Of this 6 billion tons come from burning of fuel or accidental fire and 2 billion tons from cultivation of land for agriculture. Most of this is quickly passed into the sea and is kept there stored in the form of carbonate. The interplay between atmospheric and aquatic carbon dioxide is worthnoting. The interchange between the two phases occurs through the process of diffusion, the direction of flow is dependent upon the concentration. The aquatic phase of CO_2 occurs through precipitation. A litre of rain water contains about 0·3 c.c. of CO_2. The dissolved CO_2 combines with the water of the ecosystem and forms carbonic acid (H_2CO_3), the reaction being always reversible. The carbonic acid in its turn disassociates in a reversible reaction into H^+ and HCO_3^- ions. The bicarbonate ions again disassociate reversibly into H^+ and CO_3^- ions.

$$CO_2 \text{ of atmosphere}$$
$$\downarrow\ \uparrow$$
$$CO_2 \text{ dissolved} + H_2O \rightleftharpoons H_2CO_3 \rightleftharpoons H^+ + HCO_3^-$$
$$HCO_3^- \rightleftharpoons H^+ + CO_3^-$$

As all these reactions are reversible the direction of movement is dependent upon the concentration of the components.

The carbon cycle (Fig. 3.14) in reality is a perfect one and carbon is returned

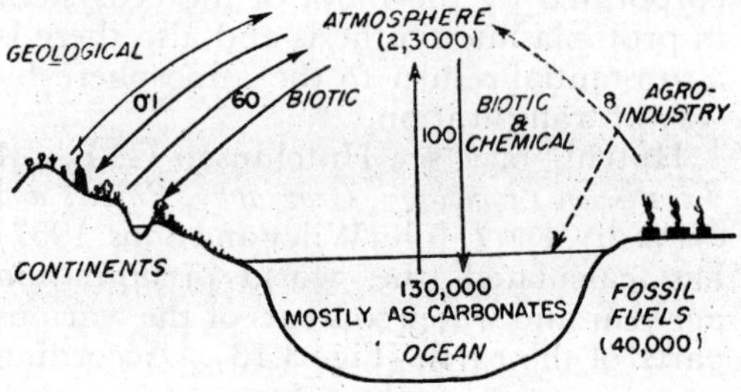

Fig. 3.14. Showing carbon dioxide cycle. Arrows indicate the circulation of carbon dioxide in major biosphere compartments. The members are in term of 10^9 tons.

to the atmosphere as fast as it is removed. The cycle also shows that there are a number of avenues by which carbon is utilized and a much larger number by which it is restored to the atmosphere.

B. **Nitrogen cycle.** The nitrogen cycle is a complex one though it is a complete and perfect cycle. The cycle exhibits marked difference from that of carbon cycle. The nitrogen cycle is extensive, complicated and at the same time ordered (Fig. 3.15).

The atmosphere contains as much as 79 per cent nitrogen. That means the organisms live in a nitrogen-rich atmosphere. But if one thinks that nitrogen reservoir is atmosphere, he is mistaken. Because most organisms are unable to use atmospheric nitrogen. The real store-house of nitrogen for the use of organisms lies in inorganic forms like ammonia, nitrite and nitrate and also in organic forms like urea, protein and nucleic acids.

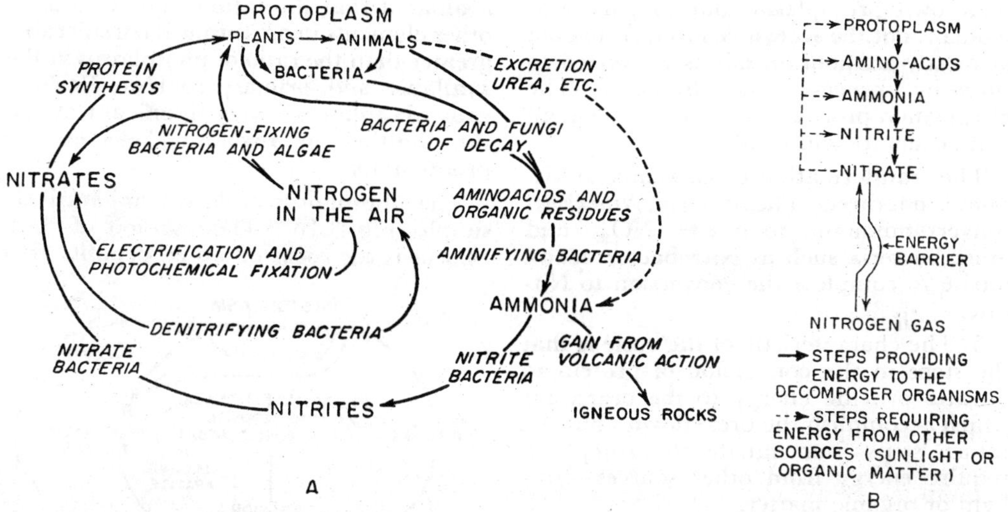

Fig. 3.15. Showing nitrogen cycle. Circulation of nitrogen between organism and environment (A). The micro-organisms are responsible for key steps. The same basic steps are shown in B.

Nitrogen fixation

Living organisms can tap nitrogen largely in the form of nitrate. So fixation of atmospheric nitrogen in nitrate form is at the helm of the affair. The fixation of nitrogen occurs through physico-chemical and biological means.

Biological means of nitrogen fixation

A few but abundant kinds of bacteria collectively called *Nitrogen-fixing bacteria* and some algae have the capacity to fix atmospheric nitrogen. At the same time there are many bacteria called *Denitrifying* bacteria which can free the nitrogen. It is by the action of this micro-organisms the nitrogen is continually entering the air and continually returning to the cycle.

The nitrogen-fixing organisms are:

1. Free-living bacteria—Ex. *Azotobactor* (aerobic) and *Clostridium* (anaerobic).
2. Symbiotic nodule bacteria on legume plant—Ex. *Rhizobium.*
3. Purple bacteria—Ex. *Rhodospirillum.*
4. Blue-green algae—Ex. *Anabaena, Nostoc.*
5. Some Lichen.

The dentrifying organisms are *Pseudomonas* and some Fungi.

Course of events in nitrogen cycle

1. The nitrogen in the air is transformed into nitrates by (1) the nitrogen-fixing bacteria and by (2) electrification and photo-chemical means. It has been estimated that electrification and photochemical nitrogen fixation is in the order of 35 mg/m²/year and biological fixation or fixation by nitrogen fixing bacteria is 140–700 mg/m²/year.

2. Nitrates and other simpler nitrogen compounds are used by plants for the synthesis of amino acid and protein which in turn go to the animals.

3. The animal metabolic wastes in the form of urea, uric acid, etc., are transformed into ammonia. The protoplasm of dead animals and plants is acted upon by bacteria, actinomycetes and fungi occurring both on land and water. These

organisms utilize the dead organisms or which are but organic nitrogen-rich substrates in their metabolism of it. They convert it to and release it in the organic form ammonia. The process is known as *Ammonification* or mineralization. A considerable amount of ammonia is also gained from volcanic action.

4. Many autotrophic and some heterotrophic bacteria only can use nitrogen occurring in ammonia form to synthesize their own protoplasm. But to most ammonia is not the accessible form. Ammonia or mostly ammonium salt is converted to *nitrate* by a process called *nitrification*. The nitrification process is slow, pH dependent and occurs in acid condition.

The transformation of ammonia to nitrates is not direct. The bacteria, *Nitromonas* convert ammonia to nitrite (NO_2) and other bacteria such as Nitrobactor act on nitrite to complete the conversion to Nitrate.

5. The characteristic of the cycle is that the steps in the conversion of protein to nitrates provide energy to the organisms which accomplish the breakdown. But the return steps from nitrite to protoplasm require energy from other sources—sunlight or organic matter.

Amount of nitrogen fixation

In 1944 it was estimated that the amount of nitrogen fixed from the air lies between 140 and 700 mg per square metre per year. Most of this fixation is biological in nature and only a small quantity (about 35 mg per square metre per year) is fixed by electrification and photo-chemical means. Recent (1970) estimates show that biological fixation of nitrogen on the earth's terrestrial surface is at least 1 gm per square metre per year and in fertile lands it may be as much as 20 gm per square metre per year.

Sedimentary nutrient cycle

In addition to oxygen, hydrogen, carbon dioxide and nitrogen living organisms require at least 13 other elements like calcium, phosphorus, potassium, sodium, chlorine, sulphur, magnesium, iron, copper, manganese, iodine, cobalt and zinc. Traces of these elements are only required by living organisms. But the requirement of calcium and phosphorus is needed a bit more than other elements. Phosphorus is a constituent of nucleoproteins, phospholipids and skeletons. Calcium is needed for skeletons, shells, antlers and other organs. These elements are obtained from food, water, salt licks and grit taken into the stomach.

Phosphorus cycle

Phosphorus is a necessary and important constituent of protoplasm. It has been evaluated that the ratio of phosphorus to other elements in organisms is considerably greater than the ratio of phosphorus in the available and primary sources. Ecologically phosphorus is very significant as it is the limiting or regulating element in productivity.

The phosphorus cycle is comparatively simple (Fig. 3.16). The reservoir of phosphorus is the rocks or other deposits that

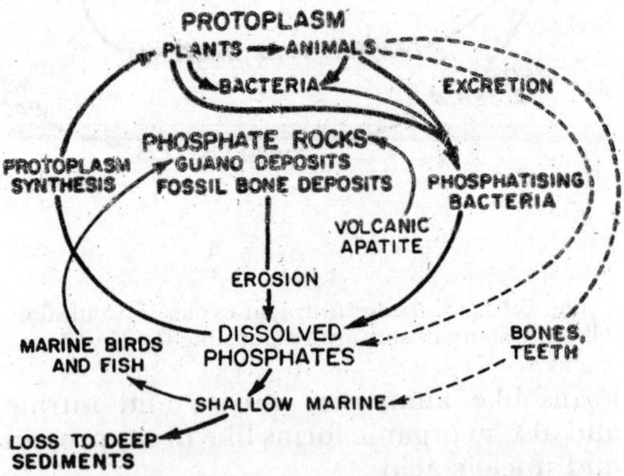

Fig. 3.16. Showing phosphorus cycle.

have been formed in past geological ages. Erosions of these reservoirs release phosphate to the ecosystems. But in the process much of the phosphate escapes into the sea, where part of it is deposited in the shallow sediments and part of it is lost to the deep sediments.

The means of return of phosphorus to the cycle is inadequate. The uplifting of sediments in most part of the world has become rare. However, sea-birds play an important role in bringing back phosphorus to the cycle through their 'guano' deposits. Man also harvests a lot of marine fish and this aids in the return of some phosphorus to the cycle.

Plants take inorganic phosphate as orthophosphate ions. This phosphorus is transferred to consumers. After death the protoplasm of plants and animals is acted upon by decomposers (phosphatizing bacteria) to make it available again as dissolved phosphate. The excreta of animals also return some phosphorus to the cycle. The bones and teeth of animals being very resistant to weathering account for some loss of phosphorus.

A study of phosphorus cycle reveals that the return of phosphate to the cycle is inadequate to compensate the loss. It is man who has hastened the rate of loss of phosphorus and has made the phosphorus cycle inadequate. It has been calculated that one to two million tons of phosphate rock are mined per year. But most of it is washed away and lost. The annual return against this is sixty thousand tons per year. The agronomists tell us that the situation is not alarming as phosphate deposits are still too large.

Sulphur cycle

Only a few organisms meet their sulphur requirements in such forms as amino acid and cystein. The source of biologically significant sulphur is inorganic sulphate.

The reservoir of sulphur lies in the soil and sedimentary rocks. The atmosphere is a minor reservoir formed by fuel combustion.

The centre wheel of the sulphur cycle (Fig. 3.17) rotates round the activity of a group of specialized micro-organisms which function as a relay team, each carrying out a particular chemical oxidation or reduction.

The sedimentary aspect of the cycle involves the precipitation of sulphur in the presence of iron in anaerobic condition. Ferrous sulphide is unsoluble in neutral or alkaline water and as a result the sulphur has the potential for being bound up under these conditions to the limits of the amount of iron present.

The biologically incorporated sulphur is mineralized by bacteria and fungi in ordinary decomposition. Some such sulphur is also reduced directly to sulphides including hydrogen sulphide by bacteria specially the *Escherichia* and *Protens*.

Inorganic sulphate (SO_4) is the source of elemental sulphur in the ecosystems. Under anaerobic condition the sulphate is reduced to elemental sulphur or to hydrogen sulphide by bacteria under the genus *Desulphovibrio*, *Escherichia* and *Aerobactor*. The presence of a large amount of hydrogen sulphide occurring in the anaerobic or deeper portion of aquatic ecosystem is inimical to animal life. The H_2S rises to shallow sediments and is acted upon by other organisms. Colourless sulphur bacteria such as species of *Beggiatoa* oxidize hydrogen sulphide to elemental sulphur. Species of *Thiobacillus* oxidize elemental sulphur to sulphate and other species of Thiobacillus oxidize sulphide to sulphur.

At the global level the regulation of sulphur cycle is dependent upon the interaction of geochemical and meteorological processes (erosion, sedimentation, leaching, rain absorption), and biological processes (production and decomposition). The

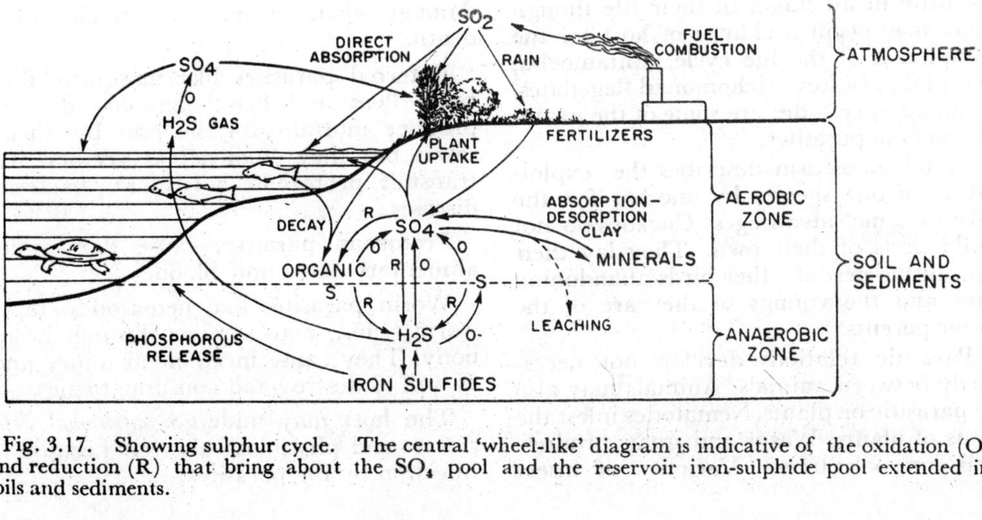

Fig. 3.17. Showing sulphur cycle. The central 'wheel-like' diagram is indicative of the oxidation (O) and reduction (R) that bring about the SO_4 pool and the reservoir iron-sulphide pool extended in soils and sediments.

interdependence of air, soil and water also aids in the regulation.

COMMUNITY DYNAMISM

Many organisms live in a habitat. These organisms establish interrelationship of various kinds amongst themselves. Co-action between organisms that are beneficial to one or more of the participants is called *co-operation*, while co-actions between species or individuals that are harmful to one or more of the participants is called *disoperation*.

Disoperations. The major disoperations present in a community are *parasitism*, *predation* and *competition*.

Parasitism. Parasitism is a relation between two individuals whereby the *parasite* receives benefit at the cost of the *host*. Parasitism is a form of disoperation which is mainly food oriented but the parasites sometimes derive shelter and protection from the host as well. For its own survival the parasite does not kill its host at least not until the parasite has completed its reproductive cycle.

Classification of parasites. Parasites are commonly classified as *Ectoparasites*, *Endoparasites* and *social parasites*.

Ectoparasites live outside the body of the host. Ectoparasites at the adult stage only are mosquito, flea and bed-bug. But the blood-sucking lice and flies, biting lice, mites and ticks are ectoparasitic both at larval and adult stages. Endoparasites are those which live in the alimentary tract, body cavities, blood or other tissues of the host. Endoparasites, in general, are parasitic in all stages of their life though there may occur a change of host for the completion of the life cycle. Entamoeba, malarial parasites, trichomonad flagellates, nematodes, cestodes are some of the examples of endoparasites.

Social parasitism describes the exploitation of one species by another for the sake of some advantages. Cuckoos do not build nests of their own. They lay their eggs in the nest of other birds abandoning eggs and the youngs to the care of the foster parents.

Parasitic relations develop not necessarily between animals. Animals may also be parasitic on plants. Nematodes infest the roots of plants. Wasps and gnats produce galls on oak trees. Varieties of insect larvae are leaf miners, wood borer, cambium feeder and fruit eaters.

Host specificity. Parasites maintain in general a host specificity. Most parasitic genera are adapted to hosts of one phylum only. The acanthocephalans *Gracilisentis* and *Tanarhamphus* are found only in the gizzards of shad fish. Each order of birds possesses its own particular species of tapeworm. This remains true even if several orders of birds live close together.

Copepods of all parasites exhibit a loose host specificity. They have been reported from various invertebrates and fish. Assignment of host specificity in case of protozoan parasites is a difficult one. Many species of parasitic protozoa have invaded more than one taxonomic host group and often several species of a single genus of parasitic protozoa invade the same host species.

Effect on parasite. Because of parasitic adaptation the parasites become restricted to special niche. Many parasites undergo degenerative changes in their organs of locomotion, sense organs and even in their alimentary canal. The reproductive structures however become well developed and some organs for attachment become new addition.

Effect on host. The parasites cause diseases of the host. By the term disease is meant a condition which affects the body or part of the body that the normal physiological functions of the body is impaired. In general, the animal parasites do not cause immediate mortality but they do cause damage to body structures and the damage when becomes too much causes death.

External parasites like mosquito, flies, ticks, fleas and lice commonly do not produce mortality of the host by themselves but they often act as vectors and transmit protozoan, virus and bacterial diseases.

Protozoan parasites cause damage of alimentary canal and blood.

Worm parasites like nematodes, tapeworm, etc., may wander through host's body. They cause mechanical injury and as well as destroy and consume tissues.

The host may undergo *nutritional deficiencies* and physiological stress because of infection by the parasites.

Evolution of parasitism. The story of the evolution of parasitism is a matter of speculation. The ancestors of parasites were all free living in water or vegetation. But how these free-living creatures became adapted to parasitic mode of life? The answer to the question is a difficult one. It may be imagined that a free-living individual living in water or vegetation got settled on the outer body surface of a larger individual accidentally. The free-living organism (the organism destined to become future parasite) in this new site found condition favourable for survival. The organism found a rich source of food in the niche and adjusted itself to become ectoparasite.

It is believed that the biting lice have evolved from psocid insects that live beneath the bark of trees. They may have migrated from this niche to the nests of birds and then to the birds themselves. Most ectoparasitic insects are probably derivatives of carnivorous, saprovorous or suckers of plant juice.

Endoparasites may in some cases have evolved from ectoparasites but more likely they come from free-living ancestors or commensals. Protozoan or flat worms might have had their first entrance into their hosts through drinking water and later invaded other organs of the body. These organisms however needed some *pre-adaptation* to live at low oxygen concentration, to avoid being digested or being carried away with the faeces. In successive generation the parasites became more suitably adapted.

Predation. Predation is a disoperation wherein one species kills and feeds on a second species. The killer individual is called *predator* while the individual that is killed is called *prey*. According to one ecologist "the difference between parasites and predators is that the predators live on the capital while the parasites live on the income."

Competition. Competition is inseparable in life. Clements has defined competition in the following way. Competition is the more or less active demand in excess of the immediate supply of material or condition exerted by two or more organisms. The materials and conditions sought by animals are food, shelter and mates. Competition may be *direct* or *indirect*. Direct competition results in active antagonism or combat between individuals. Fighting amongst Bull seals for larger harem is an instance of direct competition. In indirect competition one individual or species tries to monopolize a resource or renders a habitat unsuitable for the establishment of other organisms having identical requirements. Indirect competition is very common among plants when certain plants monopolize the water and nutrient of the soil and available light in such a way that others cannot maintain themselves. Animals often render a habitat unsuitable by their excreta for a species which otherwise would occur.

Amensalism

Amensalism is a condition in which the mere presence of a species causes hardship or difficulties of another species though the former species is neither harmed or helped. The classical example of amensalism is the Penicillium. Penicillium in cultured condition produces certain toxic substances which inhibit the growth of many kinds of bacteria. The mold of Penicillium, however, is not affected by the presence of the bacteria neither it gains by the death of the bacteria.

Co-operation

The co-operation amongst living organisms involves commensalism, proto-co-operation and mutualism.

Commensalism. Commensalism exists between two members of different species which live together and any one out of the two becomes benefited and the other is not helped or harmed. The one which derives benefit in such associations is called *commensal*. A large number of insects live as commensals in the nests of birds. Commensalism is very common in the marine habitat. The oysters keep small crabs as commensals in their mantle cavities. Some small fishes live as commensals in the cloacal respiratory trees of sea-cucumbers. Shelter and waste scraps of food are gained by the commensals.

Proto-co-operation. Proto-co-operation results when two members of different species form an association in which both gain benefit out of it. The association is not very rigid and each is capable to survive without the association. As for example the association between marine crab and hydroid co-elenterates may be cited. Hydras live on the carapace of the marine crab. In such an association the hydras gain by being moved about in the habitat and by occasionally securing food bits from the crab's diet. The crab gains by being camouflaged from the presence of hydras.

Mutualism. Mutualism occurs when two members of different species become so intimately associated that neither can survive without the other. The protozoa *Trichonympha* lives in the digestive tract of termites. Trichonympha secretes the enzymes which hydrolyse the cellulose in the termite's diet of wood. It has been seen from experiments that if the termite is freed from the presence of the protozoa, it dies of starvation. Trichonympha inside the digestive tract of termites is so adjusted that it cannot survive outside the tract excepting in cystic condition.

Intraspecific interrelationships. Only under rare and exceptional circumstances, an individual of a species becomes the sole representative of that species in a given area. The different members of a species living in a habitat form different types of association amongst themselves. These associations result in *social organisations*, *territorial segregations* and *communications* between themselves.

Needs for association. It is a common observation that animals vary in their ability of moving from place to place. Many animals are *free-living*. Free-living animals can crawl, swim, walk or fly. Some animals are *sessile*. Sessile animals remain permanently attached to rock or any other object available in their habitat. Many marine animals remain in one place for most of their lives. They are called *sedentary*. Many animals are *planktonic*. They float in water and are carried about by the action of currents and winds.

The type of locomotory ability exerts influences on the type of association that is going to be formed. Some animals are *solitary*. Solitary individuals live alone except during reproductive season. Many animals are *gregarious*. Gregarious animals form groups, flocks, herds, colonies, etc.

ACCOUNT OF SOME GREGARIOUS ANIMALS

Dependence on one another is a key character of some higher animals specially the mammals. These animals which are pressed to form a sort of 'togetherness' for their own survival have been termed *herd or gregarious* animals. Flocks of sheep, prides of lions, troops of monkeys are herd animals.

Size of herds

Herds come in all sizes. The basic unit is, however, small being constituted by not more than a dozen of participants. A herd may be confined to the members of a single family being represented by father, mother, few grown children and newly born young. This family when added with the grand parents and cousins forms a vast collection of animals of all ages and sexes. The herd may also be composed of a single male and a number of females or it may be formed by entirely male or entiretly female groups and sexes mingle only during breeding season.

Some behavioural studies on herd animals

Caribou is a herd animal in the truest sense. But the formation of the herd follows a kind of annual pattern keyed to the condition of the male's antlers. The antlers begin to grow in the spring and fall off in the next spring. A yearling bull produces only one prong on each antler but the older he becomes more *tines* (branches) he has, until he reaches the prime. At the age of four or five he gets a magnificent *rack* (set of antlers).

The antler becomes fully developed by September. The bull then scrapes off the rough hairy skin cover of the antler and sharpens the points of the antler. That is he is ready to fight for a mate. He rounds up a few yearling females and starts forming his 'harem'. He even tries to steal some other bull's cows.

At this stage he will challenge or he is challenged by some other bull and the battle starts. Each tries to gore his rival with his antlers. Sometimes the contestants, antlers become so locked that both die of starvation.

Battle over, the winner retires with his wives (about a dozen in number) and mate. The loser may attack some weaker bull or may remain isolated for the year. During winter time the bull leads his meek wives to warmer quarters and makes himself responsible for the safety of his cows, holding off wolves and other enemies. But with the advent of spring things change as the antlers fall off. The herd breaks up. The bull goes off to grow new antlers and the cows go off together to have their calves.

The common pattern of herd life here is—the members (1) split up, (2) come together for protection, shelter and reproduction, and (3) again split up. Such a pattern is followed by Bison, wild cattle and goats.

Territory of herd animals

A herd animal has no fixed home but this does not necessarily mean that it wanders everywhere. They keep themselves encompassed within a 'Territory' and stick to it. The territory may be large taking a number of days to travel from one end of the territory to the other end—the boundaries being determined by vegetation and climate.

The territory of the stags is less than a square mile. The domain of white-tailed deer is very narrow.

It is within this territory irrespective of its size the herd moves. Often it trails the same route or often it tramples out trails of its own. Some animals although sharing his territory freely with other animals of the same species, will not allow their own kind for the simple reason that they will have to compete for the same food supply and same females.

A herd of hippopotamuses make their 'head-quarter' in a gently flowing river. They spend the whole day here resting and tending the youngs. After dark they come to land in search of food and roam about in their own loop of land. This territory belongs to him and his own family and other hippos stay away. Seal and walrus herds have their favoured breeding grounds where they resort year after year. They persistently come to the same spot and put up battle against intruders.

Rules and customs of herd life

A herd may very well be compared to a society and as in a society they learn how to get along one another. Herd life has many rules and customs and consequently herd life develops many special ways of behaving.

One of these is **rank**—a system by which an individual's importance is established. Some members of the herd being physically stronger establish his authority over other members.

Peck-order amongst hens is an example of such dominance. Two hens meeting for the first time will peck at each other until one submits. The loser may not be hurt but becomes convinced that the opponent is a more important hen than she and as such considers herself lower in rank than her opponent. A stable pecking order thus becomes established in the flock. Hen A can peck all the other hens. Hen B can peck all hens in the flock except A. Hen C can peck all others except A, B and so on. High ranking hens are first to eat. They gain more weight, lay more eggs and show fewer peck scars.

Cows attain their social rank by butting and once the order is settled the discord is at an end. The head cow leads the herd to and from pasture. Sometimes the aspirants, cows second or third in rank try to outbutt the current No. 1 and when they fail they become the most discontended and ill-tempered members of the herd. The most placids, after the No. 1 are those meek members occupying the bottom of the social ladder.

Bottle-nose dolphins follow a sheep-like pattern and separate into bull herds and cow herds and get together during breeding season. Each group has its own social order. However, when the two groups unite the females drop their own system and each female takes her rank from that of her mate.

Such system of social ranking prevails among many wild animals but the system is made complicated by the presence of male members. As a rule, all males rank higher than the females.

Attributes of herd life

The important attributes of herd life are mutual protection and food procurement.

Mutual protection. Musk oxen protect themselves and their family against wolves by forming a living fortress. The musk ox has horns on its frontal bones. The horns curve down. Sensing an attack from the hungry beasts of prey the herd forms a circle. Keeping females and young ones at the centre the males occupy positions on the periphery. The deadly battle line of horns terrifies the animals of prey and very seldom they can make a meal of the musk ox.

Whales, Walruses and Dolphins offer similar group defence against the killer whales. The dolphins being surrounded by a school of killer whales form a protective circle around their youngs. As they have no organs of defence some of their members are taken by the killer whales. It has been observed that the dolphins push weaker members of the herd out of the circle offering them as a kind of sacrifice for the survival of the herd as a whole. Besides protection herd animals come to the aid of an injured fellow. A bottle-nose dolphin in distress whistles and hearing this other dolphins come to his assistance. One on either side they help the injured one to come to the surface and keep him there until he recovers. It has been observed that dolphins do the same for human swimmers.

Elephant is well known for the help he renders to his wounded comrade. Hunter after hunter has told of seeing elephants coming to the aid of a wounded mate. In one case, it has been seen that about a dozen of members of a herd struggled hard to get a mortally wounded animal on his feet. Elephants rally round a sick friend to get him back on his feet. Even the young play nurse to a sick calf.

Food procurement

Most seals and whales live in herds. They hunt together following schools of fish or octopus. The more plentiful the food the larger the herd becomes.

For lions the heavy work of finding prey is done by the lioness. She picks out a suitable position and her mate takes position upwind so that the prey gets the smell of him. The victim scenting an enemy moves away from the lion and comes closer to the lioness in ambush. The lioness kills with one swipe of her paw.

The experts in pack hunting are the cape-hunting dogs of Africa. A pack may have as many as sixty members. They chase their prey usually the antelope. The antelope often slashes back with sharp horns and hoofs and sometimes kills a dog or two. But there are many more. Sooner or later the antelope falls exhausted.

Herd animals have high I. Q.

It has been observed that herd animals are much quicker to learn tricks than solitary animals. The circus lions and tigers can only form pyramids and roar on cue. But elephants stand on their heads, play soccer, balance or balls. Dolphins toss balls through basket ball hoops. All these indicate that herd animals possess higher I. Q. than solitary animals. It is thought that herd animals regard the trainer as a member of their herd occupying the highest rank.

Social organisation. Association of individuals is the resultant outcome of random grouping that has ended in *survival advantage*. As example it may be cited that a herd of deer has a greater probability of detecting a predator. A group of insects is more apt to survive in a dry environment than a single insect because in a limited space each would lose less water as exposed surface would be less. Such co-actions have been termed *unconscious co-operation*. This type of association may be asexually gregarious combining individuals without any respect to sex or it may be sexually gregarious-combining males and females together during reproductive seasons.

Active association for mutual benefit between individuals of same species often brings forth a *social organisation*. Success of these organisations is measured in terms of the survival of the colony. Survival at the individual level is given a secondary importance. Social organisation has reached its zenith in social insects like ants, bees, wasps, termites, etc. In the social organisation of these insects there is much division of labour amongst the individuals of the colony. The king, queen, solider, worker, etc., are all morphologically and physiologically specialised. These specialisations are so rigid that the individuals can seldom survive outside the colony.

Account of colony and colonial animals

The expression of the process of evolution lies not only in the structure of the individuals but also in those associations of the individuals which are called *communities*. The communal evolution had its inception with the formation of *cell communities* between the isolated and primitive cell individuals belonging to the phylum Protozoa. The cell communities after passing through innumerable steps in the process of evolution have terminated into those enormously complicated cell communities that constitute the bodies of higher animals.

Simple-cell colony. In those protozoa which reproduce sexually, the zygote or the cell individual formed by the union of the gametes divides by fission into two or more individuals like itself and these usually separate and lead their own independent lives. Probably in a few cases the cell individuals had failed to separate and remained attatched together forming the simple or primitive colony. The characteristics of the simple type of cell colony are: (1) The colony is small being composed of relatively few cell-individuals. (2) The cell individuals of the colony are loosely attached. (3) The cell individuals are similar and more or less unspecialised.

Highly-evolved cell colony. In case of a higher animal the body is an example of highly-evolved cell colony. On analysis of the body, the following characteristics will be observed. (1) The colony is enormous, composed of myriads of cell individuals. (2) The individuals differ greatly from one another. The constituent cells are highly specialized and organised for the performance of various functions. The superficial layer of cells is specialised to form a protective skin. Certain tracts of cells in the interior are specialised to form a rigid skeleton to support and protect the softer parts, some cells are modified to form contractile muscles to move the skeleton and the movement of the body as a whole. A layer of cells has been inwardly tucked from the surface to form a tube which looks after mastication, digestion and absorption of food. Some cells have been specialized to excrete poisonous waste materials constantly being produced by the metabolic activities of the cell community. (3) Members of the cell colony are closely linked together by complex organisation so as to ensure their harmonious co-operation in the activity of the whole organism. The major part in this organisation is played by the nervous system and the blood vascular system. The nervous system serves to gain information through the sensory cells regarding the world outside, to judge the merit of the impression and to react accordingly. The blood vascular system keeps the individual linked together by supplying oxygen and food and by removing carbon dioxide and waste materials from all corners of the colony.

In highly-evolved cell community the general cells of the body have become more efficient for playing their part in the cell community.

Obelia—a lowly-organised colony

Between those two extremes we find a stage in the cell colony formation in obelia. Obelia colony consists of a family or clan of individuals, all descendants of a common ancestor and they remain more or less in complete organic continuity with one another. The individuals are fixed in relation to one another and the community as a whole remains attached to some solid objects.

On examining an obelia colony it will be seen that it is composed of a number of dissimilar individuals but each one is in continuity with the other member by a strand of living substance. The majority of the members are *Polyps* with large trumpet-shaped mouth encircled by a circle of slender tentacles. While there are some members (Blastostyles) which are devoid of tentacles and mouth.

Highly-evolved animal communities. Probably from such cell colonies evolved in course of time the highly-evolved animal communities. The characteristic of such animal communities is that the bodies of the individuals are sharply demarkated and when they live in communities the individuals remain linked together not by the continuity of their material substance but by the intangible continuity formed by social customs. Communalism even in higher forms implies division of labour and physical differentiation. True communalism is found only in two groups of organisms—the insects (Ant, Termite, Bee and Wasps) and mammals.

Social organisation in communal insects

The social organisations of two orders of insects, viz., Isoptera and Hymenoptera are well known. They indicate sufficiently advanced communal forms.

Termites. Amongst various species of termites we encounter innumerable stages of social evolution—from small communities consisting of a few unspecialized individuals to communities of several millions showing a high degree of specialization.

The termites live within nests built of soil particles or masticated and partially digested wood firmly cemented together. By habit the termites live in complete seclusion from day light. Large termite nests often form 'ant-hills' which may reach a height of 20 feet. It has been observed that termites keep in their nest domesticated beetles and other insects. These beetles and insects collectively known as 'Termitophils' have greatly modified abdomen. The abdomen bears projections and exudes food material that is licked off by the termites. In the nest of some of the old world species of termites a carefully tended 'garden' is maintained. Cultivation of the mycelium of certain mushrooms is done in these gardens. The mycelium provides food to young termites.

Members in a termite community. In a typical termite community there live five different castes of individuals.

1. **Chief males and females.** These individuals differ markedly from other individuals of the colony (Fig. 3.18). They are deeply coloured with large well-developed eyes. They are provided with wings. The chief males and females are the only sexually-matured individuals in the colony. In some species the female or the queen's abdomen shows immense increase in size for the steady production of eggs.

Workers. The workers are unpigmented, small and blind individuals. These wingless individuals may be males or females but in both, the reproductive apparatus is entirely degenerated and functionless. The workers are provided with well-formed jaws.

Soldiers. The soldiers too are unpigmented, wingless, blind and sexually undeveloped. The head of the soldiers is swollen and is provided with jaws.

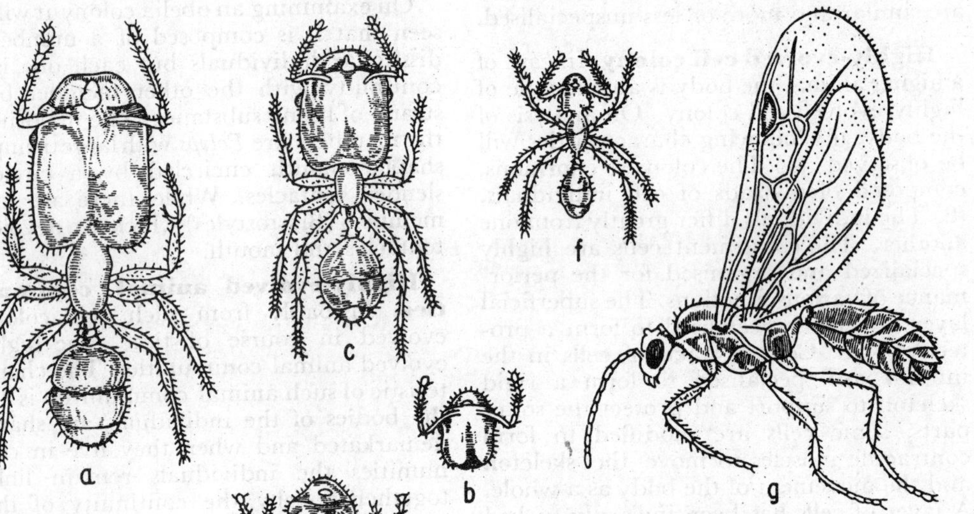

Fig. 3.18. Structural modifications in social insects. a=soldier; b–f=workers; g=male and h=female.

Young complemental males and females. Innumerable young termites are found in the fungus garden nurseries. From these grow up recruits for the various sections of the community. Those that are going to be the 'king and queen' are dark coloured and are provided with fully-developed wings and reproductive organs. While others are blind, wingless but with limited power of procreation. Their duty is to supplement the production of youngs in the event of failure on the part of the chief sexed individuals of the communities.

Formation of a new colony. The would be 'kings and queens' are produced in great number. During spring they emerge from their underground nest and swam forth on their mating flight. They are assailed by birds and other insectivorous animals at this time. However, a few become successful. The successful couple alight at a suitable spot, excavate for themselves a burrow in the earth. They go inside it, drop off wings and lay the foundation of a new colony.

Bee. On the basis of habit and physiological development three types of bees can be encountered. Some of the bees are *solitary*, some are *gregarious* and some are *communal*. The study of different species of bee depicts in a clear fashion the gradational series from solitary to communal forms. The hive bees, however, have attained the highest development.

Anthophora stanfordiana shows the beginning of gregarious life. Anthophora is technically a solitary bee. But they show a marked preference for one another's company. In a colony of this mining bees it has been seen that there are many vertical burrows. The burrows have been so dug that though they are close together there is no interference. Each burrow is the property of a single female bee. The burrow is never a store-house of food. The mother bee brings food to the larvae during the larva's entire period of helplessness.

Andrena, another small mining bee, forms similar colonies. Here the single tunnel is always branched. Within these blind branches eggs are laid and the branches are then sealed after pouring some food into it. The mother after completion of these arrangements waits at the mouth of the tunnel for the emergence of the young.

Helictus is the smallest mining bee. Each mother makes her own cell and stores food there. But the mothers like each others company. The cells of all the mothers, though retain their individuality lie close together and open to the air by common vertical passage. That means one entrance and corridor gives access to a number of cells.

Temporary colony. Bumble-bees (*Psithyrus*) form colonies but the colony lasts for one season only. These bees pair in the fall. After copulation the males die and the impregnated females pass the winter sleeping in some hole. In the following spring the females become active. They collect pollen and honey. These two are then mixed to make a pastry mass. Each female now selects a hole that of a mouse or mole or she digs one herself. She enters the hole, pours the pastry mass and lays eggs on the mass. The resultant products of development are sterile worker bees. The worker bees now take care of the nest. They enlarge it and collect food. From the egg of the mother several sexually sterile broods are formed. Males and females are produced at late summer or fall. The males and females come out, fly and pair. The males and the workers of the colony now die leaving the pregnant females only. These females become the potential founders of next year's colonies.

Permanent colony. The honey bees (*Apis mellifica*) make permanent colony. The number of individuals in such colonies may reach up to 80,000. The bee hives are made in a hollow tree (in wild state) or in deserted houses in those bees which form domestic swarms. The number of members in a colony becomes less in winter and becomes more in summer.

Social ranks and functions: The colony of honey bees is made up of three social ranks —a single *queen*, about 100 *drones* and thousands of *workers*. All these types can be anatomically distinguished. The queen and the stingless drones are *fertile*, their main functions are reproduction. The smaller-bodied workers are all *sterile* females. They build the hive, collect food, feed the queen and the drones and nurse the youngs.

A worker bee has a life span of about six weeks. A worker bee does not do the same type of work throughout its life. The age of a bee determines what work it will have to do. The house-keeping work is done by young workers while food collection is done by older workers.

The worker bees gather pollen, rich in protein and nectar, a sugary solution. The pollen is collected in 'pollen baskets'—a cavity on each hind leg and nectar is swallowed into the 'honey crop', a specialized part of the alimentary canal where the sugar part of nectar is partly digested by saliva. On returning to hive these workers unload the pollen in one chamber and the nectar in another chamber. Other members now look after the collection. Some start packing the pollen tightly in the chamber. Some start converting the nectar into honey. Nectar-filled cells are sealed with wax because it is the principle food store for the winter. The pollen being perishable is not usually stored.

Mode of development of different members

The queen lays eggs individually into honeycomb cells. Some of these eggs may escape fertilization even in a young queen. None is fertilized in an old queen whose sperm store is exhausted. Unfertilized eggs develop into drones. Fertilized eggs have the potentiality to develop either into queen or into workers. When a queen is to be raised the prospective larva is fed with 'royal jelly' containing pollen, honey and vitamins (Pantothenic acid). But new queens are not raised so long as the original queen remains in the hive, active and healthy. Larvae that are to be raised into workers are fed with pollen and honey by the nurse workers. When the queen lays more eggs that can be accommodated in the hive its food supply is cut down. Conversely when the queen slows down in egg laying more food is supplied to it.

New hive formation

When a hive becomes overcrowded, the queen, a few drones and thousands of workers come out of the hive and ultimately raise a new hive. In the older hive the workers pick up a small batch of that old queen's eggs and raise them in specially built honeycomb cells. The eggs develop into new queens. The first one that emerges from the cell immediately traces out the cells of other queens and kills them to death. In case two queens emerge at the same time they fight with each other until one dies. The victorious queen now mates with one of the drones. In the nuptial fight she receives millions of sperms and stores them in the receptacle.

Ants. Total number of species of ants are nearly 5000 and all the species are communal. Ants are found almost everywhere on land. Ant colonies are very *stable* sometimes outlasting generations of men. Such stability is due to the longivity of individual ants, since the workers live from four to seven years and the queens live for thirteen to fifteen years. Another important feature by which the ants are distinguished from all other social insects is the *diet*. The ants do not restrict their diet to cellulose as do the termites or to few specialised substances like honey or pollen as do the bees. Another characteristic of the ant colony is that a single female is never the reproductive centre of the colony. A successful species always keeps a **number of fertile females** in the colony.

Against these there is one handicap for the colony. The ants are unable to move themselves or their brood to other localities easily in case the nest is disturbed, when moisture and temperature become unfavourable or food supply fails.

Categories of social ants. On the basis of the social evolution ant colonies have been classed as:

A. **Foraging or Marauding ants.** The genus *Eciton* illustrates a state of savagery. These 'army ants' make long marches in search of food. The ants invade the nest of other ants and insects, capture larvae and pupae and use them as food. They store some of the food in their temporary nest as they are nomadic and also carry the food during the march for another insect nest.

B. **Slave-holding ants.** Some ants bring home the larvae and pupae of other ant colonies. When these larvae and pupae hatch out they are employed by the raider to do work. Some slave-holding ants (*Polyergus*) are so much dependent upon their captives that they have become unable to dig or care for their youngs. This over specialization will certainly lead to their extinction.

C. **Herding ants or Dairy ants.** Some ants keep aphids or plant lice which serve them as domestic cattle do to mankind. The brown ant, *Lasius*, collect young larvae of the corn root louse in the fall and keep them in their nests till spring. Now they are transported to the roots of certain weeds.

When corn seeds germinate the lice are taken to the roots of corn plant and herded there. In autumn these lice are allowed to pair and the offsprings are preserved as before.

D. **Agricultural ants.** Members belonging to the genus *Pogonomyrmix* make nests partly below and partly above ground in the form of large symmetrical heaps. The nests are usually located in open sunny places and where there is grass. The nests are stored with seeds and grains garnered from the surrounding grasses which are cut away in the area close to the nest. From this bare area well-worn trails diverge into the surrounding grass. It is believed that harvestor ants actually harvest their favourite grasses.

Leaf-cutting ants prepare leaves upon which fungi are grown. The fungi are systematically pruned and cared for by gardening.

E. **Honey ants.** Certain desert ants called Honeypot ants, (*Myrmecocystus*) collect nectars from the flowers and feed it to some of their fellow workers which are kept within the nest. These workers become greatly distended with honey to become the 'Living bottles'. They hang by their feet from the ceiling of the dome-shaped chambers of the nest. During dry season they dispense drops of honey to their thirsty mates.

F. **Thief ants.** Members of the genus *Soleonopsis* are small in size and live in close association with several different species of large-sized ants. These thief ants feed upon the larvae and pupae of the large ants.

G. **Commensal ants.** These ants live in a commensal way and exchange mutual benefit. Such a relation is found between *Myrmica* and *Leptothorax*. The Leptothorax ants are small in size and live in the Myrmica nests. They make one or more small chambers from which very small tunnels enter the Myrmica nests. The tunnels are so small that larger Myrmicas can not go through them. When food is needed a Leptothorax enters the Myrmica chambers and mounts a myrmica ant. Then it starts licking the body first and then the lower jaw of the Myrmica and gets a minute drop of liquid. Thus Leptothorax gets food and Myrmicas get cleaned and 'shampooed'.

Social behaviour of army ants

Colonies of Army ants (Family Dorylinae) typically consist of a single *queen*, a brood of developing *youngs*, and a large population of adult *workers*. The workers are of different sizes (Fig. 3.19). The queen is the colonies' sole agent of reproduction,

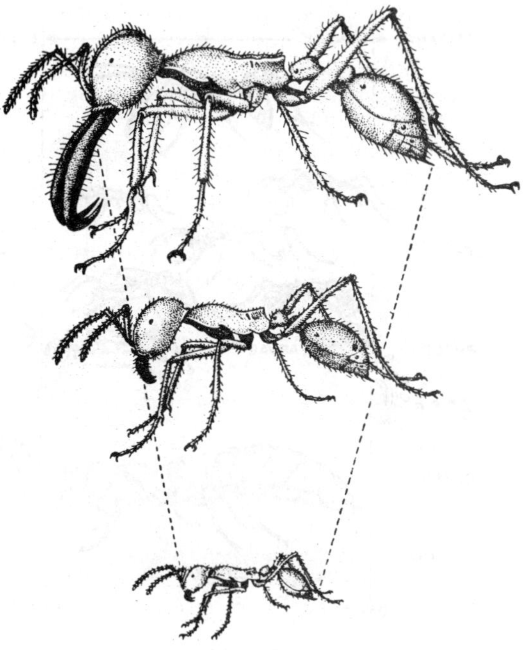

Fig. 3.19. Showing workers of different sizes of an army ant, *Eciton burchelli*.

and she is responsible to a great extent for the colonies' cohesion. The queen secretes certain substances that are attractive to the workers and therefore holds the colony together. More important, the chemical secretions of the queen actually enhance the survival of the workers. At regular intervals (about every five weeks) the queen's large abdomen swells with fatty tissue and eggs and she may lay well over 100,000 eggs in the course of a week. The eggs then give rise to the successive stages of development: larva, pupa and adult (Fig. 3.20). The newly-emerged adults are readily recognizable because of the light pigmentation on their bodies and are called 'Callows or young workers. The workers are all females but sterile with underdeveloped ovaries.

In species of *Eciton*, the workers developing from a given batch of eggs vary in size structure—a condition known as polymorphism. The developmental basis for polymorphism is not clear, although two possibilities exist. The first is that the developmental pathways leading to adult workers that differ in size and structure are determined by the amount and kind of food eaten by the larvae immediately after their hatch from the egg. This mode of development exists in Honey bees. The second possibility is that the eggs laid by the queen are so different biochemically that the subsequent development of the larvae is altered by the quantity or quality of food consumed. Whatever the differentiating mechanism may be, the adult ants that differ in size and structure also exhibit contrasting patterns of behaviour, with the result that there is a division of labour in the colony.

Small workers (size about 3 millimetres) spend most of their time in the nest feeding the larval broods. Intermedial-sized workers constitute most of the population, going out in raids as well as doing other jobs. The largest workers (more than 14 millimetres) have a huge head and long powerful jaws. These individuals are the *Soldiers*. They carry no food but customarily run along the flanks of the raiding and emigration columns. An excited 'soldier' is a formidable animal, it rears up on its hind legs, vibrates its antennae and rythmically opens and closes its jaws.

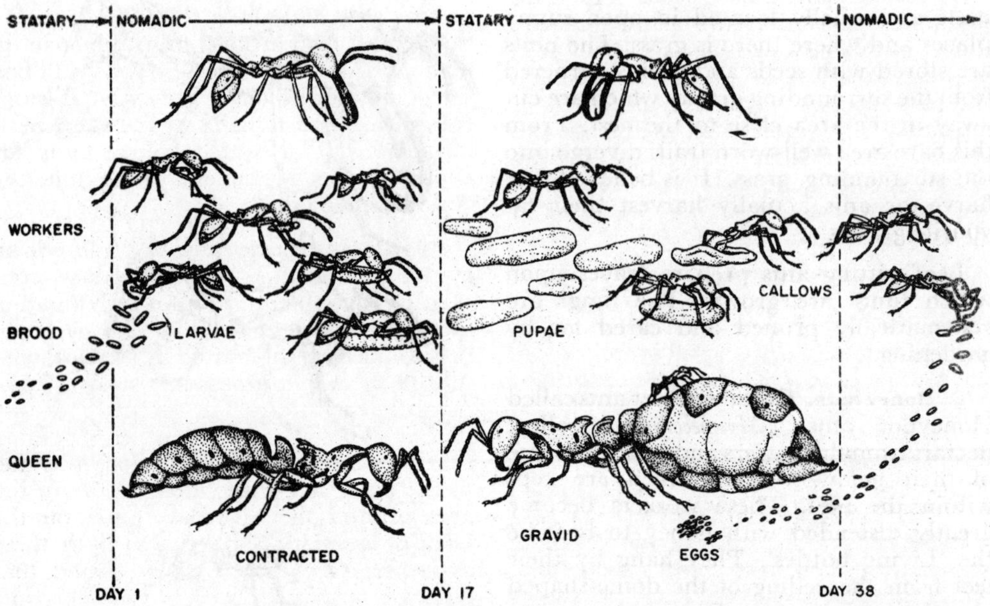

Fig. 3.20. Showing alternating phases in the behaviour of army ants. The phases of the genus, *Eciton* (a typical representative of army ants of the New World) are illustrated. Beginning with a nomadic phase (Day 1), a new generation of workers emerges after 14–17 days from the pupal stage. A colony of army ants employs large parties of raiders to get food every day. The colony shifts to a new bivouac every night to carry the larvae that will become workers in the next generation. The abdomen of the qeen remains in contracted state during the nomadic phase. The nomadic phase terminates when the larvae of the colony transform into pupal stage. With the termination of nomadic phase, the resting or statary phase begins. The statary phase prolongs for about 21 days. This time the colony stops nightly movement and the daily rading parties become smaller. The abdomen of the queen enlarges rapidly during the first week of statary life. She produces as many as 100,000 eggs in the second week. As a new generation of worker ants completes their pupal state by the end of the third week the queen's eggs start hatching into larvae to initiate a new nomadic phase.

Characteristics of the colony

(a) Large colony. The army ants live in large colonies. The Asian genus, *Aenictus* make colonies with not less than 100,000 individuals. *Eciton* of Central and South America has colony sizes approaching a million individuals. The African genus, *Dorylus* makes a colony with twenty million individuals.

(b) Periodic shifting of nesting sites. The second characteristic of the colony is the ants' periodic shifting of nesting sites. The colony movements or emigrations involve the entire colony—workers, brood and queen. At the end of each emigration the army ants settle down only in temporary bivouacs.

(c) Carnivorous in food habit. The third characteristic is that the army ants are almost exclusively carnivorous. They feed on other arthropods particularly insects and some species have been observed to eat small vertebrates such as lizards and snakes. The raiding parties of certain species are so huge that wherever they live the ants rank as major predators in the ecological community.

(d) Firm cohesiveness. The last characteristic of the army ants is their tight cohesiveness. Army ants do not forage for food individually. Instead all immigrations and raiding to new bivouacs are conducted by groups of individuals that closely follow a chemical trail deposited continuously by all the ants as they run along the ground.

Behaviour cycle of the ants

In the behavioural cycle of colonies there is a *nomadic phase* alternated by *statary phase*. In the nomadic phase a large proportion of the adult workers go out on daily raids and collect food. The raids begin at dawn. The ants come out of the bivouac and form several columns, each column later dividing into a network of branches. In running along these trails the ants do not depend much on vision. These ants (*Eciton*) have vestigial eyes consisting of only a single facet; they can detect changes in the intensity of light and almost certainly cannot reproduce an image of an object. The raiding ants stay together following a chemical trail laid on the ground by the other workers. It has been demonstrated that the substance deposited by army ants originates in the hind intestine but it is not yet established whether the substance consists of undigested food or a glandular secretion or a combination of both.

By midmorning the raiding columns overrun an area extending a considerable distance from the nest—often more than 100 metres. At the front of the raiding columns the ants attack insects and other arthropods, biting and stinging the prey, pulling it apart and carrying the softer pieces back to the nest. Thus the column is actually a two-way stream, with some ants advancing and others returning with their prey. At night fall the entire colony moves to a new bivouac typically immigrating along one of the principle raiding trails of that day. It may take the colony most of the night to complete the move to the new nesting site. The daily routine of massive raids and emigrations from bivouac to bivouac is followed for 14–17 days. Then the colony settles down abruptly, it stops emigrating and remains at the same nesting place. This statary phase lasts for about 3 weeks. At the end of the time the cycle begins again.

Regulating factors in the behaviour cycle. At one time there were two contrasting hypotheses about the regulating factors. One suggested that the cycles of behaviour were influenced by physical conditions of the environment, such as temperature, humidity, air pressure or the phases of the moon. The other suggested that the stimulus for emigrations was due to the depletion of the food supply in the area around the bivouac. Later on the work of Schneirla showed that both these hypotheses are incorrect. He found that generally within a given environment some of the colonies were in the nomadic phase and some were in the statary phase—that rules out environmental conditions as the determinant of whether or not a colony would make nightly emigrations. Schneirla disposed of the second hypothesis by observing that a colony of army ants would sometimes move into a nesting site that had just been vacated by another colony and the new-comers would remain at this site even for a three-week statary period—clearly indicating that the food supply around the bivouac had not been exhausted.

The actual regulator in the ants' nomadic and statary behaviour was not some external influence but the breeding cycle within the colony. It has been observed that the nomadic phase always coincides with the period when a larval brood is developing in the colony and the statary phase begins when the larvae start to spin cocoons and go into the pupal stage. When these ants emerge from the pupal cocoon as young workers, the nest stirs with high activity.

Vertebrate versus Invertebrate societies. A society has been defined as a grouping of organisms within a population. The vertebrate society, as it appears, is more successful than the invertebrate society.

The success of vertebrate societies lies in the functional versatility of the individuals. In the insect society the reproductive ability of the majority of the members is suppressed. The reproduction of the minority results not only toward the formation of new individuals but also toward new formation of the whole society. It is evident that amongst social insects, the fate of social institutions is hinged to the fate of a single female and it is her genes that provide the continuity from one social generation to the next. In contrast to that all adult members of a vertebrate society are capable of reproduction and as such the foundation of social continuity is more broad based.

Society has been given an advanced evolutionary status. The insects and vertebrates occupy the zenith of their respective evolutionary lines. Although these socities are different in origin and organisation, similar pattern of social behaviour is evident. Ants and man are unique among animals in making war, practicing slavery and agriculture and in domesticating other animals.

Social hierarchies. Study of animal behaviour has established the presence of *social hierarchies* in many species. It has been seen that in a herd of deer an old and experienced female usually acts as leader of the herd. During periods of danger the female leader guides the whole herd.

Dominance—subordination behaviour. Classical studies of peck order in a flock of chicken has revealed the dominance subordination behaviour that occurs in many associations. In a flock of chicken the peck order determines the position of a chicken in that flock. Lowest in that order stands the individual who is pecked at by all. The highest in that order stands the individual who is not pecked at all. The second highest in the scale is the one who is pecked at by the highest placed individual only. It has been seen that if food, water or a specific standing space is desired by two individuals of the flock, the higher in the peck order simply pecks at the other. The lower on the scale immediately makes room for the dominant one.

Territory. In most vertebrates there is a tendency to defend certain area against the intrusion by the other individuals of the same species. The dimension of the territory may be a small area like the nest or shelter or may be an extensive one. In many animals territorial behaviour appears only during breeding season. In some birds the males set up a territory and defend it against the intrusion of other males. Similar behaviour is exhibited by many fishes and mammals.

Communication. Communication may be defined as an act which influences the activity of one individual by some behaviour of another. The evolution of the techniques of communication in the animal kingdom is progressive but complex in nature. The complex nervous and hormonal systems help in communications between cells and tissues. In man, communication system ranges from simple body positions and facial expressions like smile, frown to complex and symbolic verbal language. A survey of the communication system in the animals points out that several sensory receptors are involved in the process. They are:

(*a*) *Odour*: In all probability odour is the primitivemost means of communication. Its role amongst the invertebrates is little known. The role of *scent organs* in mammals has been worked out in some details. The musk or other scent glands are useful in marking territories, identifying the members of the same group and also as sex attractant during reproductive seasons. The secretions of these glands are steroid or steroid-like compounds that might have evolved from the excreted forms of sex hormones.

(*b*) *Tactile communication*: The practice of nose rubbing in many mammals and the rubbing of antennae in many insects is a common feature. By such an act the members belonging to the same society recognise each other.

(*c*) *Visual communication*: Some animals with well-developed vision practise visual communications amongst themselves. Workers on animal behaviour have observed that the various positions of the ear and tail of one wolf have different meanings to other wolves.

(*d*) *Auditory signalling*: The most highly developed means of communication is auditory signalling. Sound production is done in a variety of ways. Some insects produce sounds by rubbing legs or wings. Swim bladder in many fishes is used as sound-producing organ. Vocal cords are well developed in male frogs and mammals. The membrana semilunaris inside the syringeal cavity subserve the function of vocal cord in most birds.

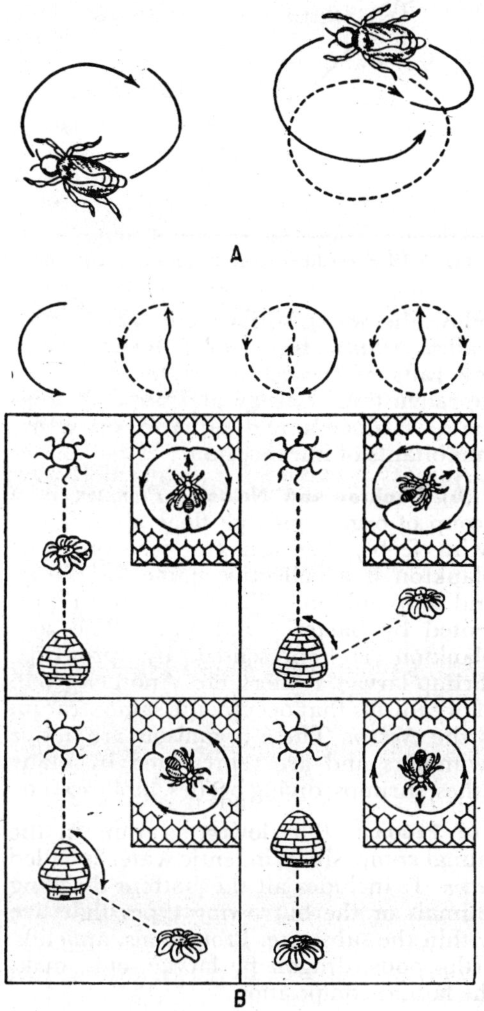

Fig. 3.21. Language of Bee. A. Round dance and B. Wagging dance.

(*e*) *Language of the bees*: Amongst the invertebrates the bees have developed a complex language through dancing (Fig. 3.21). Frisch and others have made exhaustive studies on bee communication. In general the bees adopt two mechanisms to communicate to other bees the information about the source of nectar. They make *round dance* and *wagging dance*. A round dance indicates that the food lies near-by. The round dance consists of small circles made first in one direction and then in the reverse direction. A wagging dance indicates that the food source is further away and it informs other bees of both distance and direction. It is done on a vertical surface inside the hive. It consists of three distinct phases. In the first phase it makes a *semicircular walk*, the second phase consists of a *straight rush* and the final phase consists of another *semicircular walk* to complete the circle. The straight rush indicates the direction of the food. Up the side of the hive means towards the sun, down the side of the hive indicates away from the sun and diagonals indicate the angles towards or away from the sun. The frequency of the dance indicates the distance. It has been observed that rythms of four dances per minute indicate a distance of about 10 km and that 16 per minute refers to a distance of 1 km.

STRATIFICATIONS AND ZONATIONS OF HABITATS

In ecology each habitat has been divided into several strata and zones. The term *stratification* means a vertical layering of organisms and environmental conditions within a biotic community. The term *zonation*, on the other hand, refers to the horizontal arrangements of both the biotic and abiotic factors. Stratifications of several habitats are given below:

Stratification of a typical forest

A forest community is a typical example of terrestrial stratification because here a number of strata both above and below the soil can be recognised.

A forest bears an *overstory stratum*. This stratum is composed of trees which are forty or more feet in height. These trees often form a *canopy*. Relative proximity of the trees determines whether the canopy will be open or thick-set. Below this lies the *understory stratum* extending from twenty feet in height to a little below the overstory. The trees of this stratum are usually shade-tolerant and never grow to the height of the overstory. Below this stratum lies the *transgressive* stratum extending from four feet in height to a short distance below the understory stratum. The trees of this stratum are also shade-tolerant and have the potentiality to grow to increase the density or to replace those species making up the overstory or understory strata. A *seedling stratum* ranges between the soil level and the lower limit of the transgressive stratum. These seedlings are hardwood species. Finally, there is a *herbaceous stratum* containing non-woody plants (Fig. 3.22).

Fig. 3.22. Stratification in a typical forest.

Animals residing in the wood are also stratified but the stratification is not so rigid because of the mobility of the animals. Animals move from one stratum to another stratum in search of food or in response to other abiotic factors. Insects, birds, squirrels and many other animals travel between seedling to upper strata. In the herbaceous stratum rabbit, deer, wolves and foxes are found. Spring tails, beetles, fly larvae, millipeds, centipeds, mites, spiders, annelids, snails, etc., are found on the soil surface.

Stratification in lentic water

Quiet water has been termed *lentic* water and flowing water is called *lotic* water. Ponds, lakes and shallow pools are examples of lentic water. The strata in a lentic water fall under the following categories (Fig. 3.23):

(a) Neuston. It is the faunal group found on the surface film covering a body of water. Those animals which are found on the upper surface of the film are called *supraneuston* while those that are found

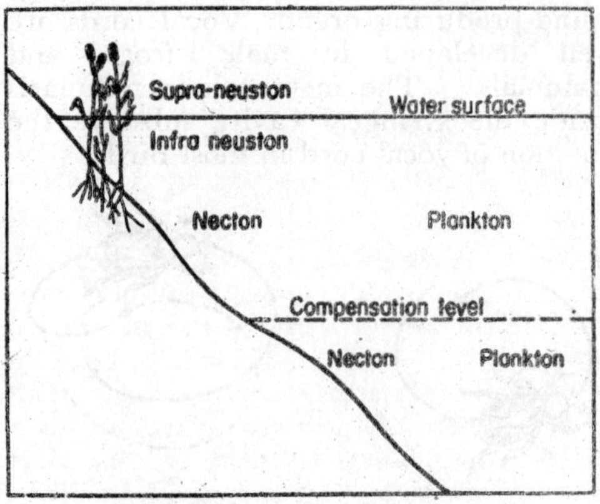

Fig. 3.23. Stratification in a fresh water pond.

below the water surface are called *infraneuston*. Aquatic birds, water boatman and egg rafts of mosquito are the fauna of supraneuston. Larvae and pupa of mosquito, cladocera and aquatic snails are inhabitants of infraneuston.

(b) Plankton and Nekton. *Plankton* is a group of organisms that float aimlessly or swim feebly beneath the neuston. The plankton is a collective name for phyto- and zooplankton. Phytoplankton is represented by bacteria and algae. The zooplankton is represented by protozoa, shrimp larvae, rotifers, etc. Another group of organisms that occupy the same stratum is the *Nekton*. These organisms are active swimmers and are represented by many fishes, shrimps, diving beetles, molluscs, etc.

(c) Pedon. The lowest stratum in the faunal composition in lentic water is called *pedon*. It includes all the bottom dwelling animals or the burrowing types that live within the substrate. Protozoans, annelids, arthropods, dragon fly larvae, etc., make the faunal composition.

Fresh-water zonation. Zonation of lakes and ponds is made on the basis of water depth and the type of vegetation. The following ones are recognised:

(a) Supralittoral zone. It is the zone lying just above the edge of standing water. This zone is not submerged but is exposed to splash and current action. Life is sparse in this zone and is represented by bulrushes, sedges, insects and earthworms.

(b) Littoral zone. The littoral zone extends between water's edge and to a depth of six metres. The rooted hydrophytes are limited in this zone. Animal representatives are directly or indirectly dependent upon the vegetation present. The vegetation present serves as anchorage for egg masses of snails and insects.

(c) Sublittoral zone. The sublittoral zone extends between the lower limit of the rooted vegetation to a depth of ten metres. The lower limit of this zone marks the *compensation level*. It is the level where the rate of photosynthesis is equal to the rate of respiration of the flora. The sublittoral zone is often called *shell zone* because large number of molluscs live here (Fig. 3.24).

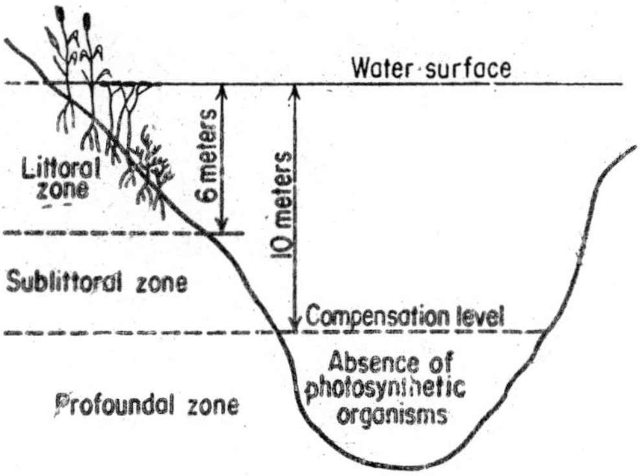

Fig. 3.24. Zonation in a fresh-water pond.

(d) Profoundal zone. It is the deepest zone in any fresh-water habitat. Light penetration is very low in this zone. Water in this region contains high hydrogen sulphide and is acidic in nature. Molluscs and larvae of insects belonging to the orders Diptera, Odonata, Ephemeroptera, etc., are found here.

Marine zonation. The sea and ocean are subdivided into two basic zones—a *neritic province* and an *oceanic province*. The neritic province is of limited extent and extends between high-tide marks and the outer mark of continental shelf (a depth of 200 metres). The oceanic province is of considerable magnitude. The neritic zone is further subdivided in a number of more restricted zones.

Subdivisions of neritic zone

(a) Intertidal zone. The zone is limited between high and low tide marks. As this zone is readily accessible to ecologists, it has been worked out thoroughly. This is a complex one and the complexity is due to the different types of organisms and to the diverse nature of the habitat. Tides periodically cover and uncover this zone twice daily. This makes the temperature, light, salinity and rate of evaporation extremely variable.

The intertidal zone may be made up of loose or stable rocks or may be sandy. The biota in the loose rock area are Collembola, *Gammarus*, *Orchestia*, *Nereis*, Hermit crabs, Starfish and Sea-urchin. Large and stable rocks are inhabited by Boring mollusks, Barnacles, Annelids, Sea-urchins. In sandy zones Tellina (Lamellibranch), burrowing copepods and mole crabs form the principle biota.

(b) Sublittoral zone. The sublittoral zone extends from the low-tide mark down to a depth of 200 metres. The substrate of this zone is of soft consistency being formed by sand, clay, small rocks and broken shells. Numerous animals are found in this habitat. The important ones are Foraminifera, *Clione* (sponge), Sipunculida, *Chaetopterus*, Nemertine worms, Sea-fan, Sea-pen, *Maia* (spider crab) *Palincerus*, *Pecten*, *Dentalium* and many echinoderms.

Subdivisions of the oceanic zone

The oceanic zone is subdivided into the following zones (Fig. 3.25).

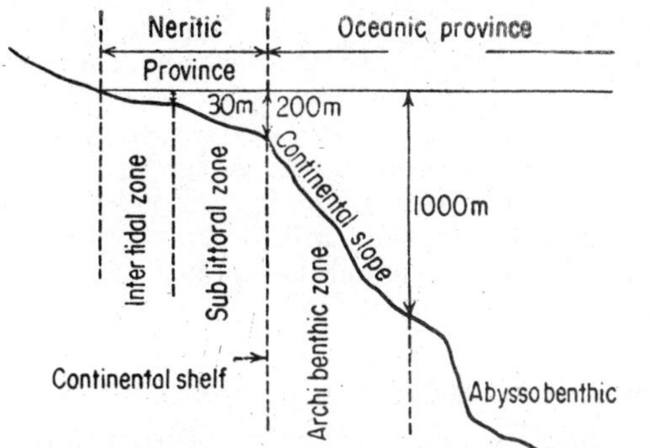

Fig. 3.25. Zonation in ocean.

(a) Archibenthic zone. The zone ranges from the continental shelf to a depth of

1000 metres. The drop off of this zone is very sharp. At the lower level of this zone the temperature of water remains fairly constant throughout the year. Pressure of water on an average is 400 lbs. per sq. inch and a little amount of blue-violet light is available in the zone. Animal biota in this zone is sparse and is represented by Sponge (*Geodia*), Sea-cucumber, Starfish, Brittle star, Brachiopods, etc.

(b) Abyssobenthic zone. From the lower level of the archibenthic zone down to the very depth of the ocean extends the abyssobenthic zone. This zone lies beyond the continental shelf. The characteristics of this zone are: no seasonal change in temperature, little dissolved oxygen, no light, pressure more than a ton per square inch and little gross movement of water. The substrate of the abyssobenthic zone is loose and is made up of deposits, oozes and clays. Animal life is sparse with one or two organisms per meter. Scattered forests of sponges (*Cladorhiza*) are found between the depths from 1600 to 2000 metres. As the substrate is of loose consistency, the bottom dwellers are mostly burrowers or are with long stalks. Annelids, long stemmed sea-pen, crinoid, bryozoans and tunicates are the sessile forms found in this zone. The active forms are represented by broad-footed *Neopallina*, sea-cucumber, pycnogonids and the largest of the crustacean crab, *Kaempfferia* with a leg expanse of more than five meters. Animals living below 1000 metres depth lose their body pigmentation. All the deep-sea fishes and invertebrates in general possess wide mouth opening. Bioluminescent organisms are very common. The light emitted by these organisms is used to lure the prey or to recognise the individuals of the species.

PART FOUR

ADAPTATION

Adaptation to environment is one of the basic characteristics of the living organisms. Living organisms are plastic and possess the inherent properties to respond to a particular environment. Adaptation to environmental dynamics is a biological process eternally operating in nature. It is a facet of evolution and involves structural diversities amongst living organisms that are heritable. Organisms exhibit numerous structural and functional adaptations that help them to survive as species and to overcome the tremendous competition in nature. This part deals with the adaptation of animals to environment.

ADAPTATION—a biological process

Living organisms show two basic properties (i) adaptability leading to (ii) adaptation. The term adaptability is applied to the power of orientation of the organisms to new environmental condition. All organisms possess the power of adaptability to a limited extent to various environmental changes. Mammals are adaptable to various climatic conditions. Adaptability and adaptation are two quite separate biological processes. Adaptation is defined as the permanent moulding in organisms to such an extent that they can live in a particular environment suitably. It is a characteristic of living forms which develop over a length of time, when certain morphological and physiological modifications are initiated that enable them to survive within the jurisdiction of a particular environmental condition.

A survey of the biological world reveals that all animals are harmoniously living in different ecological conditions. Every region has a peculiar and characteristic physical condition, which indirectly gives rise to varied forms. The island fauna represent the typical examples. The evolutionary history of the animals also gives a large number of instances of adaptive changes. Fishes are the primary aquatic vertebrates which show all the basic adaptations to their primal aquatic home. From fishes the amphibia, the first tetrapod evolved. Having appeared on land they had to modify to live in a completely different environment. The amphibians show adaptive duality. They show modifications for aquatic medium as well as for terrestrial living. The reproductive system in amphibia is not adjusted for terrestrial life, they had to come back to the watery home for the purpose of reproduction. The reptiles are the true land adapted forms in the phylogenetic history of vertebrates. The reptiles hold the pivotal position from which both birds and mammals evolved and underwent parallel evolution. These living organisms have shown all the possible ways of living and furnish extensive adaptive radiations.

Adaptive convergence and divergence. As a result of living in a similar environment, organisms of quite distant and unrelated groups show close structural and functional convergence. The reverse is also true in nature, where organisms, originating from the same stock exhibit adaptive divergence as a response to living in completely different environment. All secondary aquatic vertebrates show adaptive convergence (Fig. 4.1).

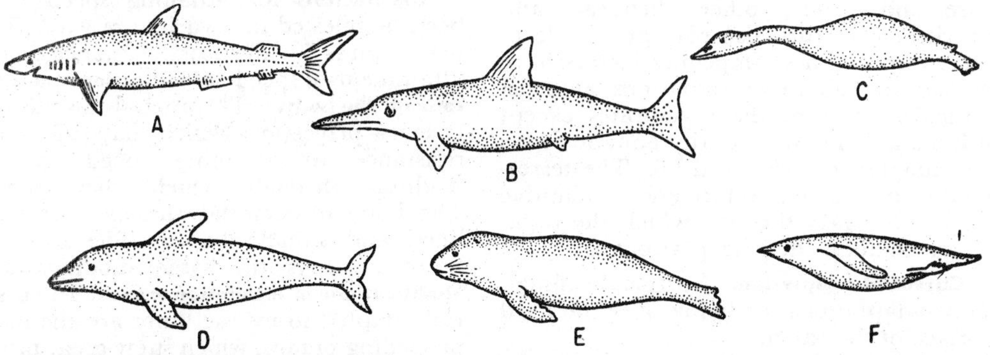

Fig. 4.1. Showing adaptive convergence of body form in aquatic vertebrates. A. Shark. B. Ichthyosaur. C. *Hesperornis*. D. Dolphin. E. Seal. F. Penguin.

Structural and functional adaptations. The adaptations exhibited by organisms to a particular environment are both structural as well as functional in nature. In a particular environment the structural and functional adaptations are almost inseparable. The structural adaptations are more obvious than functional ones. The instances of structural adaptations are plenty in nature. The typical case of functional adaptation is the modifications of the gastro-intestinal tract in vertebrate body which is due to adaptation to different types of food. But both these modifications work harmoniously in an organism to suit a particular environment. So in our present discussion on adaptations of animals to different environments, the terminology 'adaptation' is used to mean both morphological and physiological modifications.

Organisms in relation to environments. The different adaptations exhibited by the organisms depend upon their surroundings. The environment of an organism is not only the physical environment, but also includes the biogeochemical (biological, geological and chemical) and biotic environments. Of all these, light, temperature and water are the three main factors. Light causes modifications of the eyes in vertebrates, temperature plays a very important role and the water seems to be a very important physical factor, especially from the ecological point of view. Some animals are adapted entirely within watery medium and others are adapted to land. Structural adaptations for living in these environments are quite obvious and contrasting. Besides these important factors such as light, temperature and water, other chemical and nutritional factors also play great role in adaptations. These impelling factors have caused the adaptive radiations amongst animals. Most of the vertebrates except fishes and some secondarily aquatic forms are adapted to terrestrial life. The terrestrial forms also exhibit divergent adaptive lines. The ways through which the organisms manifest their adaptations are:

Cursorial adaptation. Cursorial animals show adaptations for living over the hard surface of the earth.

Fossorial adaptation. Fossorial animals exhibit profound adaptations for living beneath the surface of the earth and lead subterranean life.

Scansorial or arboreal adaptation. The organisms have chosen to lead life above the surface of the earth onto the trees and become adapted accordingly.

Desert adaptation. The organisms living in deserts show, in addition to primary cursorial adaptations, the special adaptations against extreme temperature, lack of moisture, lack of vegetation characterising the deserts.

Volant adaptation. The organisms show extreme adaptations for aerial life.

Aquatic adaptation. This adaptation made the animals suited for living in watery medium.

Cave adaptation. Adaptations for living in caves.

Deep-sea adaptation. The organisms, in addition to aquatic adaptations, show adaptations for living at the extreme depth of the sea.

In all adapted animals, the structures of the body which are in direct influence of the environment become extremely modified, while the internal structures which are more conservative become less modified.

PHYSICAL ENVIRONMENT AND ADAPTATION

CURSORIAL ADAPTATION

Cursorial adaptation signifies the modifications for attaining speed on hard surface of the earth. In cursorial animals, speed is the prime requisite.

Modifications for attaining speed

Adaptations for attaining speed have been manifested in a variety of ways. The main aim is to offer least resistance in the attainment of speed. **Modifications of contour of the body.** The animals are devoid of extra projections which may offer least resistance in attaining speed in the medium through which they move. The body of cursorial animals is streamlined and spindle-shaped. The body of horse portrays the same modifications. **Modifications of locomotor organs.** In cursorial adapted forms the limbs are the main propelling organs, which show great modifications for speed. Of the limbs, the hindlimbs show greater modifications than the forelimbs. This fact is amply recorded in the phylogenetic history of the horses. The ancestral horse, *Eohippus* of Eocene period had four digits in the forelimbs and the hindlimbs had only three digits. The forelimbs are modified towards getting the food. **Attainment of digitigrade condition.** Primarily the terrestrial forms were plantigrade because they used palm or sole for locomotion. But in cursorial animals the digitigrade condition becomes more perfected. The speediest representatives are digitigrade in nature. Highest perfection is reached in ungulates where special solepads in the form of hoofs are developed at the digital tips to absorb shock and to reduce mechanical friction during fast movement on the hard surface. The ungulates show all the stages in the conversion of the plantigrade condition to the digitigrade condition. **Elongation of the limbs.** The phenomenon of lengthening of the limbs seems to be the important step towards the attainment of speed. The long bones are conservative, but the digits together with the carpals or tarsals become very much elongated. The bones of the sole and palm become fused together to form the '*cannon bone*' which increases the length of the limbs. Such fusion of small bones reduces the friction between small bony pieces during locomotion. The lengthening of the limbs is thus caused by the growth of the distal sectors of the limbs only. The hindlimbs of birds also show speed adaptation and show the fusion of metatarsals. **Reduction of number of digits.** The presence of five digits is the usual occurrence in plantigrade animals. But with the attainment of digitigrade condition, there is always the reduction in the number of digits. The climax is observed in horses where the limbs have but one digit. **Reduction of ulna and fibula.** There is always a tendency towards the reduction of ulna and fibula. In horse the fibula is reduced to a small vestige. **Restriction of movement in one plane.** There is always a restriction of movement at one plane excepting the shoulder and the hip. The articulation of bones prevents universal movement but permits movement in one plane only. The main aim of such adaptation is to increase the locomotory power.

Attainment of bipedality. Attainment of bipedality is the other attribute of cursorial adaptation. Mode of progression by two feet is repeated in many vertebrates. The human feet become secondarily readapted for bipedal movement. Better development of hindlimbs and the reduction of the forelimbs are noted. In bipedal forms the hindlimbs are exclusively the locomotory agent and become cursorially adapted. The forelimbs become reduced to a considerable extent. Due to disparity of function, the limbs become unequally developed. The forelimbs in cursorial animals, show reduction and usually serve as the prehensive organs. **Maintenance of balance.** In bipedal animals maintenance of equilibrium is essential. In semi-erect bipedal animals, the tail helps to counterpoise the body. Kangaroos have and the Dinosaurs had a very powerful tail which acts as the 'third leg'. **Reduction of the length of neck.** In bipedal animals, particularly in mammals, there is always a tendency towards the shortening of the neck to bring the head near to the shoulder. The number of the cervical vertebrae is usually seven but the cervical vertebrae coalesce into a rigid body.

Modifications for food-getting

The lengthening of the limbs in cursorial animals takes the head sufficiently away from the ground. So to get food and drink, the neck and the skull become elongated. The elongation of the neck and the skull bear a definite ratio with the lengthening of the limbs, which is clear from the phylogenetic stages of the horses already discussed in Part One of this volume.

FOSSORIAL ADAPTATION

To get more food and safe shelter, animals tried to go beneath the surface of the earth to lead the subterranean life. Fossorial adaptation shows different gradations and as a consequence different degrees of structural modifications are encountered to tunnel the soil. Animals living above ground may dig the soil for search of food. In these forms the structural modifications are limited to the specialisation of the digging device only. The tusks of the elephants and the snout of the swine serve as digging machine. Although structural alterations are not so great, the associated structures become modified. In elephants the development of the incisors into tusks causes tremendous modifications of the skull and jaws. But the animals which dig the soil for retreat exhibit greater modifications. In this category we find two gradations: (1) Fossorial animals dig for retreat but had to come over the surface for food. (2) Fossorial animals dig for retreat and find their food also beneath the surface. The animals which come under this group are wholly fossorial and show greatest specialisations.

Modifications for fossorial life

Fossorial animals have moulded their body to offer least resistance to dig the soil. The body becomes either spindle-shaped or fusiform which helps to penetrate the medium very easily. The greatest diameter of the body lies near the shoulder. The shoulder girdle is strongly built. In

Echidna the greatest diameter of the body lies further backwards. Extreme modifications are observed in snakes, Gymnophiona and in some lizards where the body becomes very much elongated, cylindrical and limbless. The tail becomes invariably short in fossorial animals. In extreme cases the tail is vestigial which may serve as a tactile organ. The sense organs have the general tendency towards reduction. The eyes and the external ears tend to become vestigial. The reduction of these structures correlates with the degree of fossorial adaptation. Complete subterranean life causes total visual reduction. The eyes are very small and insignificant in the members of Bathyergidae and Geomyidae. In *Spalax* the visual apparatus is a complete structure being represented by small specks. The eyes are imperfectly developed and nonfunctional in marsupial mole (*Notoryctes*), vestigial in common mole (*Talpa*) and remain covered by integument in *Chrysochloris*. The pinna in burrowing animals may cause obstruction, so they tend to disappear. The external ears are very small in Geomyidae, extremely reduced as fringe of skin in Bathyergidae. In most of the fully fossorial animals the external ear is entirely absent.

Digging apparatus

Fossorial animals must possess efficient digging apparatus. **Snout—as the digging machine.** The snout comprises an important device for digging in *Heterodon* (hog-nosed snake) and swine. The skull becomes subconical and lacks the expanded zygomatic arch. An extra bone (*prenasal*) may be developed at the tip of the nasal cartilage to aid in digging as observed in *Talpa* and swine. **Teeth—as the digging machine.** The teeth, particularly the incisors and the canines in many animals, help in digging. The canines act as digging apparatus in swine. The incisors are usually forwardly protruded to facilitate digging. The tusks of the elephants are the effective digging instruments. **Forelimbs—as the digging machine.** Extreme modifications and specialisations of the forelimbs are observed in the fossorial animals. These limbs constitute the most effective digging machine. The limbs show the tendency towards the reduction in length and have become strongly built, because shorter limbs have positive advantage during burrowing. The forelimbs become not only short and strongly built but also they become very much broadened. They also differ materially from the hindlimbs. They have undergone divergent specialisation for the purpose of digging (Fig. 4.2). The hindlimbs remain less modified and help to

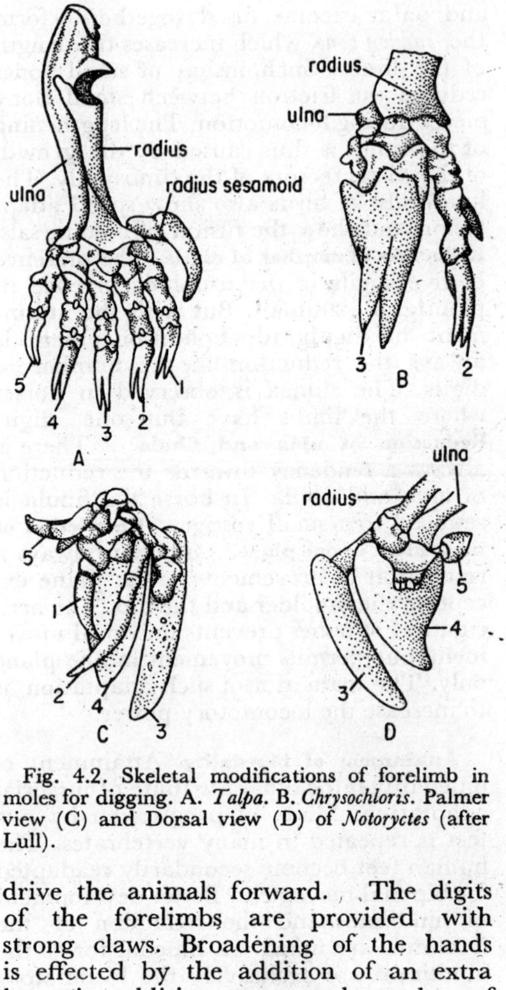

Fig. 4.2. Skeletal modifications of forelimb in moles for digging. A. *Talpa*. B. *Chrysochloris*. Palmer view (C) and Dorsal view (D) of *Notoryctes* (after Lull).

drive the animals forward. The digits of the forelimbs are provided with strong claws. Broadening of the hands is effected by the addition of an extra bone in addition to normal number of five. This extra piece of bone is called the *radius sesamoid* or *os falciforme*. This increases the breadth of the palm in some moles and *Echidna*. In *Chrysochloris*, the hand has four digits but its effectiveness is increased by the elongation of two middle fingers which are furnished with powerful claws. The long bones of the forelimbs are very strong and are provided with prominent ridges for the insertion of the muscles. The proximal ridges of the humerus are highly developed to give attachment to the powerful shoulder muscle. The olecranon process is usually large to give attachment of the powerful triceps muscles.

Strengthening of the shoulder. Powerful shoulder is an asset for fossorial forms, because it has to withstand great pressure during digging. Narrow shoulder is another modification of the fossorial forms. In monotremes and northern moles, the entire shoulder girdle is shifted forward to the neck (Fig. 4.3). In Chrysochloris the sternum is convex inward. In true mole the shifting of the shoulder is also caused by the elongation of the first sternal segment, which bears very short clavicle. In case of monotremes, the shoulder girdle is supported by a very large T-shaped episternum.

SCANSORIAL (ARBOREAL) ADAPTATION

Arboreal life is sought for safety, retreat and for abundance and easy procurement of food. From the phylogenetic history of the vertebrates, arboreal life is very significant, because the flying vertebrates, excepting flying fishes have been derived from scansorial vertebrate forms. Like fossorial forms, the scansorial types also furnish gradations of specialisations. The first category includes the animals who are usually rock and wall climbers. Geckoes and house lizards are the typical representatives where the adaptation is limited to the development of climbing organs at the bases of digits. The flying squirrels, included in this group show also scansorial adaptation in addition

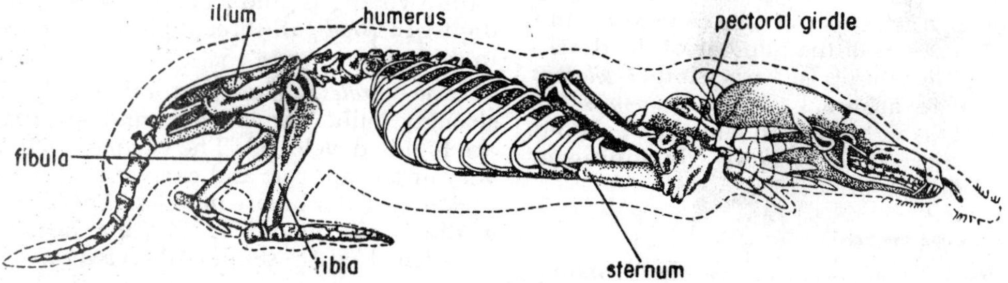

Fig. 4.3. Skeletal framework of a mole, *Talpa* (after Lull). Note the shifting of shoulder girdle towards head.

Hindlimbs as driver. The hindlimbs remain less modified. The femur is not stout as the humerus and the ridges are not so prominent. Only alteration is observed in the partial fusion of the tibia and fibula, and the greater development of the calcaneum. In *Talpa*, a large bone is attached to the side of the tibia which can be compared with the *os falciforme* of the forelimb. The only function of the hindlimbs is to push the animal forward during digging. The ilium and ischium are situated parallel to the vertebral column and are greatly elongated.

Modifications of the vertebral column. The vertebral column is rigid. The sacral vertebrae show the tendency of fusion. In *Talpa* and the hedgehogs, the vertebral column is strengthened by the development of intercentral ossicles. Several cervical vertebrae become either more elongated or show fusion.

to their volant modifications. The second category comprises of many insectivores, rodents and carnivores which can climb the trees very easily as well as they are quite at home on the ground. They actually lead double life, in trees and on land. The third category includes the animals who are fully arboreal and make the trees their home and may in rare occasions descend to the ground. These forms have undergone greatest modifications for living on trees. The animals belonging to this category exhibit different modes of progression: (1) Progression on the upper surface of the trees by the help of the fore- and hindlimbs. Most of the arboreal forms move by this method. (2) Progression on the under surface of the trees by both the pairs of the limbs. The sloths are the typical examples. They possess large, recurved and powerful claws by which they can remain suspended from the branches of the tree.

They are also able to move upon the branches of the tree. Other forms, such as the bats and *Galeopithecus* may also be included under this category. (3) Progression by swinging by the forelimbs is observed only in Primates. They can move with speed and accuracy by the forelimbs from tree to tree.

Modifications for scansorial life

The general contour of the body and the limbs are extremely modified to lead perfect scansorial life.

Body architecture. For climbing, the chest along with ribs, the shoulder and the pelvic region are very strongly built. In fully arboreal types, the thorax is subcircular and the ribs are extremely curved. In Sloths, the ribs are many to afford the weight of the viscera in an inverted state. Another modification is the elongation and the increment of the number of the dorsolumbar vertebrae. In a tree-sloth, *Choloepus* and in an arboreal rodent, *Capromys* the number is increased from nineteen to about twenty-five. In *Dendrohyrax*, six additional vertebrae are present.

Locomotor organs

Limbs. In arboreal forms, the proximal portion of the limb becomes extremely elongated and bears a definite ratio with the climbing ability. The greatest modification is seen in *Hylobates* where the arms are very long and can even touch the ground while standing erect. The feet may also be modified as prehensile or grasping organs. The digits in this form may be opposable. In those cases where the feet are not prehensile, the digits are furnished with powerful claws as seen in Squirrels. The claws may be curved. In *Erethizon*, a tree porcupine, in addition to recurved claws, the sole possesses spines which help in climbing. Presence of adhesive discs at the tip of the digits or on soles is climbing organs. These are observed in tree-frogs, Geckoes and *Dendrohyrax*. In tree-frogs the sticky secretion of the adhesive discs helps in adhesion. In Geckoes, adhesion is caused by creating numerous vacua on the adhesive digits. In *Dendrohyrax*, adhesion is done by creating partial vacuum on the soles. The feet are modified as prehensile organs in marsupials and primates. In marsupials the hallux and the fourth digit are opposable, while the second and the third being bounded by a common integumentary sheath, thus showing *syndactyly*. In Primates, the first digit is opposable. Another remarkable climbing modification is observed in *Koala* where each foot has long clawless great toes and clawed syndactylous second and third toes. The fourth and fifth toes are also provided with powerful claws. In *Chamaeleon*, syndactyly is observed in both the limbs which are powerful grasping organs. In scansorial birds the hallux opposes the second and third toes and thus offers a very firm grip during perching. In Woodpeckers, Parrots and Cuckoos, the toes are provided with strong claws. Although the digits are very vital in arboreal life, occasional reduction in the number of digits is observed in many forms. Greatest modifications are observed in *Choloepus* where the hand is provided with two digits and the foot has three digits. In *Bradypus*, three digits are present in each limb.

Limb girdles. The pectoral girdle is strongly built. The clavicles and scapulae are highly developed. The clavicles play a very important role and can withstand the pressure of the breast muscles. The pelvic girdle is less modified. The ilia become broadened out to support the viscera.

Modification of the tail

In most arboreal forms, the tail is prehensile. In Chamaeleon the tail is prehensile. In nonprehensile forms, the ectodermal scales (as in *Anomalurus*, a flying squirrel) prevent the animal from slipping. One of the most remarkable examples is *Ateles*, the spider monkey, where the tail is highly prehensile.

DESERT ADAPTATION

Desert is a specialised terrestrial environment which has its quota of animals and plants living their life quite suitably by making numerous morphological and physiological adjustments. It is extremely difficult to define a desert, because various types of desert are encountered. To name a few, these are: (1) low-rainfall desert (sandy deserts); (2) cold deserts (such as Polar and Alpine deserts); (3) low-nutrient deserts caused by the excessive use of the soil; (4) deserts due to toxication such as the adjoining areas near the volcanoes and (5) high-salt deserts. Of these types of deserts, the low-rainfall desert is usually regarded as the most typical one. The common feature of all these deserts is the presence of dryness where rainfall is usually less than 10 inches in a year.

Characteristics of the low-rainfall desert. It is generally regarded as a vast stretch of drifting sand with the following physical conditions: (i) low rainfall, (ii) extreme temperature, (iii) low water vapour in atmosphere resulted from excessive solar radiation, (iv) increased air movement which results into sand and dust storms and (v) poor vegetation excepting in rare cases, e.g., oasis. In this portion of the land evaporation of water exceeds its condensation.

Deserts have their own inhabitants which show peculiar modifications for desert life. To the desert forms two primary necessities stand on the way. They are moisture-getting and moisture-conservation. Defence against physical and organic environments is also a very important factor in desert life.

Modifications for moisture-getting

One of the primary requisites for desert life is to get adequate moisture. The desert forms have developed a number of devices to obtain moisture which is really a rarity in desert. In desert, moisture suspended in atmospheric air is not showered as rain but it may form only dews which are utilized by diverse means. For animals living in desert the main source of water is the juices of plant or may be the blood of the prey. Desert animals have developed the habit of occasional drinking of water and in some cases the collected dews taken along with plant food is sufficient to quench the thirst. The skin of *Moloch horridus* is hygroscopic which has the property of absorbing moisture from the atmosphere like that of blotting paper.

Modifications for moisture-conservation

Conservation by storage. The primary method of conservation of moisture is done by utilizing little quantity of water in metabolism. It was previously believed that camel, the ship of desert, possesses numerous pouch like diverticula, called the *water cells* in the stomach which store water for the emergency use in desert. But this belief is not acceptable now (*vide* J. E. Hill, 1946, *Natural History*, **55:** 387). These pouches are not sufficient to hold as much water as the animal would require during a prolonged journey through sandy desert where no water is available. The water cells may sometimes be filled with water which are generally metabolic water collected from the different parts of body to moisten the food to facilitate digestion. Some quantity of water is stored in the connective tissues and muscles. This quantity of water usually comes from the breakdown of glycogen stored in muscles and of fat present in the hump. It has been estimated by physiologists that every 100 g of fat oxidised in the body yielded 110 g of water. An estimated quantity of 8 gallons of water may be produced from the breaking down of fat stored in the hump alone.

Conservation by avoidance of evaporation. The desert animals can conserve moisture by being non-perspiring. Perspiration or sweating is a physiological process in mammals to keep the body cool, but in desert non-perspiration is a special adaptation.

Modifications for defense

Protection from hostile environment is necessary for desert animals. One of the notable unfavourable physical conditions prevalent in desert is the extreme temperature.

Protection from extreme temperature. In sandy desert the day temperature is extremely high and the mercury column may touch 182° Fahrenheit, but the night is cold and the hot wind does not blow. The desert animals usually seek the shadow of great rock in daytime or they can burrow the sand to escape the extreme cold or heat. The desert insects can withstand considerably high temperature. They can sit or move comfortably on hot stones which are unbearable for other animals.

Protection from sand. Protection against the harmful effect of sand on the vulnerable parts of the body is of primary importance. So the eyes, nostrils, ears are extremely modified to fight the sand storm.

Eyes. For security and for searching suitable place for shelter, acute vision is necessary. In camels, the eyes are large and are guarded by plenty of long eyelashes. In a burrowing desert snake, *Typhlops*, the eyes are small and are overhanged by headshields. In some lizards, *Agama* and *Phrynocephalus* the edges of the eyelids are broadened into plates beset with scales. The upper eyelid is enlarged in some cases as observed in *Teratoscincus*. In some reptiles the eyelids are extremely modified. In *Mabuia* the lower eyelid is greatly enlarged with a transparent window, through which the animals can see when the eyes are closed. Extreme specialisation is found in *Ablepharus* where the whole of the lower eyelid is transformed into a transparent membrane which is united with the upper eyelid.

Nostrils. The nostrils are directed upwards in most desert reptiles. In many desert snakes the nostrils are protected by complicated valves and may be reduced to small pinholes. In camels the nostrils can be closed. *Saiga tartarica* possesses a large inflated nose and the narial apertures are placed far backward.

Ears. In desert reptiles the ear opening is very small and may be protected by scales. In extreme cases the opening may be absent. In camels the openings of the ears are protected by hairs.

The greatest modifications are observed in camel and ostrich where the head is placed on a very long neck. This brings the vital sense organs as far as possible above the sand level.

Protection against enemies

Desert animals have adopted various devices to protect against the predators. These are: **Colouration.** The hues of the desert animals are such that they can harmonize their body colour with the colour of the sand. One of the typical examples is *Gazelle* which matches the surrounding sands and stones to such an extent that it becomes impossible for the predators to detect the animals when at rest. The prominent yellow and black colouration of *Gila monster*, the only venomous lizard, shows warning colouration. **Modifications of body surface.** The dry, hard and spinous surface is the characteristic of many desert animals. The spinous scales of the body of *Moloch horridus* and the large spines on the body of the horned toad, *Phrynosoma* (Fig. 4.4) are the two typical examples. **Venom.** Possession of the venom is an attribute of desert adaptation. Many desert animals are venomous. The desert snakes, *Gila monster*, Ants and Spiders are some of the poisonous

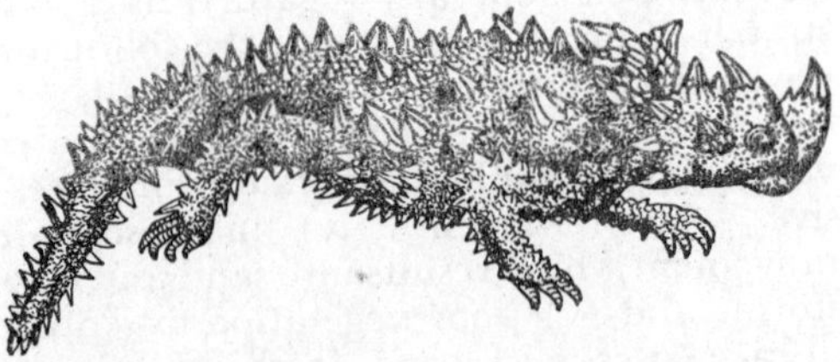

Fig. 4.4. A horned toad, *Phrynosoma*. Note the presence of spinny skin.

desert animals. The common red desert ants, the rattlesnakes and all the spiders in desert are dreadly venomous. *Tarantula* is a venomous and dreadly spider. The Scorpions are also highly venomous. **Odour.** Repulsive odour is also an adaptive feature in many desert forms.

Modifications for speed

The desert animals usually move with high speed, because they have to travel far off in search of food, drink and safety. The sand-adapted feet of camel serve as a very effective locomotor organ.

VOLANT ADAPTATION

Volant animals constitute a very interesting and highly specialised group of vertebrates. They have undergone extensive adaptive radiations and have beautifully modified their body to lead a perfect aerial life.

Duality in volant adaptation. Volant adapted forms are not such as the term denotes, but they must return to the surface of the earth for rest. All the volant forms exhibit two-fold adaptations. (1) Volant adaptation for flying in air and (2) Cursorial or Arboreal or aquatic adaptation for the surface life. As a consequence two directional structural adaptations have occurred. The most illustrative volant forms are the birds, where the forelimbs have modified into wings for volant life and the hindlimbs are specialised for cursorial and/or arboreal or aquatic life. In true volant forms, the modifications for flight in most cases have superseded other adaptations.

Modifications for volant life

In discussing the modifications for volant life, the discussion is restricted only to the structural adaptations for flight as the other adaptations such as the cursorial or arboreal, have been discussed previously. In volant forms the cursorial or arboreal adaptation is quite typical. In aquatic forms the digits of the hindlimbs have become webbed to aid in swimming.

Prerequisites for volant life

For living in air an organism must possess the following unavoidable requisites: (1) Organs for flight; (2) Lightness and rigidity; (3) Energy and power; (4) Speed; (5) Balancing and (6) Controlling. These are the trump cards upon which the avian organisation is based.

Structural modifications

The degree of structural modifications in aerial forms depends upon the ability of flying. In all the flying forms the contour of the body tends to modify in such a way as to offer least possible resistance in air. Amongst the vertebrates two types of flight are encountered.

(1) Passive flight. This type of flight is also called the gliding flight, when the animal takes a leap from a high point which is either the tree or the peak of mountain. After taking the initial leap, the animals can glide to a lower region and thus cover some horizontal distance. During such gliding the main role is played by the gravity and in most cases is helped by sustaining structures.

(2) True flight. This type of flight is caused by the activity of the organs for flight and the organisms have the power of sustained movement through the air.

Organs for flight

Expanded pectoral fin. The inclusion of the flying fishes as the true flier is questional. There are several genera of fishes which can fly. The most illustrative example is the *Exocoetus* where the pectoral fins become greatly expanded as the organs for flight. The pelvic fins are comparatively small. The lower lobe of the tail fin is much longer and can accelerate the speed. The length of the flight can exceed 200 to 300 yards. The other flying fishes are the various species of *Dactylopterus* with beautifully coloured wings, *Pantodon* of Africa, *Gastropelecus* of British Guiana, *Pegasus volitans* of Japan, India, China and Australia.

Webbed feet as the sustaining surfaces. The only volant adapted amphibia is the genus *Rhacophorus* whose digits of the limbs are webbed. The expanded web helps the animal to have a prolonged leap although the rudiments of patagia are present in front and behind the arms. Amongst the tree-frogs, *Rachophorus pardalis* has rather greater power of gliding.

Patagia as the sustaining surfaces. Patagia are primarily folds of skin expanding from the body and may be supported by various skeletal elements. In a lizard, *Draco* (or 'flying dragon') the sides of body become extended outward to form two large membranous surfaces supported by five to six elongated ribs. The patagia can be folded against the body when not in use. In another form, *Ptychozoon*, the patagia are present along the sides of the neck, body, tail and limbs. In birds, traces of patagia may be present (prepatagia and postpatagia) in front and behind the arm. The development of feathers in birds has led to the degeneration of the patagia. In mammals, particularly in *Galeopithecus* (or 'flying lemur') the patagium extends not only between the fore- and hindlimbs but is also present between the neck and the forelimbs, between the tail and the hindlimbs and even between the digits as webs.

Wings as active organs for flight. Wings are the greatest anatomical deviation recorded in the vertebrates for flying. Wings are complex organs consisting of several structural elements. Three types of wings are observed in vertebrates (Fig. 4.5).

(i) *Wings in bats.* The humerus is well formed, the ulna is vestigial but the radius is very long and curved. The pollex is free and clawed. The other digits are extremely elongated and support the patagium. In *Megachiroptera* the second digit is independent and bears a claw. But in *Microchiroptera* the second and the third digits are attached distally and are devoid of claws.

(ii) *Wings of Pterodactyl.* The radius and ulna are almost equally developed. The fourth metacarpal is heavily built and bears the large wing-finger. The other three metacarpals are slender and support the other three digits which are independent and clawed. The wing-finger supports the entire anterior region of the patagium.

(*iii*) *Avian wings*. The forelimbs are modified into true wings which are composed of bones, muscles, nerves, blood vessels and

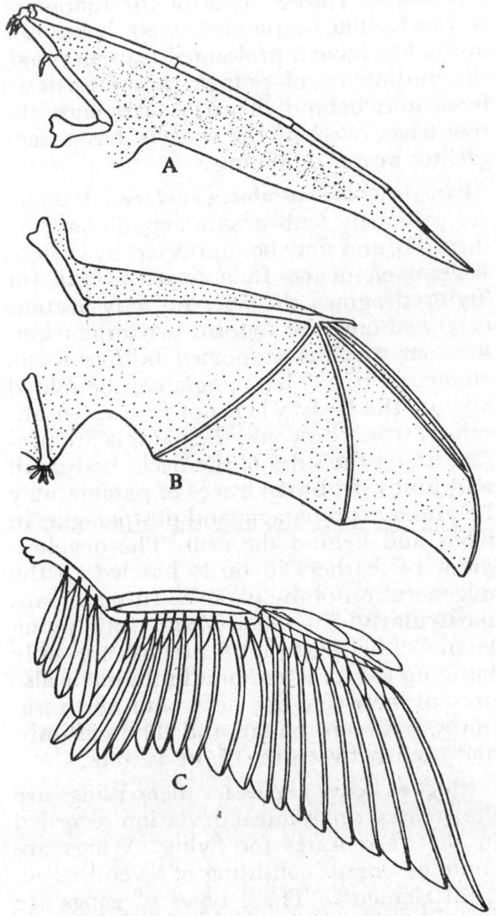

Fig. 4.5. Wings in vertebrates. A. Pterodactyl. B. Bat. C. Bird.

feathers. In modern birds, the wings have the framework of bones. The humerus is large, strongly built and possesses prominent ridges for the attachment of flight muscles. The radius and ulna are slender, stout and slightly curved. Only three phalanges are present and there is no trace of the fourth and fifth digits. The bones of the wrist, hand and fingers are extremely specialised both by the loss of bones or by the fusion of bony units. In modern birds, the claws are lacking in the digits, but in fossil birds and in some ratites claws may be present. The patagia are rudimentary as their functions are taken over by the feathers. The feathers are borne by the wings. The feathers are called the remiges. Overlying the basal part of the remiges there are several rows of coverts which close the interstices between the remiges to make the wing a continuous surface. The muscles are highly developed. Several flight muscles such as the *pectoralis major*, *pectoralis minor*, *coraco-brachialis*, *tensores patagii* and many others are intimately associated with the wing and regulate the activities of the wings.

Modifications for lightness and rigidity. Many elements combine to make the volant animals the models of mechanical perfection. Bones are stout to withstand the pressure of flight muscles during flight. Bones of the pterodactyls and birds are hollow and pneumatic. The climax is reached in birds, especially in carinates, where the development of air-sacs; fusion of the vertebrae; loss of internal organs such as the gall bladder, urinary bladder, loss of right ovary and the vestigial right oviduct in many birds makes the body much lighter. In birds the skull bones are paper-like thin and are sutureless. Although in birds the bones are pneumatic and thin, the bones have a secondary plastering to make them rigid.

Modifications for obtaining extra-energy and power. The capacity to convert chemical energy into mechanical motion through the combustion of fuel is the secret of attaining sustained power. Greatest modifications are observed in birds. The lungs are proportionately smaller in size but the efficiency is increased by the development of the air-sacs which also help to send oxygen directly to many tissues. In birds the body temperature is rather high which hastens the combustion. Non-conducting coat of feathers over the body prevents surface loss of heat. Crude power as food is stored inside the body in the crop. The heart is proportionately larger in size and the circulatory system as a whole is very efficient. The physiological efficiency of respiration is increased by the development of air-sacs in birds and chameleon.

Modifications for obtaining speed. Speed is a must for volant life. The body is trim-built or fusiform and without any extra resisting projection. In carinates the existence of air-sacs between the flight muscles help in flight by reducing friction.

Modifications for balancing and steering. To equalise the air-pressure on the wings, the surface of the wing can be decreased or increased. In carinates the air-sacs are so arranged that the proper centre of gravity can be maintained by shifting the air from one side of the body to the other. The cerebellum in bird is very well-formed. The tail is also provided with the rectrices which as a whole helps steering during flight.

Modifications for control. Especially in case of birds, the brain is well developed with well-formed corpus striata. The optic lobes are well-formed and show the hint of quadrigemina. The eyes are provided with pecten which increases the acuity of visual perception. The eyes in case of pterodactyls were also very well developed and large. The eyes were provided with sclerotic plates. Bats have good vision, especially in twilight. They have well-developed tactile sense. The ear and the facial appendages are the main sensory seats to prevent the crush with the unwanted approaching objects. The patagia in bats are also very sensitive.

In birds, the neck is usually long. Birds have a wide range of vision all around, they can rotate the head to 180° due to the possession of heterocoelous cervical vertebrae. The jugular veins are also very long and form a loop at the anterior end.

AQUATIC ADAPTATION

Almost all the classes of the vertebrates have representatives who lead aquatic life. Two types of aquatic vertebrates are encountered, the primary aquatic vertebrates and the secondary aquatic vertebrates. The primary aquatic vertebrates are the fishes that have evolved from aquatic progenitors. The fishes are perfectly adapted to aquatic environment. Besides the fishes, many lung-breathers have gone back to the primal aquatic home for food and safety and exhibit extreme modifications for aquatic life.

Requisites of primary aquatic adaptation

The adaptation of fishes to a watery medium is perfect. So the structural adaptations of the fishes may be taken as the requisites for aquatic adaptation.

Body contour. The body is compressed into a stream-lined form with no extra protuberances that may cause obstruction for swift movement through water.

Locomotor devices. Locomotion is primarily effected by the lateral undulations of the body. The fins also help in the process and are regarded as the accessory locomotor organs. Of the fins, the tail fin is the most important prop elling organ. In addition to the unpaired fins, the paired fins, comparable to tetrapod limbs, are present and serve, in addition to normal function, as stabiliser.

Hydrostatic device. Some sorts of hydrostatic mechanism are necessary for aquatic life. Almost all the fishes excepting the elasmobranchs have a gas-filled swim-bladder.

Structural modifications in secondary aquatic vertebrates

The lung-breathers, due to the stress of circumstances, have been compelled to go back to the water. As a result of living in watery home they exhibit extensive modifications. The greatest difficulty which the lung-breathers had to confront was the inability to breathe in water. Modifications for aquatic life involve many anatomical and physiological alterations. The entire modifications can be marshalled into two perspectives, securing of food and the attainment of swift passage through water. Internal and external alterations are so profound in quality and quantity that one may be at a loss to interprete them properly. The amphibians constitute a transitional group to furnish partial aquatic adaptation. The structural modifications in them are limited to the possession of webbed feet, laterally compressed swimming tail which may have fin-like outgrowth along the upper border. They can respire by gills in adult stages, but the lung-breathers are numerous. But the greatest modifications are observed in aquatic reptiles, birds and mammals. They furnish typical examples of adaptive convergence.

Modifications for locomotion

Adaptations connected with the body forms. There is always a tendency to eliminate any unevenness of the body contour that would offer resistance during progression through water. Reduction or absence of neck in whales, dolphins and manatee brings the head near to the thorax. In cetacea, sirenia and pinnipedia the greatest breadth lies in the anterior one-third of the body. The body becomes stream-lined. The tail becomes enlarged and the mobility of neck is lost in all the fully secondary aquatic forms. In Protomogale, Limnogale and River-otters the body is fusiform, but in Zeuglodonts the body is not fusiform but is largely anguilliform. The chest tends to become cylindrical and is modified to bring the visceral cavity higher towards the back and thus insuring greater stability in floatation and also increases the lung capacity. The ribs are highly arched dorsally and move upward.

Modifications in 'Oar propulsion'. In case of oar propulsion a gradation of modifications of both the limbs is encountered. In Turtle, Plesiosaurs, Walrus, the limbs are equivalent in size.

Forelimbs and pectoral girdle. In highly-specialised aquatic mammals the humerus has become shortened. The clavicle is also absent in Pinnipedia, Sirenia and Cetacea. The scapula is elongated for the attachment of muscles. The digits are webbed in Pinnipedia and *Ornithorhynchus* and the degree of webbing varies greatly. In Walrus, the digits are folded and are covered by skin, but the individualities are retained. In Sirenia, the development of paddle is seen, where the mobility of the various joints is lost and the entire skeleton is enclosed by skin showing no external division of phalanges. In Cetacea, the forelimbs are modified into flippers and show both hyperphalangy as well as hyperdactyly. In Dolphin, the length of two or three digits is increased while the others are reduced.

Hindlimbs and pelvic girdle. The hindlimbs in water opossum (*Chironectes*), Protomogale, Hippopotomus, Phocidae, Odobaenidae are the chief and/or exclusive swimming organs. In *Ornithorhynchus* and Otariidae the hindlimbs act as equilibrators.

The general tendency of the hindlimbs and the pelvic girdle is to disappear. In rare cases, small bony vestiges representing the pelvis and limbs may be seen buried deeply in the muscle.

Modifications for 'Caudal Propulsion'. The tail is used as a rudder and for the purpose of keeping the head downward when the animals search food in the mud. The tail is modified for two fundamentally different principles, especially in aquatic mammals. The tail becomes narrow horizontally and broadened in vertical plane. This condition is observed in Hippopotomus, Protomogale, Nectogale where the tail is narrow. The tail is rounded in Chironectes. The tail is flattened vertically and broadened in horizontal plane. It is single-lobed and rounded in Manatee. The tail may be notched in whales, Dugong and Sea cow (*Rhytina*). In Cetacea and Sirenia, the tail has two flukes. The caudal fin of marine reptiles is quite different from that of aquatic mammals. It is horizontal and the terminal part of the vertebral column equally supports the flukes. In secondary aquatic vertebrates, the unpaired fins, if developed, are not supported by skeleton. The dorsal fin is triangular in *Ichthyosaur*. In many whales, the dorsal fin is entirely lacking, but in *Balaenoptera musculus* it is small and situated well off upon the tail.

Modifications for physiological regulation

Modifications of the integument. In aquatic vertebrates the integument is extremely modified. Loss of armour has occurred in *Ichthyosaur*. In majority of aquatic mammals, loss of hair is seen. In Cetacea and Sirenia, there is no trace of hair excepting few bristles around the muzzle. The hairs may be absent even in foetus in White whales and Narwhales. Acquisition of a subcutaneous layer of fat (*blubber*) in aquatic mammals compensates the loss of hairs by helping in the retention of body heat. The sweat and sebaceous glands are also absent.

Modifications for the protection of sense organs

Olfactory sense organ. In case of aquatic vertebrates, the external nostrils have shifted towards the apex of the head. The nostrils are closed involuntarily and opened voluntarily as in Cetacea and Sirenea. The closure is done by the action of the *maxillonasolabialis* muscles and the apertures are further complicated by the development of pads and valves. The aperture is crescentic in Sirenea. The valves guarding the narial apertures close at the end of inspiration and open at the initiation of expiration in Cetacea. In Odontoceti the external narial aperture is single and medially placed. The whole olfactory apparatus is appreciably reduced in Pinnipedia.

Acoustic sense organ. In case of aquatic mammals, there is a general tendency for the elimination of the external ear. The opening of the ear may be closed which is usually accomplished by pulling into the orifice of a valvular plug. In Phocidae the external auditory aperture is small and pinhole-like. In Cetacea the ear bones are extremely hardened.

Visual sense organs. The eyelids show the tendency of degeneration in aquatic vertebrates, especially in mammals. In Cetacea the eyelids are absent and the nictitating membrane is present over the eye-ball. The eyes become adapted to aquatic vision. The visual organs exhibit great optical adaptation. The lens is spherical and the refractive index is high. Unusual number of rods are connected with a single cell and tapetum lucidum is extensively developed. The other adaptations are the reduction in the size of cornea, the unusual development of choroid and perichoroidal lymph spaces. The cornea is laterally thickened and the sclera is also greatly thickened. The glands are comparatively well developed to produce oily secretion.

Modifications of soft parts

Digestive system. In case of aquatic forms there is a general tendency for the teeth to become simplified and become sharply pointed for grasping the slippery prey. The teeth may be absent in one jaw as in Sperm-whale, but in Baleen Whale both the jaws are devoid of teeth. The function of the teeth is taken up by the baleen plate. The molars of Platypus are shed at a very tender age and their places are taken up by the horny plate. Although well-toothed as a rule, Ichthyosaur showed reduction of teeth in some cases. The buccal cavity in Cetacea is very large and the stomach becomes complicated by the formation of chambers. The salivary glands are absent or reduced in aquatic mammals.

Respiratory system. In Porpoise, the lungs are provided with cartilaginous armature to give unusual strength and incompressibility. In Cetacea and Pinnipedia, the intrinsic musculatures of the lungs and the disposition of the diaphragm enable to emtify the lungs more completely. The lungs are spacious. In Cetacea and Sirenea, the diaphragm has become horizontal in position.

Circulatory system. In most of the aquatic mammals, especially in Cetacea and Sirenea, the blood is comparatively rich in haemoglobin. Another peculiar feature is the presence of retia mirabilia in Cetacea, Sirenia and Phocidae.

Urinogenital system. Particularly in case of aquatic mammals, there is a tendency to eliminate the scrotum. The testes are not actually abdominal but are situated in a pouch near the inguinal ring. The offsprings are precocious.

Disposition of mammae. In Cetacea and Sirenia there is one pair of mammae situated inguinally. In some Sirenea, the paired mammae are axillary and situated practically upon the posterior border of the flippers. In *Myocastor*, there are two pairs of mammae situated almost upon the back, one pair is located upon the shoulder and the other pair is situated near the haunches.

Modifications of hard parts

Skull. The floatation and the degree of water pressure have exerted tremendous influence in modelling the shape of the Cetacean skull. In aquatic mammals, there is a tendency of shortening the cranium and the facial portion of the skull tends to become elongated. The zygomatic arch is greatly reduced to vestige as seen in Cetacea. In Dolphins and whales, the posterior portion of the skull is globular and the anterior portion is prolonged into a rostrum-like structure. In Sirenia the bones are compact and are heavily built for getting submarine vegetation. But in Cetacea and other swimming forms the bones are made light with spaces filled with fats. In Mystacoceti the bones of the skull are telescoped at the occipital region.

Vertebral column. The vertebral column tends to become simplified. The cervical vertebrae are ill-developed and become condensed. In whales, the number of cervical vertebrae is seven but becomes fused into a compact bony piece. The articulating processes of the vertebrae in the trunk region become reduced but become elongated in the tail region for the attachment of the tail muscles. The sacrum is undifferentiated in aquatic mammals and the sacral vertebrae lose their identities.

The morphological as well as the physiological alterations are most striking in aquatic animals. Moulding force is so severe in case of aquatic adapted forms that the identity may be lost and one finds perfect covergence of bodily forms between diverse groups. All the secondary aquatic forms have evolved from terrestrial ancestors. Although this fact is not supported by palaeontology, still the subsequent stages can be reconstructed with considerable accuracy. The terrestrial forms may have been forced to come to water for food and for enemies that pursue them. On coming to water they may again be confronted with aquatic enemies and as a result they are forced back to land. Some of them may encounter favourable condition in water and relinquish the terrestrial home and eventually become highly specialised aquatic animals.

ADAPTATIONS IN CAVE-DWELLERS

Caves are essentially the abandoned channels of underground rivers in the limestone regions. The caves also provide homes for animals. Denudation of the surface by external agencies may produce caverns where ingress of cave-dwellers occurred. On coming to caves, the animals became isolated from others with suitable adaptations for living in caves. Like the deep-sea forms, the cave fauna are relatively recent in origin and the adaptive changes undergone by them are mainly degenerative specialisations.

Physical conditions in caves

Absence of sun-light and uniformity in temperature are the two striking physical characteristics existing in caves. The caves may be divided into the following zones depending on variable physical conditions. *Twilight zone.* This zone comprises of the mouth of the cave and a little amount of light may penetrate. This zone constitutes the transitional region of the cave. *Fluctuating temperature zone.* Seasonal or diurnal variation of temperature is felt in this zone. *Inner cave region.* In this zone light is absent and the temperature remains more or less constant.

Structural modifications

The structural modifications of the cave-dwellers are mainly due to lack of light, scarcity of food and changelessness physical conditions. Two categories of cave-dwellers are seen, the temporary visitors to the caves and the permanent inhabitants. Amongst the temporary visitors to the caves the instance of brown bat (*Myotis lucifugus*) is worth mentioning. The adaptations in permanent cave-animals are quite striking. The principal modifications are: the bleaching of pigmentation of the skin; the reduction of eyes and development of highly sensitive tactile organs, the scarcity of food has modified the organs of digestion, as a result the body becomes slender with delicate appendages.

Characteristics of permanent cave-dwellers

Cave-dwelling fishes. Many fishes inhabit

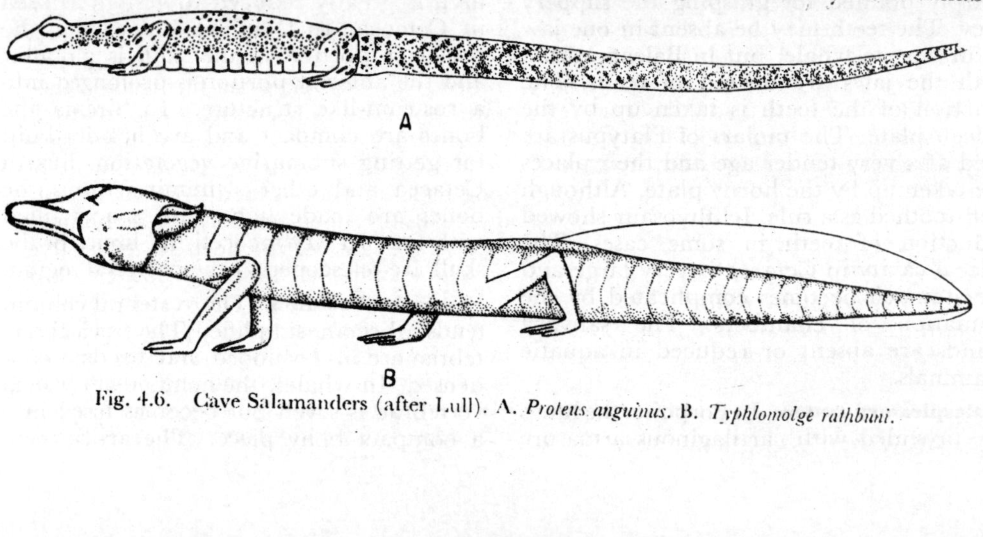

Fig. 4.6. Cave Salamanders (after Lull). A. *Proteus anguinus*. B. *Typhlomolge rathbuni*.

the caves and show structural modifications. The catfish (*Gronias nigrilabris*) is almost blind. Majority of the cave fishes show sign of degeneration of eyes. In a small cave fish called the *Typhlichthys subterraneus*, the eyes are present in young stage but become useless in adult. *Amblyopsis spelaeus* is totally blind but possesses well-developed tactile sense organs.

Cave-dwelling amphibia. Amongst Amphibia, a large number of salamanders live in caves permanently and have undergone degeneration (Fig. 4.6). *Typhlotriton spelaeus* has normal eyes in larval condition but become degenerated in adult. The eyelids are fused and the retinal layer lacks rods and cones. In *Typhlomolge rathbuni*, the eyes are devoid of muscles. The eyes are entirely covered by skin and non-functional. The limbs are extremely elongated and slender. *Proteus anguinus* is another peculiar cave amphibian where the body is white and lives in complete darkness. It can exist for a long time without food. It is completely blind but the body is experimentally seen to be very sensitive.

Cave-dwelling amniotes. Amongst the amniotes there is no true cave-dweller. Although the structural modifications in some burrowing lizards and snakes are same as the cave-dwellers, these cannot be due to cave-dwelling. Amongst the birds there is no permanent cave-dwelling form. Only one mammal, *Peromyscus leucopus* (white-footed mouse) exhibits a sort of transitional adaptation between epigean and cave-dwelling forms. They possess protruded eyes and long tactile whiskers.

As already discussed the environment in caves is nonfluctuating and is more or less uniform. This changelessness has made all the cave-dwellers to look alike. The cave-dwellers are weak and degenerated forms. To escape from the epigean enemies, they were forced to take the resort in caves for retreat. Absence of light and lack of food have caused the cave-dwellers towards degeneracy.

DEEP-SEA ADAPTATION

Of the bathymetric realms, the most important and significant one is the Holobiotic or Marine realm. This realm is again divided into four characteristic subrealms, such a Strand, Shallow sea, Pelagic and Abyssal zones. The abyssal realm includes all the waters below the depth of 100 fathoms. The abyssal realm is divided again into the abysso-pelagic and the abysso-benthonic zones. The abysso-pelagic zone has no substratum and the animals have pelagic characteristics, while the abysso-benthonic zone has a substratum and the animals inhabiting in this region are extremely modified and peculiar.

Physical characteristics of deep sea

Four remarkable characteristics prevail in this area. They are as follows:

(1) Absence of sunlight. The limit of penetration of sun's rays is about 200 fathoms. Beyond that depth there is no sunlight.

(2) Quiescence. Because of depth the movement of water is almost absent and the movement is exceedingly slow.

(3) Cold temperature. Beyond a certain depth the temperature is nearing to freezing point and the temperature remains constant. Diurnal and seasonal fluctuations of temperature cease.

(4) Pressure. The pressure of the water column is enormous and increases with the depth.

(5) Lack of green vegetation. Besides the above four factors, another factor is the total absence of green vegetation due to absence of sunlight.

General trends of adaptation in deep-sea animals

For leading deep-sea life, the animals had to confront the adverse physical conditions stated above. As a result they exhibit certain structural variations. The general trends are summarised as: (*i*) The deep-sea animals are weak and delicate. (*ii*) The hue of the body is generally simplified. (*iii*) The deep-sea forms are either with powerful telescoping eyes to catch maximum possible volume of light rays or are totally blind. (*iv*) Development of long feelers to act as tactile organs. (*v*) Almost all the deep-sea forms are luminescent. (*vi*) Most of the deep-sea fishes live on the exudes of decaying matters that have led to the loss of masticatory power. Others are seen to possess powerful jaws. (*vii*) Most of them have wonderful devices for caring the youngs and the others produce enormous number of youngs to overcome the hostile environment. (*viii*) Small size is the another characteristic of deep-sea living.

Structural modifications of deep-sea invertebrates

Almost all the phyla have representatives who lead deep-sea life. Amongst the invertebrates, sponges, corals, hydroids, few echinoderms such as brittle-stars and stalked crinoids, tube-dwelling annelid worms, some brachiopods, cephalopods, some arthropods like the barnacles are some of the deep-sea forms. The modifications of the invertebrates are quite diverse and the description will be limited to the modifications of the vertebrates.

Vertebrates of the deep sea

Amongst the elasmobranchs, the true sharks do not exhibit deep-sea characteristics excepting the luminous sharks (*Spinax niger*). The silver sharks or Chimaeroids show deep-sea appearance in having huge eyes and a long attenuated body and tail.

Of the teleost fishes, the typical deep-sea forms is the *Cetomimus* which has a large mouth, small teeth, very small eyes and scaleless body. But in another form, *Ipnops* there are no eyes and two large luminous organs are present on the head. The family Stomiatidae is characterised by the possession of scaleless body and well-developed luminescent organs. The scales may sometimes be present but are extensively delicate. In *Gastrostomus*, the body is long, slender and rows of luminous organs are present on the lateral sides of the body. The mouth is bounded by very large jaws. The cod-like forms (Gadiformes) have reduced mouth and dentition. The eyes are extremely large, the trunk is reduced and has a filamentous tapering tail.

The most remarkable group of fishes is the anglers which show typical deep-sea characteristics. The paired fins are adapted for crawling on the bottom of the sea and the anterior fin-rays of the dorsal fin function as a lure. In *Linophryne* the fin-rays are provided with luminous organs to attract the prey. Another species, *Oneirodes*, is blind but has luminous organs to compensate the loss of eyes. In another deep-sea fish, *Protostomias*, specialised light producing organs are present as rows on the lateral side of the body.

Another striking feature is encountered in deep-sea fishes. Due to enormous pressure of water column, the body in flat fish becomes flattened and the mouth is shifted to the lateral side of the body.

The structural modifications in deep-sea forms are due to the peculiar physical conditions of the deep sea. The deep-sea forms are geologically very recent in origin. These forms were originally the inhabitants of the pelagic or littoral regions which migrated to the deep sea and became adapted there. Due to the changelessness of the physical conditions in deep sea the evolutionary possibilities are very low.

BIOLOGICAL ENVIRONMENT AND ADAPTATION

Organisms are greatly influenced by other organisms of the same species or belonging to other species. All the living organisms have prolific reproductive rate and as such the reproductive potential of a species tends to go far ahead of food supply and available space. This results into competition for survival and the inevitable consequence is that some members survive at the expense of others and develop new adaptations which their genotypes permit. Such relationships existing between different species for food and shelter may result into keen competition. The main source is the food chain which produces a large number of forms.

DEVELOPMENT OF PREDATION

Competition for food may result into **predation (a mode of living** that involves **feeding on other organisms)**. Predators are **all animals and few primary** types of predators **are encountered:** *Carnivores*—the **animals under this category** are adapted to feed on other animals. *Herbivores*—the animals that are adapted to feed on plants. *Omnivores*—the animals that are adapted on mixed diet.

DEVELOPMENT OF ANIMAL ASSOCIATIONS

In course of environmental adaptations, sometimes different species of organisms become intimately associated with one another. The relationships are quite essential for protection, transportation, food, etc. The relationships are as follows:

Mutualism. Organisms of different species remain intimately associated and derive benefit from living together. The association of cellulose-digesting protozoa with the termites may be cited as a typical case of mutualistic partnerships.

Commensalism. Organisms of different species are intimately associated in such a way that one derives benefit while the other is neither helped nor harmed. The example of barnacles may be taken as a typical case. The barnacles are usually attached to the shell of the molluscs.

Parasitism. This is a specialised relationship between organisms of different species, in which one (parasite) derives full benefit at the expense of the other (host). Two types of parasites are seen: (*i*) **Ectoparasites.** The parasites living upon the bodies of the organisms such as lice, ticks and leeches. (*ii*) **Endoparasites**—the parasites that live within the body of the host. The examples of the endoparasites are plenty in animal kingdom, such as Entamoebae, Malarial parasites, Tapeworms, Flukes, etc.

PARASITES AND PARASITIC ADAPTATIONS

Parasitism in wider sense is a specialised mode of living and the parasites in their efforts to survive have made innumerable adaptations and compromises. From the evolutionary point of view, parasitic existence is not actually a phenomenon of degeneration, but offers a chance of living in this crowded and inhospital world. The parasites in course of time have established a unique harmonious co-existence with the host by the process of evolutionary adjustment. The ecological aspect of parasitism has been dealt in detail in Part Three of this book.

Factors necessary for survival of the parasites

The parasitic existence of the parasites depends upon finding a suitable environment in which they can live and propagate. Of other factors, the following four seem to be most vital.

(1) Provisions for successful admission to the body of the host.

(2) Suitable environmental conditions in the host.

(3) Protection against the normal metabolic processes of the host.

(4) Absence of reaction on the part of the host that may interfere with the normal metabolic activities of the parasites. The host should be tolerant and resistant.

Adaptations of the parasites and the host

Parasitism leads to adaptations in the parasites as well as in the hosts that house them. Profound structural and functional modifications are encountered due to parasitism.

Adaptations in the parasites. The modifications in the parasites show two contrasting features, some structures become simplified while others become highly specialised. The modifications in parasites depend on the degree of parasitism. The greater the degree of parasitism the more pronounced is the departure from the normal pattern. Gradations of adaptations are observed in parasites. Structures are slightly altered in the ectoparasites but they become highly modified in endoparasites.

Loss of unused structures. Some unused structures are totally or partially eliminated. They are the organs of locomotion which become reduced or atrophied; reduction of the central nervous system and sense organs; gradual obliteration of the alimentary canal; reduction of the excretory system, etc.

Modifications of the epidermis. The epidermis is modified for protection and for absorbing nutrition as seen in parasitic helminths. Exoskeleton is totally lost. The organs of feeding often become more specialised and are better developed to suit the purpose. Food storage devices may be better developed in ectoparasites and the posterior portion of the alimentary canal may atrophy due to lack of residual matters. In tapeworms, the alimentary canal entirely disappears. Most of the helminths lack alimentary canal, because they are constantly bathed in a medium of digested or semidigested medium. Simplification and eventual loss of digestive system indicate the increasing dependence on the host for food and nutrition. Development of organs for attachment and penetration to the hosts are highly developed. The modifications for the purpose of attachment are the hooks, spines, acetabula or sucking organs of the helminths. Modifications for the purpose of penetration are the glands in the larval forms of trematodes, in the vicinity of mouth of adult helminths are the devices for penetration into the body of the hosts. Modification of the reproductive organs is aimed to produce prolific number of the offsprings to overcome the obstructions. The eggs are usually produced in astronomical number. Occasionally self-fertilization occurs. In extreme cases completion of life-history needs intermediate host. The chance of survival is increased by the development of resistant envelope to seal the content of the egg from heat, drying and injurious influences. Such phenomenon also helps in the dispersal of the progenies.

Adaptations in the host. The hosts that provide shelter to the parasites, also exhibited certain adaptations. The reproductive glands are in most cases affected and in extreme cases may lead to degeneration. The reversal of sexes occurs in many hosts. In some extreme cases maturation and completion of the life-cycle is not possible.

Parasitism—a case of degeneration or specialisation?

Parasitism furnishes one of the most significant examples of the power of adaptation in organisms to live parasitically. They exhibit all the possible far-reaching adaptations for the specialised environment prevailing within the body of the hosts. In some cases the parasitic adaptations are so severe that they diverge widely from the normal types of the families they belong. As a method of adaptation for survival, parasitism must be regarded as a remarkably successful way of living, although the evolutionary trends may often culminate into retrogression and resulted into degeneracy. The degree of degeneration may often depend on the degree of parasitism. Although the parasites in their efforts to overcome obstacles may become degenerated, parasitism has got great survival value in this congested world.

Evolutionary value of parasitism

Nothing is definitely known about the origin and evolution of parasitism. But it can be said beyond any doubt that parasitism is more of recent origin than the free-living forms since free-living forms must have existed before the parasites could obtain hosts to live. Parasites, like other organisms, have been subjected to the evolutionary dynamics and their study can furnish valuable information about organic evolution.

Ectoparasites to come first. Ectoparasites were the first instances of parasitism and exhibit different degrees of gradations. (*i*) The larvae are parasites but the adults are free-living as seen in the Glochidium larva of bivalved molluscs. (*ii*) The adults are ectoparasites but the larvae are free-living such as the leeches and lampreys. (*iii*) Both the adults as well as the larvae are ectoparasites as seen in Mallophaga.

Endoparasites reach the pinnacle of parasitism. The endoparasitic existence is the extreme case of parasitism. In the endoparasitic forms almost same types of gradations are observed. (*i*) The adults are free-living but the larvae are parasites as seen in certain forms of hymenoptera. (*ii*) The adults are parasites but the larvae are free-living as seen in certain copepods and cirripeds. (*iii*) Endoamoebae, malarial parasites, tapeworms are parasites all through their life.

Ectoparasites evolved before the endoparasites because the change from a free living existence to ectoparasitism does not appear to be so difficult as to what we see in endoparasitism. The relationships between ectoparasitism and endoparasitism furnish much material for speculation in the evolution of the parasites.

Host-specificity and survival value

A parasite (essentially the endoparasites) needs a definite condition in the host. In almost all cases, the physiological demands of the parasites remain more plastic than its environment. But there is always a limit to such plasticity, as a result most of the parasitic species become host-specific. Host-specificity arises because a particular parasite is adapted to take specific type of food especially the protein which produces specific antigen and antibody reaction. The degree of host-specificity bears a close parallelism with the evolution of parasitism. The more host-specific a parasite is, more divergent it appears to be from its original phylogenetic stem.

Reproductive faculty and survival value of parasitism

The truth of evolutionary progression is that the mutation offers the species new variations for the Natural Selection to work with. The main source of mutation lies in the process of reproduction. In parasites the possibilities of mutation are lowered because most of the parasites are hermaphroditic and their genotypes are practically devoid of heterozygocity. But this is compensated by the possession of specialised methods such as the precautionary measures in egg lying taken by the parasites to have good chance of survival. Prolific rate of egg production gives ample scope for mutation. The phenomenon may be illustrated with the examples of *Ascaris* which produces about 200,000 fertilized eggs in a day. Fish tapeworm lays about 36,000 to 1,000,000 eggs per day. It is known that the mutation is the failure of faithfully copying the genes. So when there is prolific rate of egg production, the rate of frequency of mutation increases. By this way of rapid duplication, numerous mutations are produced which provide the raw materials for Natural Selection to work with to reach the evolutionary goal. While any mutant gene arising in the host may be favoured by Natural Selection as far as the elimination of the parasites is concerned by creating a different internal environment for that species of the parasite. The mutations in the parasite also may be fovoured by Natural Selection to adjust the parasite in the altered environment or establish it in a new species of host, the sum total effect of this battle between the host and the parasite is the possibility for emergence of new species both in the parasite and the host.

Amongst the possible modes of living in nature, the parasitic existence, although regarded to be a case of degeneration, is certainly a most specialised type of living and has great survival value. Now question arises how the parasites became parasitic. When a parasite from the generalised environment comes under the shadow of parasitism, it becomes confined in a specialised environment and becomes isolated. In the realm of parasitism Natural Selection works in a most contradictory way. The survival of a parasite depends on the host-parasite relationship. Parasites cannot to be too much parasitic. The host must live for providing house for the parasites to live. Natural Selection works in two ways—in one hand it insists the parasites to be parasitic and on the other hand it checks them not to become too much parasitic.

COLOURATION IN ANIMALS

All living organisms are adapted to survive in the particular environments they inhabit. For living suitably in a particular environment, living organisms exhibit numerous adaptations. Most of the adaptive features encountered in nature are protective in a variety of ways. Animal colouration is generally regarded as a sort of protective adaptation, where many animals resemble the backgrounds against which they live. While discussing the colouration in animals, emphasis is given to the ways whereby the colours are produced and then we will pass on to the significance of such colouration in the biological world with suitable examples. Instances are numerous in the animal kingdom, but to restrict the discussion only typical cases are cited.

Colour production in animals. Colouration in animals is produced in animals both by physical as well as chemical means. Colour is usually produced by a sort of interference with the beams of white light. The physical colouration is caused by surface sculpturing of the body. The beam of light falling upon this surface refracts and breaks up into its component rays. The chemical colouration is primarily due to the presence of pigments. Presence of pigment is very common in animals. Pigment may be present in the interior of the body as well as in the outer surface. Chemical colouration may be produced by the absorptive power of certain pigment which by oxidation may cause various colours. Haemoglobin and haemocyanin are the pigments of the blood. Cells containing the pigment are usually differentiated into two broad categories—first cells containing the pigment give an unchangeable colour to the animals and the second category is the *chromatophores* or changeable pigmented spots which produce flushes of colour over the skin. Various types of chromatophores are seen in animals. They are *melanophores* containing black or brown pigment; *erythrophores* containing red pigments; *xanthophores* containing yellow pigment and *lipophores* containing a combination of red, yellow and orange pigments. The ability to change colour is called as *metachrosis* and is controlled by two agencies—nervous and endocrine systems. The chromatophores are influenced by the skin and eyes through the pituitary gland and us produce changes in the hue of the body. The activities of the chromatophores are greatly enhanced if a reflecting layer of guanin is present beneath them.

Types of colouration in animals

Colouration in animals is usually classified under the following categories depending upon its biological significance.

Sympathetic or cryptic or concealing colouration. The colour of the animals harmonises with its surroundings to such an extent that it completely blends into the background. By this way the animals lose their conspicuousness. This tactic may help them to escape the eyes of the enemies (Protective colouration) or to wait in order to catch hold of the prey (Aggressive colouration). Hares, weasels, ptarmigan of far North moult into white coat of hair when the surrounding is covered with snow. Many marine invertebrates and the larvae of some fishes are transparent. Bark-inhabiting insects usually fit with the colour of the background on which they live.

The colour of the same animal may vary. Species having a wide range of distribution, vary their body colour to harmonise the different colours of the surroundings. Such colour is designated as the *local colour*. Several grasshoppers belonging to the family *Acrididae* exhibit such local colouration. The body colour of the desert adapted gazelle, varies from white (in sandy plains) to dark gray (in the volcanic districts). Another phenomenon is called as the *seasonal colour*, where the body colour of the same species varies in different seasons of the year. A number of birds and mammals exhibit such change of colour. The Artic fox, some hare, the weasel change their body colour from summer to winter to harmonise the changing surroundings.

As a consequence of adaptation all animals harmonise the backgrounds they live. The desert forms are usually dull or gray, such as the gazelle, the camel, the lion, etc. Most of the forest insects are green in colour to harmonise the green colour of the plants. Arctic animals are mostly white to harmonise the snowy background. In many sea birds the dorsal side of the body is blue or gray and the ventral side is white. When viewed from above, these birds harmonize the background of the sea and to the sky when viewed from below. Generally, the upper surface of the body is darker than the lower. This phenomenon of countershading is effective in camouflage because it minimises the contrast between the shaded and non-shaded parts in light.

Warning or revealing colouration. The colour of the body of some animals is conspicuously red and yellow and they also advertise their unpalatability. Possession of stout sting and conspicuous yellow and black colours of the body in bees and wasps are the typical examples. Other instances are the coral snakes, tiger salamander, *Gila monster*. These creatures are almost immune from attack, because of the conspicuous exhibition of a dangerous nature. Prominent markings on the dorsal side of the hood of Cobra is another instance. The large eye-like spots on the hind wings of a moth, *Smerinthus* are displayed when they are annoyed, but are ordinarily kept concealed under the front pair.

Recognition marks or signal. These markings are used by some species to help in identifying from other species and to warn them in time of danger. These are of great use particularly in gregarious animals when mutual help in time of danger is necessary. The conspicuous whitely coloured under surface of the tail of virginia deer acts as signal at times of danger. The white rump patch on *Antilocapra*, the cottony-white tail of *Lepus sylvaticus* are all flashed into prominence when they are frightened. The recognisation marks are usually used by individuals of the same species and are borne by many insects.

Alluring colouration. Certain animals exhibit some colouration which may act as lure for others. Certain carnivorous spiders resemble the orchid blossom very closely in form and colour which is an alluring adaptation. Similar characteristic is present in some African lizards which are protectively coloured excepting a brilliantly coloured patch at the corners of the mouth which usually acts as a lure.

Mimetic colouration. The colour of the body of the animals may resemble very closely between the animal and any other objects. This colouration may be protective in concealing their conspicuousness or may be warning as well.

Confusing colouration. Many butterflies and moths by exhibiting recognisation marks make their position very conspicuous but at the end of flight, they cover the bright area very quickly. This sudden change from extreme conspicuousness to extreme inconspicuousness is very confusing to the predators.

Sexual colouration. Colouration in different sexes has significant value in sexual selection. The males of many birds and some other forms are more conspicuously coloured than their females. The actual cause and significance of such colour difference seem to be confusing, but it is a fact that such colour differences exist.

The biological significance of colouration in animal kingdom is quite interesting and has always played an important role in determining the chances of survival in this hostile world. Whatever the cause of animal colouration may be, the adaptive perfection for whatever purpose may they be used, is the result of Natural Selection.

MIMICRY

One of the most wonderful protective devices in nature is mimicry, a phenomenon which organisms bear to other animate or inanimate objects in form, colour and attitude. By imitating others, these particular groups of organisms can escape the eyes of the observers or can exhibit an apparent aggressiveness which is not at all real. The former case is classified as *protective mimicry* while the last one is the *aggressive mimicry.*

Protective mimicry. Protective mimicry is divisible into two categories: Concealing mimicry and warning mimicry.

Concealing mimicry. Most of the protectively mimic forms exhibit devices of concealing themselves in order to escape the clutches of enemies. Some insects exhibit protective resemblances to objects of the surroundings in forms and colours. A crab, *Cryptolithodes* with its smooth rounded form, texture and colour of the body perfectly harmonises the white pebbles of the beach to such an extent, it becomes very difficult to detect them. Some geometrid caterpillars rest in a position which mimic a twig of the tree on which they live. In another geometrid moth, *Schizura nucronis* such mimicry is reached to the peak where not only does the rough-barked caterpillar resemble the twig perfectly but the adult also does (Fig. 4.7A). Some walkingstick insects of the family *Phasmidae,* with their body, slow movement, attenuated appendages and matching colouration resemble dead or green twigs and leaves. Most perfect example is the leaf insect called *Phyllium,* where the wings along with the flattened and expanded body and limbs are mostly green (Fig. 4.7B). An Indian butterfly (*Kallima*) resembles a dead leaf when it rests with folded wings (Fig. 4.7C).

Warning mimicry. Many remarkably mimic forms exhibit warning mimicry. These forms advertise coloured texture which are either distasteful or poisonous if devoured. Some of the deadly poisonous snakes belonging to the family *Elapidae* are brilliantly coloured. Many poisonous coral snakes are mimicked by other harmless snakes of different genera. A host of imitators are present amongst the insects. Amongst the butterflies there are many forms which mimic other distasteful forms. The most remarkable example is *Basilarchia archippus,* which are highly palatable for the insectivorous birds, imitate the *Anosia plexippus* an inedible form.

Thus a relatively palatable species gets advantage from the predators, if it resembles a warningly coloured inedible species. This 'false warning colouration' is termed as the *warning mimicry of Batesian type.* If two of unpalatable species resemble each other, is called *Mullerian mimicry.* Mullerian mimicry is advantageous for both but Batesian mimicry is good only to the edible species.

Aggressive mimicry. This is the characteristic of some carnivorous forms. Spiders are the typical examples. In most cases the yellow body harmonises so perfectly with the colour of the flower upon which they rest that they become invisible for the visiting insects which constitute the spider's prey. Some spiders resemble an orchid blossom very closely in colour and forms which act as a lure. Similar phenomenon is observed in some African lizards where a patch of brilliant colour in the corner of the mouth is an alluring one.

Mimicked forms of animals are usually defenseless and this mechanism of imitating others is a highly adapted feature and at the same time has great survival value. Opinions differ as regards the origin of mimicry in animal kingdom. It has been suggested by many that mimitic forms may have arisen as drastic mutations. Whatever the causes may be, it is quite suggestive to regard that Natural Selection plays the greatest role in the production of mimicry in animal kingdom.

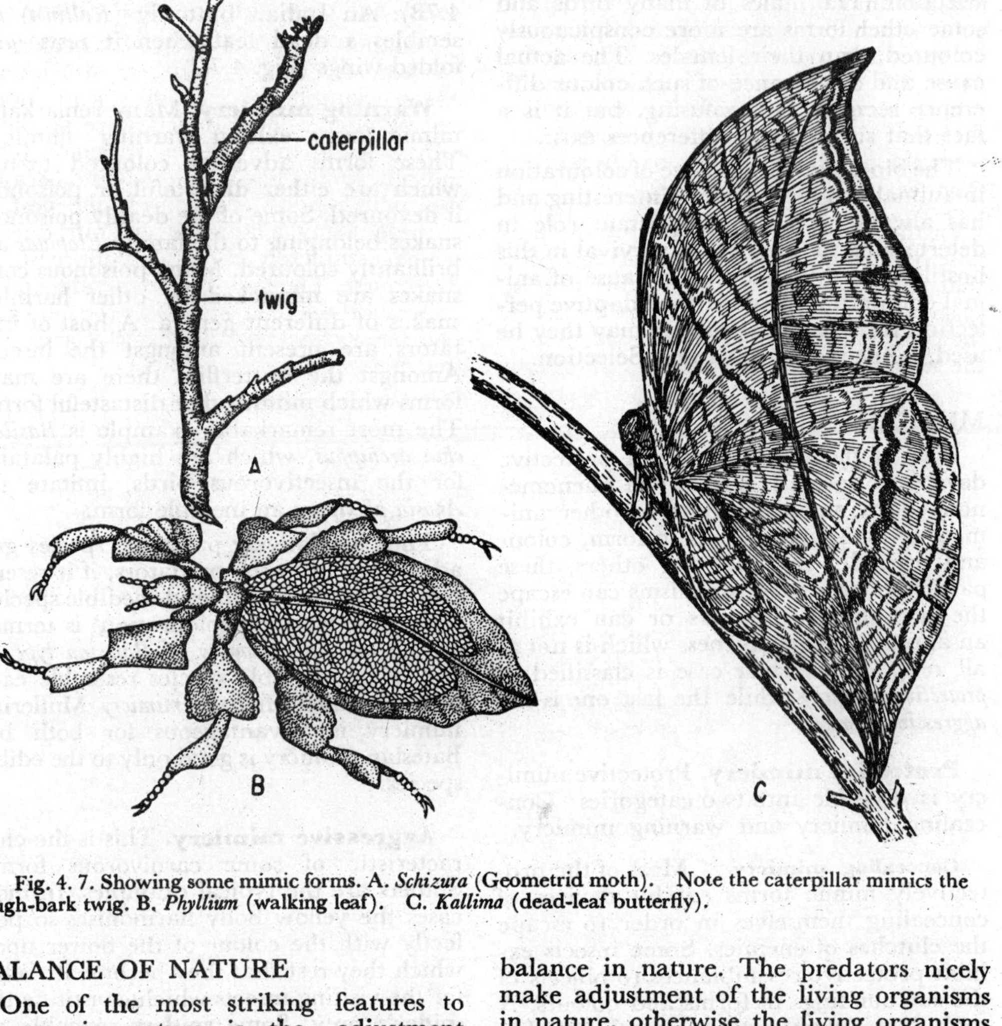

Fig. 4. 7. Showing some mimic forms. A. *Schizura* (Geometrid moth). Note the caterpillar mimics the rough-bark twig. B. *Phyllium* (walking leaf). C. *Kallima* (dead-leaf butterfly).

BALANCE OF NATURE

One of the most striking features to observe in nature is the adjustment and interrelationship between organisms. As already discussed, the reproductive potential of the organisms always runs far ahead of food and space available in this earth. So to survive in this world, competition between organisms has become the rule. This competition ultimately gives rise to the evolution of *predation* in the animal kingdom. Profound modifications in animals are observed to reach the ultimate goal. A survey of the colouration and mimicry in animal kingdom gives an idea of the wonderful ways how the hunters and the hunted make biological adjustment to overcome the situation.

Predation is a remarkable phenomenon which helps in the maintenance of the balance in nature. The predators nicely make adjustment of the living organisms in nature, otherwise the living organisms will overproduce and disturb the balance. This phenomenon of maintaining the equilibrium in nature is designated as the *balance of nature*. If the balance is not maintained the normal population ratios will be upset and thus result into imbalance.

The balance of nature is very delicately

adjusted and can very easily be disturbed. There are many notable examples to show how such balance of nature can be deranged easily.

In the western part of United States, Coyote was a harmful predator, because it eats important livestock animals. The Coyotes were extensively hunted by man so that their numbers were reduced to a minimum. With the decline of Coyotes, smaller animals, particularly the rabbits upon which they prey, increase in a prolific rate and become more detrimental. After the disappearance of Coyotes, the rabbits increase to such an extent that they eat away all the green vegetables that supports the livestock. The destruction of the Coyotes brings worst result.

In the city of Connecticut, for the purpose of eradiction of the measuring worms that destroy the elm trees, the English sparrows were introduced. But owing to the quarrelsome disposition they drove out the native insectivorous birds and offered in exchange a widespread pest and abundance of most unattractive sparrows.

To eradicate rats which brought plague in Jamaica, the mongoose (*Herpestes griseus*) was introduced. The mongoose destroyed the rats and the disease was ultimately controlled. Due to non-availability of rats, the mongoose started taking the poultry and ground birds and insect-eating lizards as food. The reduction of these animals caused the rapid multiplication of destructive insects and ticks which become widespread.

Instances of partial restoration of natural balance after being upset are also many. As an example the case of a scale insect can be cited. These insects were accidentally brought in contact with lemon trees in California. They multiplied and became destructive pests. To control the pest, a natural enemy, a lady-bug was introduced, which completely eliminated the pests in due course. The lady-bugs depend specifically upon the scale insects as the food and became extinct simultaneously with the destruction of the scale insects.

PART FIVE

EMBRYOLOGY

Living organisms have an inherent tendency to increase in number. Such increase enhances the frequency of emergence of more variations, though the living forms have a general trend '*like begets like*'. This phenomenon involves self-duplication, but so many complicated processes operate in establishing and maintenance of pattern that it cannot be called as mere multiplication. This part of the book discloses the events involved in the propagation of diversities which may briefly be termed as Developmental Events.

THE SCIENCE OF DEVELOPMENT

The science of embryology is usually defined as the process of development of an individual from the zygote to a stage resembling the adult form. But in recent years, the science of embryology has achieved a new meaning. It not only involves the changes through which the young passes but also it includes such events as growth, repair, maintenance of pattern, aging and death. Thus the science of development includes all the events, which begin from the initiation of germ cells to the death of the individual. The old term "*Embryology*" has thus got a new name, "*Developmental Biology*".

Historical review. The origin of the complex adult organism from a zygote has drawn the keenest attention of exponents of natural sciences since earliest times. The earliest philosophers tried to reason out the dynamics of development without performing any experiment. The science of development is as old as Biology itself. The correct interpretation of developmental dynamics rests on the cell theory advocated by M. J. Schleiden (1804–1881) and T. Schwann (1810–1882) in the year 1839. Most of the embryological observations prior to 1839 were erroneous and grotesque.

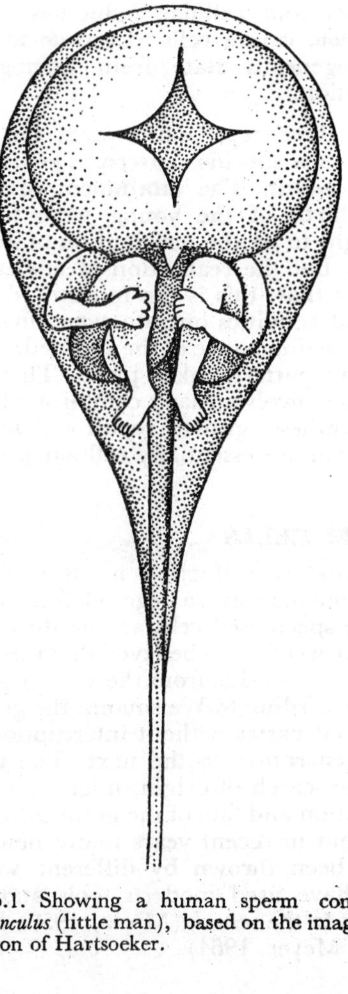

Fig. 5.1. Showing a human sperm containing the *homunculus* (little man), based on the imaginative illustration of Hartsoeker.

Two distinct schools of developmental Biologists approached to solve the problems of development with eagerness. They established (*i*) the THEORY OF PREFORMATION and (*ii*) the THEORY OF EPIGENESIS.

Theory of Preformation: This theory postulated that the gamete (ovum) contains a more or less perfect miniature of an adult animal in its substance and development involves mere growth and unfolding of predetermined pattern, i.e., growth and unfolding of the miniature form into an adult stage. The miniature human form was designated as *homunculus* (little man). Every ovum in the ovary was assumed to contain such homunculus. The dealers of this concept (Preformationists) speculated that an ovum in the ovary of a female must contain a homunculus, the homunculus itself must contain a secondary homunculus in the ovum of the homunculus and this again tertiary homunculus, and so on until the homunculi become too small to exist. The preformationists have estimated that mother Eve carried about two hundred million generations of homunculi. It was also beleived that when the homunculi become used up, no more human beings could be born.

With the discovery of human spermatozoon by Antonj van Leeuwenhoek (1632-1723), an additional complication was added to the preformation concept. Leeuwenhoek named the spermatozoon as the animalcule. With the establishment of significance of fertilization, the question arose as to whether the ovum or animalcule contained the homunculus. A microscopist, Hartsoeker described human spermatozoon by newly invented microscope and showed the existence of *homunculus* within the human spermatozoon (Fig. 5.1). Since then the preformationists became divided into two rival groups—*Animalcultists* (*Spermists*) and *Ovulists*. August Weismann (1834–1914), the discoverer of the continuity of germplasm, was the champion amongst ovulists.

Theory of Epigenesis: The theory of Epigenesis relates that development involves differentiation which originates in a homogeneous mass of living material. There is no preformed tissue or organ at the beginning. The developmental changes are epigenetic, which mean that the determination is established subsequently. This was first advanced by Aristotle who was aware about the transformation of chick embryo. William Harvey was an ardent supporter of epigenetic concept along Aristotelian lines. Harvey and Kaspar F. Wolff (1733–1794) added additional concepts to this theory. Besides his physiological contributions, Harvey worked on the development of mammalian and avian embryos. To Harvey we owe the famous revolutionary aphorism in embryology, "Omne vivum ex ovo". Although the epigenetic theory was initiated by Aristotle, K. F. Wolff (1733–1794) and Johann Friedrich Meckel (1781–1833) placed the epigenetic concept on a logical platform. We accept the theory of Epigenesis as true but we do not accept it in the form advanced by Wolff. With the advancement of biological science the epigenetic theory also underwent modifications. With the advent of twentieth century, numerous experimental methods have been applied to find the causal mechanism of development. Wilhelm Roux (1850–1924), one of Hæckel's students, interpreted development through experimental methods. His experimentations and the experiments and work of Endres (1895), Spemann (1901–12) proved that there is no preformation in development, on the contrary, the developmental changes are *epigenetic*, which mean that at the beginning, cells are indetermintate in nature and determination is established subsequently. Spemann's discovery of 'organizer' and its role in embryonic development added new dimension to this field. Spemann was awarded the Nobel Prize in 1935 for his discovery. Since then all eyes were focused to search the answer of a simple question, how determination occurs. But unfortunately, despite extensive experimentations by men of different disciplines, a clear-cut answer was not available. But this search placed Embryology to the most important position in Biology, where number of disciplines reconciled together to realise the question of development. The answer to this question will not only solve the mystery of determination but also will help to understand various other problems related to life itself.

Mechanism of development. The prerequisite of development is *reproduction*. Asexual or sexual reproduction, whatever it may be, it denotes the beginning of development. This is followed by three distinct steps, *growth*, *interaction* and *differentiation*. An important realisation of developmental biology is that all developmental steps are essentially cellular processes which include: **Growth,** involves cellular synthesis and cellular division. **Interaction,** includes cell movement, cell contact and cellular induction. **Differentiation,** causes cells to transform into a heterogeneous state from homogeneous condition.

All these steps lead towards the attainment of a specific pattern. This is called *morphogenesis*. The attainment of pattern and birth of the young were formerly thought to be the termination of development, but the realisation of modern biology is that, this is the midway, developmental activities operate even after birth, when maintenance of the attained morphogenetic pattern takes place. This maintenance involves many complicated events like synthesis, growth, repair and death, all of which are essentially cellular processes.

GERM CELLS

Sexual reproduction needs one sperm cell from male and an egg cell from female. These specialised cells are produced from the germ cells. It is believed that these germ cells are set aside from the very beginning and according to Weismann, the germinal material passes without interruption from one generation to the next. The inquest for the search of origin, migration, transformation and fate of the germ cells, is old one, but in recent years many new lights have been thrown by different workers, who have used modern tools and techniques to the search (Mintz, 1959; Simon, 1960; Meyer, 1964).

Development of germ cells. Development begins with the union of sperm and egg. These reproductive cells originate from the germ cells. The available information shows that series of interesting events happen in the life of germ cells from their origin to transformation. These events are the preparatory phases on which the development of individual depends. These developmental events begin long before the development of the individual.

Origin of germ cell is distinct in organisms where chromosome number differs in germ and somatic cells. In the roundworm, *Parascaris equorum*, the four blastomeres produced by the second cleavage exhibit differences in nuclear behaviour. The blastomeres are designated as A, B, C and P. A large portion of each chromosome in blastomeres A, B and C passes into cytoplasm and breaks up in the substance of cytoplasm. This phenomenon is regarded as *chromatin diminution*. The blastomere P remains unaltered. At the next cleavage division, one of the daughter blastomeres of P shows chromatin diminution, while the other remains unchanged. The same phenomenon occurs at the fourth cleavage division, as a result only one blastomere, P_4, retains the full number of chromosomes. All the germ cells eventually develop from P_4. In recent years, it has been described that in different animals the factor which guides the cell to be germ cell resides in cytoplasm. This area is called the "*area of germ cell determinant*". Any nucleus passing through this area develops the potency to be the nucleus of germ cell. Followings are the brief survey of literature on the origin of germ cells:

Waldeyer (1870) regarded that the germ cells originate from the coelomic epithelium around the gonad. Nussbaum (1880) advocated that the germ cells are formed outside the gonad and from the site of origin, the germ cells migrate within the gonad. Weismann (1885, 1892) established that germplasm is segregated completely independent of somatoplasm and results into the differentiation of germ cells. Boveri (1892) stated that cytoplasm of germ cell plays important part in determining the germplasm. Witschi (1914), Gatenby (1916) stated that in frog, a cytoplasmic substance in the yolk near the vegetal pole and on the roof of the gut is marked as site of germ cell formation. Swift (1914, 1916) stated that in chick, germ cells originate extra-embryonically and are first seen in 16-somite stage. Mintz (1959) established that germ cells originate in the yolk sac of 8-day-old mouse embryo.

The above findings indicate clearly that germ cells are segregated at the very onset of development of the individual and germ cells originate at a site far from the reproductive organ. This finding has raised the most interesting question about the migration of germ cells.

Migration of germ cells. Three distinct views are held regarding the migration of germ cells. These views are: (1) The germ cells migrate by performing amoeboid movement to the region of developing gonad. (2) Germ cells, from their site of origin, are carried by the blood stream and after travelling through different organs settle in the reproductive organ. (3) Various foldings which occur in embryonic development bring about the shifting of germ cells from their site of origin to the gonad.

Fate of germ cells. Two different opinions are forwarded to explain the fate of germ cells: (1) The germ cells which migrate into the reproductive organs are not destined to be the gametes at all; they degenerate and are replaced by new cells which are formed by the reproductive organs themselves. (2) The germ cells after migrating within gonad persist and transform into functional gametes. Recent experiments with improved techniques endorse the second view.

Transformation of germ cells. Within the gonad, the primordial germ cells are known as *primary gametogonia* which multiply rapidly by mitosis. This state is called the *phase of multiplication* and the resultant cells are then known as *secondary gametogonia*. The gametogonia enter into the *phase of growth* which is more pronounced in oogenesis. In male the gametogonia produce sperm cells and are called *spermatogonia*, while in female, gametogonia giving rise to ova or egg cells are known as *oogonia*. Sperm and ovum are responsible for (1) bringing together of hereditary factors in the new individual from the parents and (2) to provide material substance from which the new individual will arise. The gametogonia in both the sexes transform to fulfil these two purposes. The development of male gametogonium or spermatogonium to sperm is called *spermatogenesis* and change of female oogonium to ovum is known as *oogenesis*. The gametogonia then transform into the specific gametes and in both the cases, the transformation involves (1) reduction of diploid to haploid number of chromosomes and (2) considerable preparation in different components of the cell. It may be mentioned here that germ cells carry the potentiality of becoming both sperm and egg. It is the influence of the cells of the testis or ovary, which determines the final fate of it. Spermatogenesis and oogenesis are discussed below (Fig. 5.2).

Spermatogenesis. With the differentiation of testes, the primordial germ cells give rise to spermatogonia. Further

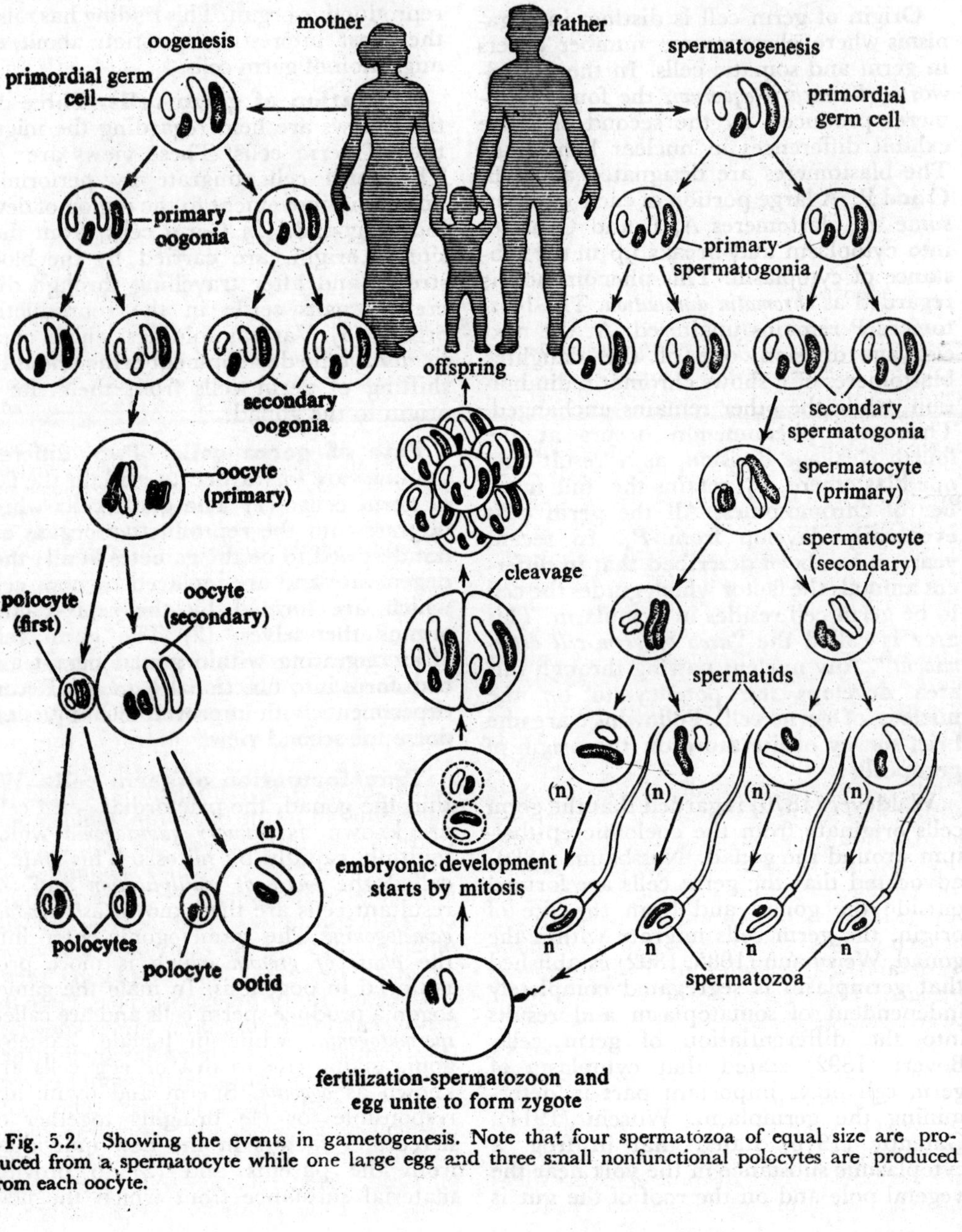

Fig. 5.2. Showing the events in gametogenesis. Note that four spermatozoa of equal size are produced from a spermatocyte while one large egg and three small nonfunctional polocytes are produced from each oocyte.

multiplication results in transformation of gonial cells into primary spermatocytes. The primary spermatocytes undergo the first meiotic division to produce secondary spermatocytes which give spermatids by the second meiotic division. Without further division, each spermatid differentiates into a spermatozoon—the functional

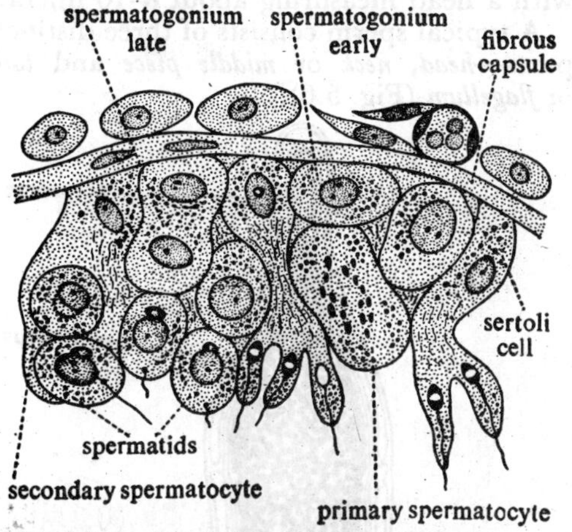

Fig. 5.3. Sectional view of a part of human testis showing spermatozoa in all stages of development in the seminiferous tubule.

male gamete (Fig. 5.3). As a result of meiotic cell-division four spermatids are produced from one primary spermatocyte.

Spermateleosis. The spermatids possess haploid set of chromosomes but are unable to function as male gametes. A spermatid transforms into a spermatozoon which involves a series of cellular changes. This process of differentiation is called *spermateleosis*. The events in spermateleosis are illustrated diagrammatically in Fig. 5.4.

The acrosome is a derivative of Golgi bodies. Numerous small vesicles appear within Golgi apparatus, each vesicle contains a homogeneous granule. These vesicles coalesce to form a single large vesicle containing numerous granules. This large vesicle ultimately comes to attach with the surface of nucleus. It is known as *acrosome vesicle* and the granules are called *acrosome granules*. These granules fuse to form a large *acrosomal granule*.

The nucleus with acrosome vesicle moves towards the peripheral end of the cell. The vesicle spreads over the nucleus, thus the vesicle becomes entirely occupied by the granules. The centrosome consists of two centrioles after second meiotic division. At the time of formation of acrosome vesicle, the centrioles move towards the *plasmalemma* and one of the centrioles attaches itself with the membrane. From the point of attachment a fine thread

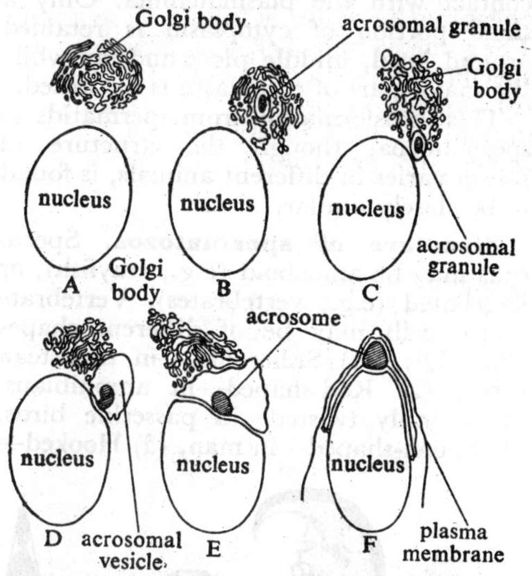

Fig. 5.4. Sequence of development of acrosome from Golgi body in spermatogenesis. A. Golgi body consists of a series of concentric membranes around a collection of small vacuoles. B. The vacuoles start enlarging and enclose a dense body (acrosomal granule). C. The granule becomes enlarged. D–E. The vacuole with its acrosome granule becomes closely applied to the anterior end of the nucleus. F. Showing the formation of the cap of the spermatozoon from the wall of the vacuole.

grows out and extends from the cell as a small flagellum, this is the *axial filament* of the future tail.

The centrioles then return to the nucleus and one of them fixes itself to a notch on the nucleus at the opposite pole of *acrosome vesicle*. During inward journey, the centriole which is in contact with the membrane draws the plasmalemma around the flagellum. Thus the flagellum, which apparently seems to be inside the cell, actually remains outside the cell and is bounded by double plasma membrane. The proximal centriole continues to remain attached with nucleus, but the distal centriole comes down again to bring the fold of plasma membrane back to the general surface. Here the distal centriole forms a ring and is known as ring centriole. Mitochondrial bodies then associate around the axial filament between proximal and ring centrioles. The axial filament extends outside the cell as a long flagellum and fine fibres originating from proximal centriole encircle the axial filament, thus establish the structure of the tail of sperm. The nucleus of a spermatid during its transformation into a spermatozoon loses water and becomes elongated to keep the acrosome vesicle in close contact with the plasmalemma. Only a small portion of cytoplasm is retained around head, middle piece and tail while the major part of cytoplasm is discarded.

This transformation from spermatids to spermatozoa, though the structure of sperm varies in different animals, is found to be closely similar.

Structure of spermatozoa. Sperm cells may be amoeboid (e.g., crayfish), or flagellated (e.g., vertebrates). Vertebrate sperm cells may be of different shapes (Fig. 5.5): (1) Spheroidal—in teleostean fishes, (2) Rod-shaped—in amphibians, (3) Spirally twisted—in passerine birds, (4) Spoon-shaped—in man, (5) Hooked—

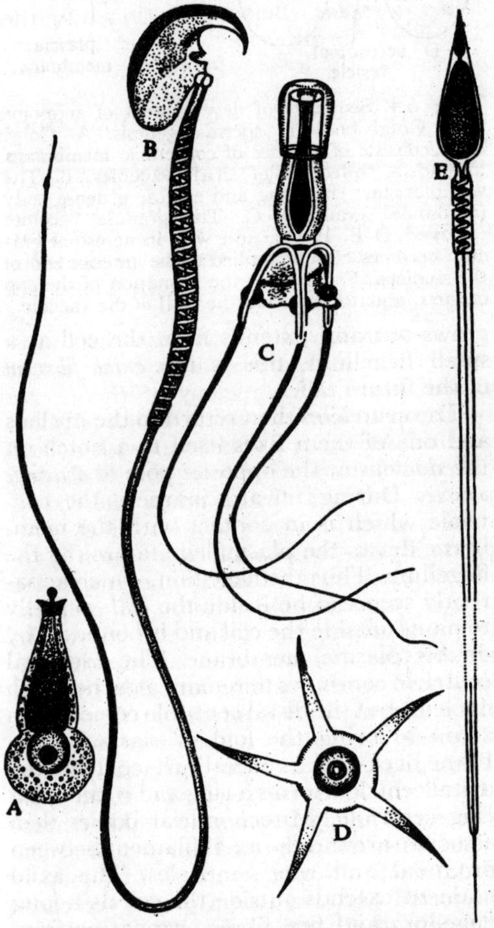

Fig. 5.5. Showing various types of spermatozoa. A. *Ascaris*. B. *Mus*. C. *Nephrops*. D. *Maia*. E. Man.

in mouse, etc. But in all cases, the sperm is built on same structural plan. The description of sperm cell given below is based on the structure of the sperm of rabbit (*Oryctolagus cuniculus*) which has been extensively studied by both light and electron microscopes. A spermatozoon of rabbit measures about 60–70 micra long with a head measuring about 8–10 micra.

A typical sperm consists of three distinct parts—*head*, *neck* or *middle piece* and *tail* or *flagellum* (Fig. 5.6).

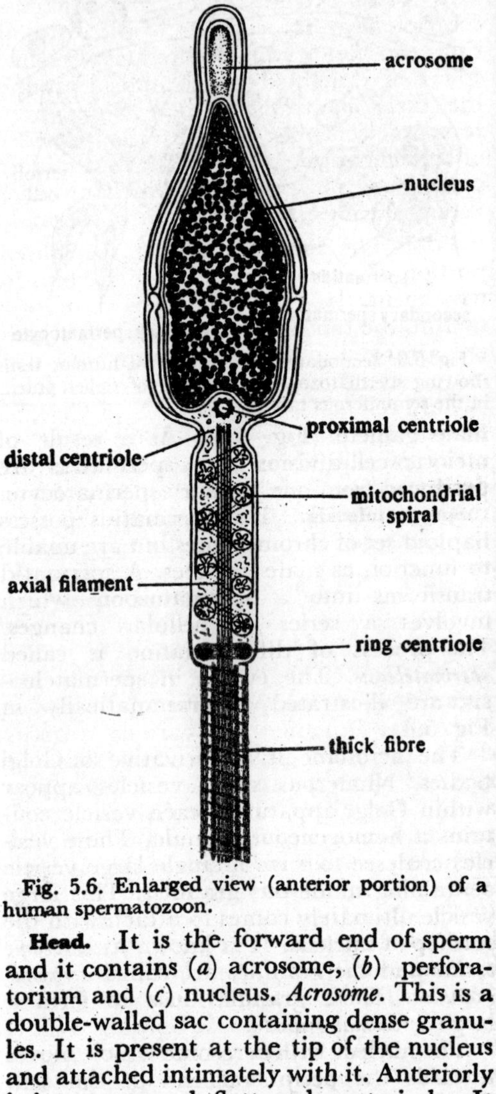

Fig. 5.6. Enlarged view (anterior portion) of a human spermatozoon.

Head. It is the forward end of sperm and it contains (*a*) acrosome, (*b*) perforatorium and (*c*) nucleus. *Acrosome*. This is a double-walled sac containing dense granules. It is present at the tip of the nucleus and attached intimately with it. Anteriorly it is convex and flattened posteriorly. It contains enzymes for breaking down the egg membrane during fertilization. *Perforatorium*. This structure is not present in rabbit and other mammals. But invertebrate sperm (e.g., Sea-urchin) contains perforatorium between acrosome and nucleus. It forms acrosome filaments and plays important role in fertilization. *Nucleus.* It forms the major part of the head and has a groove at its posterior end. It contains densely packed material within its nuclear membrane. The nucleus contains *DNA* and a protein called *histone*.

Neck or middle piece. The midpiece varies greatly in structure in different animals. This piece extends from the proximal centriole near the nucleus to the distal or ring centriole. It is formed by a firmly coiled spiral of elongated mitochondria around the axial filament and nine fibres from proximal centriole. Due to the presence of mitochondria, it is thought that middle piece is a 'power plant' which provides energy during the locomotion of sperm.

Tail. The sperm tail is a specialised portion of a spermatozoon which helps in movement. In the spermatozoon of sea-urchin, the tail is made up of fibres which are arranged in a circle of nine with two more extending down the middle. This pattern resembles that of cilia and flagella. In the spermatozoon of rabbit such an array of tail fibres is present which is designated as the *axial filament*. The fibres in the axial filament are actually the contractile units responsible for the movement of sperm tail. The sperm tail of rabbit has in addition another outer ring of nine more fibres. The fibres of the outer ring are much thicker than the inner fibres. The coarse outer fibres are not present up to the posterior tip, but end one after another at various regions down the tail. But the axial filament remains up to the tip. The axial filament and outer fibres are formed from the proximal centriole which remains attached with the posterior indention of the nucleus. The pattern of arrangement of fibres (9+9+2) is best developed in mammalian spermatozoa. The said pattern is encountered in other forms, viz. Honeybee (*Apis mellifera*), Fruitfly (*Drosophila melanogaster*), Grasshopper (*Gelastorrhinus bicolor*), Snail (*Helix pomatia*), Sparrow (*Passer montanus saturatus*), Snake (*Lampropeltis getulus*) and many others.

Characteristics of sperm cell. A sperm cell has the following characteristics:

1. It has a highly specialised structural organisation.
2. The major part is the nucleus and cytoplasm has modified to form special locomotory devices.
3. It is devoid of stored food and protective envelope.
4. Size is much smaller than egg, but is produced in greater number than the egg.
5. Its function is to seek the egg and to stimulate it to develop.

Oogenesis. The egg or ovum serves two functions—it carries the maternal chromosomes and at the same time it provides the ground substance on which the future embryo develops. To be prepared for these double purposes, the egg during its development undergoes complicated modifications which result—(1) production of haploid nucleus, (2) acquisition of food reserves and (3) preliminary organisation of the cytoplasm. The entire process is called *oogenesis*. Before entering into the details of the process it must be remembered that the process of oogenesis begins at a stage when the individual itself is in embryonic condition and ends either shortly before or immediately after fertilization.

Stages of oogenesis. The process of oogenesis has been worked out in different organisms and the results show close similarity in the process. In the present discussion the oogenesis of amphibians will be followed.

The gametogonium (here known as oogonium) which is destined to form egg is called *oocyte*. The nucleus of the oocyte swells up to form a vesicle which is called *germinal vesicle*. It is filled up with a fluid called *nuclear sap*, which contains sulphydryl protein. Meiosis begins when the individual is in embryonic state, but till the prophase stage of first meiotic division it remains suspended. During the suspended period, considerable amount of preparatory work happens within the oocyte. In the nucleus, chromosomes become dispiralised and appear as long and thin threads. Each chromosome contains at first single filament but later it becomes doubled. These filaments at short intervals contain paired thick, dense swellings called *chromomeres*. The chromomeres roughly correspond with the expected number of genes. From each chromomere a slender loop is projected outward to unite with the chromomere of other filament. The loop is enclosed by a matrix. The shape and size of the loop and also the matrix vary in different chromosomes. These loops are known as puffs and the chromosomes containing puffs are called lamp-brush chromosomes. Callan and Gall have demonstrated that puffs are constant for each pair of chromomere and are always found in the same position on the chromosome. The filament part of the puff is made up of DNA whereas the matrix contains RNA and protein. At the cytoplasm, two kinds of events occur. Establishment of physico-chemical differences occurs in the various parts of the egg cell. This is done by aggregation of different substances within the cytoplasm in graded fashion and mapping out of the different fields of development of the future embryo. Considerable amount of synthesis continues within the egg for making the reserves of food materials. The most important reserve is the yolk. This inert substance is composed of a phosphoprotein (called vitellin) and lipids (different fats).

Types of Egg. According to the quantity of yolk, the egg may be:

(1) *Alecithal*—Eggs having no yolk, e.g., mammals.

(2) *Microlecithal*—Relatively yolk-free eggs are called the microlecithal types. Eggs of coelenterates are of microlecithal types.

(3) *Megalecithal*—Large yolky eggs are called the megalecithal types. Examples: Eggs of reptiles, birds and monotremes.

(4) *Telolecithal*—Eggs are large in size and contain much yolk like megalecithal egg but the yolk pushes the cytoplasm at one end. Examples: Chick.

Factors controlling the cytoplasmic events. It is believed that the accumulation of different substances which happens within the egg is controlled by external and internal influences. **External influences.** (1) Immunological studies have shown that there exist similarities between blood protein and the cytoplasmic content of the egg (including yolk). (2) In the rapid growth phase, vitellin (which is characteristic of yolk) increases rapidly in the blood of the individual. It indicates that external factors are responsible for providing nutritive materials to the egg. (3) Cells investing the developing oocyte transfer complex molecules to the developing egg. These substances are either synthesized by them or transferred from the general circulation. **Internal influences.** Simple molecules enter within the egg and these are converted into more complex substances by the nucleus and cytoplasm. Callan and Gall have shown that RNA produced at the lateral loop of lamp-brush chromosome comes first to remain free in the nuclear sap and later passes through porous nuclear membrane into the cytoplasm. Considerable amount of protein is synthesized under the instruction of this RNA.

Regarding the events of growth during oogenesis, it may be said that the entire event is divisible into a *slow phase of growth* and a *rapid phase of growth.* The slow phase continues for a longer period during which synthesis of different substances (including yolk) continues within the egg. Rapid phase begins after the attainment of sexual maturity, when rapid accumulation of different substances occurs. It has been shown that during this phase many substances are contributed to the egg from general circulation. In the oocyte of almost all the vertebrates, the microvilli and pinocytotic vesicles are concerned mainly with the transportation of substances into the oocyte from the follicle cells (Fig. 5.7). The differentiation of cortical

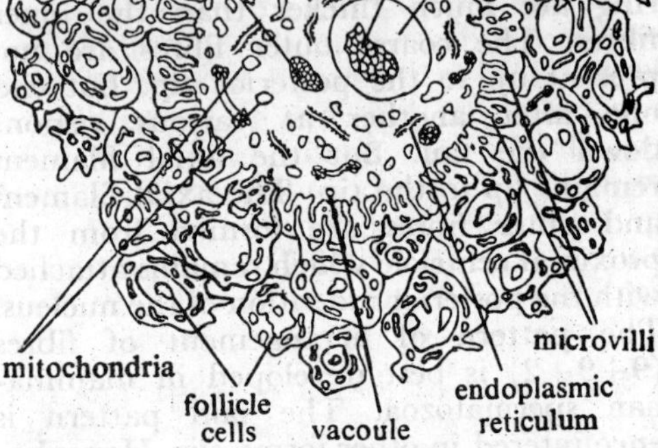

Fig. 5.7. Schematic diagram of a young mammalian oocyte and its associated follicle cells (after Bodemer).

granules (spherical bodies surrounded by a membrane) is also a significant event in egg development. The cortical granules are composed of acid mucopolysaccharides and are arranged in a layer adjacent to the plasma membrane. The cortical granules play vital role in fertilization.

Theories regarding the organisation of ovum. During organisation of egg, different substances get oriented in specific pattern, which plays important role in establishing the spatial arrangement of the developing embryo. Three different theories are forwarded to explain this establishment of determination.

(1) **Quantitative or gradient theory.** According to this view, one end of the egg possesses highest metabolic activity and controls the activity of other regions. This activity gradually reduces at the other end.

(2) **Qualitative theory.** According to this concept, there are actually different substances in the egg which control the formation of different parts of the embryo.

(3) **Two-factor theory.** Quantitative gradient exists within the egg in the beginning but after some time different parts of the egg become qualitatively different. Thus this view is a compromise between the first two theories.

In amphibians, when egg is fully matured the first meiotic division ends—this results into the formation of two haploid cells. But contrary to sperm cell formation the cytoplasmic division is unequal, i.e., a large cell called *secondary oocyte* and a very small cell called *first polocyte* are produced. The first polocyte remains attached with the outer surface of the oocyte. At this stage the egg is liberated from the body. After the entry of sperm cells, second division of meiosis ends. At this time the secondary oocyte divides unequally producing one large cell called *egg* and a small cell called *second polocyte*. The first polocyte divides equally. To sum up the result, it may be said that during oogenesis, single oocyte produces one large cell called egg (or ovum) and three small cells called polocytes. The egg is the functional reproductive cell while the polocytes are abortive in nature.

Structure of ovum or egg. The size of egg varies greatly in different animals. The size is chiefly attributable to the quantity of yolk present. The eggs of reptiles, birds and monotremes have, in addition, a coating of albumen. The albumen subserves both protective and nutritive functions. The egg is surrounded by a plasma membrane which encloses cytoplasmic part called vitellus. Within the vitellus, most prominent is nucleus which is 200–300 times larger than an ordinary cell nucleus. Nucleus contains a single large nucleolus rich in RNA and enveloping chromatin aggregates. Immediately beneath the plasma membrane, there is a thin layer of cytoplasm called cortex. In many eggs, the cortex contains special granules, called *cortical granules* which play important role in fertilization. The substance which constitutes the cortex is principally protein which has much influence on the future development. The cytoplasm contains various organelles: mitochondria, Golgi apparatus and reserve food material in the form of yolk. In some forms (e.g., Sea-urchin), the maturation of egg begins after the release from ovary and in others (e.g., Rabbit) maturation begins within ovary. Ovulation (i.e., release of ovum from the ovary) takes place at the metaphase stage of second meiotic division and completion of meiosis occurs immediately with the entry of sperm cell. The egg or ovum contains one or more membranes which differ in different animals. These egg membranes may be:

Primary membrane. It is formed by the ovum itself. It includes (*a*) *vitelline membrane*. It is a constant thin membrane which remains closely apposed with the plasma membrane before fertilization. (*b*) *Chorion*. In *Stylea*, a stiff chorion is found to be originated from the ovum itself. (*c*) *Zona radiata*. Generally a striated and less constant layer which—(*i*) In shark, bony fishes, amphibians and reptiles are present as a layer between vitelline membrane and plasma membrane. (*ii*) In birds it is difficult to differentiate. (*iii*) It is unstriated in mammals; here it is known as *zona pellucida* and it is debatable whether it is primary or secondary in nature.

Secondary membrane. It is formed by the cells of the ovary which remain immediately around the ovum. Examples: (*i*) chitinous shells of different eggs, (*ii*) membranes which are present immediately outside the frog's egg, (*iii*) cells of the graafian follicle in mammals. The secondary membranes are not seen in urodeles, reptiles and birds.

Tertiary membrane. These are formed by oviducts and special glands associated with the oviducts. These are always formed after the discharge of egg from the ovary and are of following kinds:

(*i*) *Albumen or Horny capsule* of the elasmobranch eggs. Albumen in birds and shell-membrane and calcareous shell of reptilian and avian eggs. It is to be noted that reptilian eggs contain little albuminous covering except in snake. (*ii*) *Tertiary membranes* are absent in mammals excepting prototheria where leathery shell is present.

Significance of egg membranes

(1) Primary membranes. Vitelline membrane plays important role in development. After the event of fertilization it prevents the entrance of more sperm cells and at the same time helps the egg to assume bilateral symmetry from a state of radial symmetry in unfertilized state. (2) Secondary membranes are chiefly protective and are also responsible for processing nutritive materials from the surrounding area. (3) Tertiary membranes meet one of the two requirements, (*i*) provision of organic food and (*ii*) adequate water supply. Eutherian and metatherian mammals have solved both the problems by the attainment of complete or incomplete viviparity. In other vertebrates, the first requirement is met by the storage of yolk or some other devices, but the second need gives a serious threat to the life of oviparous group. Amphibians, to avoid the threat of drought, return either to water for egg laying or developed certain devices, viz., (*i*) eggs deposited in a foam, (*ii*) formation of oviducal jelly and (*iii*) keeping the eggs in urinary bladder. Reptiles, birds and monotremes have developed waterproof shell for the purpose.

The transformation of the primordial germ cells into gametes occurs in three distinct steps—phase of multiplication, phase of growth and phase of maturation. The phase of multiplication is closely similar in both spermatogenesis and oogenesis, but the other two phases are quite different. The differences are summarised in Table Embryology–1.

ANALYSIS OF DEVELOPMENTAL EVENTS—FERTILIZATION

Fertilization is the process which marks the beginning of an individual. It involves the union of two gametes—sperm and ovum. The union involves a number of physical and chemical events which result into: (*i*) the *activation* of the egg and (*ii*) *amphimixis*, i.e., intermingling of parental hereditary traits in the offspring.

Prerequisite of fertilization

The essential prerequisites for fertilization are sperm from male and ovum from female. The ensurement of fertilization demands—(1) maturity of sperm and ovum and (2) coming together of these cells. The first demand is fulfilled during *gametogenesis* and second one is facilitated by the act of *sexual reproduction*. In many animals, i.e., Amphioxus, fishes and amphibians, the fertilization is external, in others like, reptiles, birds and mammals, the process is internal. In animals where internal fertilization occurs, number of devices are seen for the transfer of sperm cells within the gonoduct of the female.

TABLE EMBRYOLOGY–1

Spermatogenesis	Oogenesis
Growth phenomena in nucleus and cytoplasm are not pronounced in spermatocytes.	Growth phenomena are much pronounced in the oocytes. It involves both nucleus and cytoplasm.
Each spermatocyte produces four functional spermatids of equal size.	Each oocyte produces one large functional ootid and three small abortive polocytes.
During maturation, (*i*) Nucleus remains small which may contract during spermiogenesis into a compact elongated mass. (*ii*) Nuclear fluid is small in amount. (*iii*) Nuclear maturation takes place before cytoplasmic differentiation.	(*i*) Nucleus enlarges greatly. (*ii*) Nuclear fluid increases considerably. (*iii*) Nuclear maturation takes place simultaneously with cytoplasmic differentiation.

TABLE EMBRYOLOGY–1 (*cont'd.*)

Spermatogenesis	Oogenesis
Nucleus and cytoplasm become polarised along a long anteroposterior axis. Nucleus occupies almost the anterior end.	**Usually the nucleus is shifted towards the animal pole, while the cytoplasmic materials become polarised towards vegetal pole.**
Little quantity of food substances is stored within cytoplasm.	Considerable quantity of food substances (deutoplasm) is stored which is composed of carbohydrates, fats and proteins.
Considerable quantity of cytoplasm is discarded and some Golgi materials, centrioles and mitochondria are retained.	Cytoplasm is increased with the formation of basis cytoplasmic stuffs.
Specific membranes are not formed around the sperm.	Specific membranes are formed around the egg.
Capable of producing androgammic substances helping in fertilization.	Capable of producing gynogammic substances which help the fertilization process.
Active metabolic state is assumed before fertilization	Metabolic processes are slowed down before fertilization event.

State of gametes before fertilization

Ovum. After maturity, the egg is fully prepared with organised membranes, cytoplasmic particles and reserve materials. Physiologically the egg lies with suspended metabolic state and reduced glycogen content. Revival of metabolic activity occurs only after fertilization, which must happen within a specific period of time. If no fertilization occurs within that period, the egg undergoes degeneration.

Sperm. Sperm cells are adapted for locomotion. They are metabolically active and obtain their energy from both exogenous and endogenous pathways. Recently, it has been shown that synthesis of some protein takes place within the sperm.

Events in fertilization

The entire events in fertilization may be divided into three phases: A. *Penetration*, B. *Activation* and C. *Fusion of gametic nuclei*.

Penetration. During this event the sperm and ovum come in contact. The act involves a number of *physical* and *chemical* processes.

Physical process. Leeuwenhoek, for the first time, described the structure of sperm and inferred that development begins with the impregnation of ovum by sperm. F. R. Lillie (1919) advanced the view that chemical events accompany the physical events of sperm and egg contact. He showed that in sea-urchin, penetration is possible due to the interaction of two chemical substances—*fertilizin* by the egg and *antifertilizin* by the sperm. The fertilizin is a glycoprotein with a molecular weight of 82,000 or more. It contains a number of amino acids and a polysaccharide. The antifertilizin is composed of acid proteins. The fertilizins and antifertilizins combine in a specific way to establish an initial bond to facilitate penetration of the spermatozoon into an egg. The fertilizin-antifertilizin reaction is species specific and resembles antigen and antibody reaction. Tyler (1959) forwarded that during fertilization, sperm head is engulfed by the surface of the egg. He termed this phenomenon as specific pinocytosis. Colwin and Colwin (1961) made detailed study of events during penetration in the Enteropneust, *Saccoglossus* and a polychaete worm, *Hydroides*, under electron microscope. According to their observations (Fig. 5.8):

(*i*) Sperm head comes in contact with the vitelline membrane.

(*ii*) The plasmalemma of the sperm head

and the membrane of acrosome vesicle unite and split in the middle.

(*iii*) The acrosome granules come in contact with the vitelline membrane and produce lesion to it.

(*iv*) The acrosomal plasma membrane everts into one (in *Saccoglossus*) and many (in *Hydroides*) tube-like projections which approach the plasma membrane of egg through the gap of vitelline membranes.

(*v*) The plasma membrane of the egg sends one or several finger-like villi.

(*vi*) The opposing tubes—from egg and sperm unite and the acrosomal plasma membranes of sperm and egg become continuous, thus a connection is established between sperm and egg.

(*vii*) Sperm nucleus and centriole pass through these tubes inside the egg cytoplasm.

(*viii*) Later, plasma membrane of egg and acrosomal plasma membrane of sperm become a continuous membrane of zygote.

Similar events have also been recorded in toad (Fawcett and Burges, 1956), in cat (Fawcett and Burges, 1964), in chick (Nagano, 1962) and in man (Hartman, 1961).

It is to be noted that these studies have given a new insight to the meaning of fertilization. Formerly, it was believed that fertilization means only the fusion of nuclei, but now it is understood that it involves fusion of two cells.

Chemical process. Penetration is largely a chemical mechanism. Since Lillie forwarded the concept of fertilizin-antifertilizin, many workers provided more information about these chemical substances. Tyler regarded that the spermatozoon contains an enzymatic substance (*sperm lysin*) which causes local dissolution of egg membrane to make the path clear for sperm entry. But other workers like Hartman, Rothschild and Runnstrom have collectively called the substances as *gamones;* these gamones are regarded as the *active principles* involved in the process of fertilization. The gamones from sperm are called *androgamones* and those from the egg are known as *gynogamones.*

Precursor of the active principles. According to Brachet, RNA compound of low specific gravity which is present in high concentration at the outer side of the gametes are the precursors of the active principles.

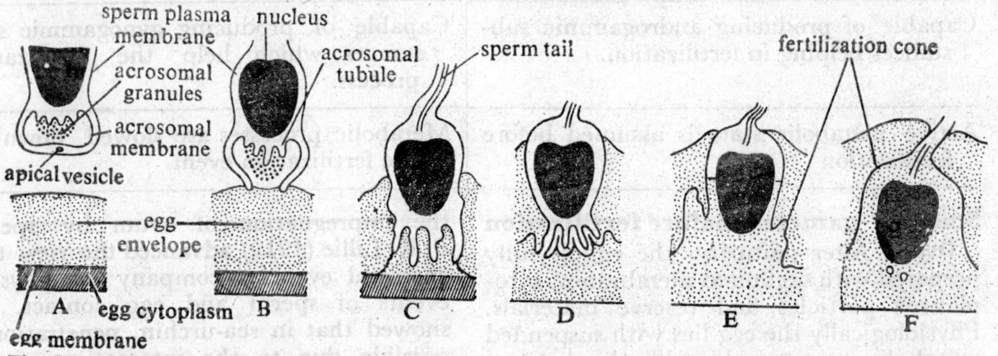

Fig. 5.8. Showing the events of the fusion of sperm and egg cells during fertilization. A. Approach of a sperm towards egg. B. Opening of the sperm tip after its contact with egg envelope. The acrosomal and sperm plasma membranes become continuous. C. The acrosomal wall is everted. D. Sperm and egg plasma membranes interdigitate and become fused (E) and fertilization cone arises. F. Sperm parts mingle with egg cytoplasm.

Chemical nature of active principles

(*a*) *Gynogamone* (I). It contains a bluish-red pigment called "*echinochrome—A*", which remains in association with proteins. It is responsible for liberating Gynogamone (II).

(*b*) *Gynogamone* (II). This is a non-dialisable lipase containing 5% nitrogen but without protein. It can be precipitated by ammonium sulphate, and if heated for two hours at 95°C., its property is lost.

(*c*) *Androgamone* (I). It is a colourless, nondialisable colloid, which is resistant to boiling and cannot be precipitated by ammonium sulphate. It gives a colour reaction with protein and is quickly inactivated by trypsin.

(*d*) *Androgamone* (II). This substance is yellowish in colour and soluble in sea-water. It is also resistant to boiling.

(*e*) *Androgamone* (III). It is dialisable, trypsin resistant and soluble in methanol.

Mechanism of actions of active principles. The success of fertilization depends upon the equilibrium of two sets of gamones. In addition to gamones, certain other substances are necessary for the success of fertilization. In case of mammals, these are: (1) *Hyaluronidase* and (2) *Glutathione*. Hyaluronidase, produced by the plasma membrane of the spermatozoa, helps to break the cells of enveloping layers and Glutathione maintains the metabolic stability of sperm cells. Successful fertilization thus depends upon the presence of specific quantity of sperm cells which provide the requisite quantity of these chemicals.

The sperm cells immediately after liberation, release androgamone (I), the production of it continues till the arrival of sperm to the attraction zone of the egg. It is responsible for retarding the excess activity of the sperm and thereby results into the conservation of energy. Almost at the same time the egg liberates Gynogamone (I). It excites the sperm movement and attracts them to the egg, by positive chemotaxis, the result is the oriented movement of sperms to the egg. Immediately after the touch of the first sperm with the egg the latter produces Gynogamone (II). It binds or agglutinates the sperm in contact and at the same time provokes the production of Androgamone (II) in other sperms which make them nonagglutinated. The Gynogamone (II) also initiates the restoration of metabolic activity in egg itself which is marked by the production of second polocyte. The Androgamone (II) produced in sperm helps it in agglutination with egg to penetrate by dissolving the mucous jelly, in others it neutralises the Gynogamone (II) and thus negates agglutination. Another gamone from the sperm, called Androgamone (III), obtained only from the sperms of sea-urchin, is believed to have the power of liquefying the egg membranes.

Activation. It means the awakening of a sleeping egg to a state of activity. Though many changes in the penetrated egg are well known, yet clear understanding of specific biochemical events which give the necessary stimuli of activation is abegging.

The first event of activation involves the cortical reaction and formation of fertilization membrane (Fig. 5.9). The vitelline membrane lifts itself from the cortical surface of the egg to form the perivitelline space. Subsequently dense lamellar materials (derivatives of cortical granules) are accumulated in the inner side of the vitelline membrane which is now called the *fertilization membrane*.

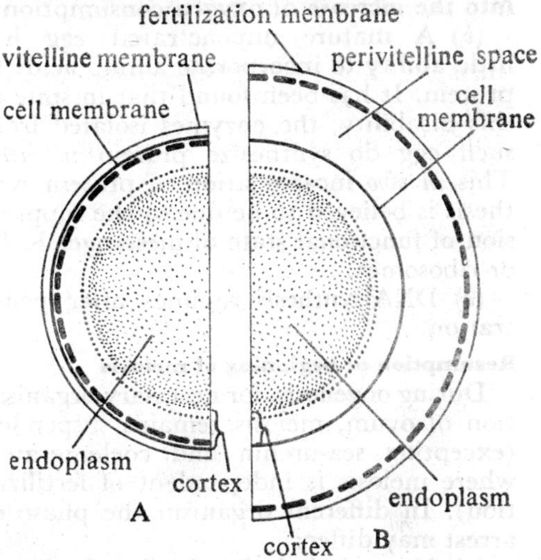

Fig. 5.9. Showing the cortical reaction and development of fertilization membrane (after Bodemer). A. Prefertilization stage and B. Postfertilization stage. Note that the large cortical granules (shown as dark broken lines beneath the cell membrane in A and beneath the fertilization membrane in B) become associated with the fertilization membrane while the smaller ones are associated with cell membrane.

Structural change of penetrated egg

(*a*) Cytoplasm moves in streamlined fashion towards the point of sperm attachment.

(*b*) Different components of the cytoplasm are reshuffled and reorganised.

(*c*) Some irregular amoeboid movements result into the final return to spherical shape.

(*d*) The refractive index of the cortex alters.

(*e*) Thickness of the membrane increases.

(*f*) The cytoplasmic part or vitellus reduces in volume.

(*g*) There is redistribution of calcium in cytoplasm which results into the change of viscosity in the cytoplasmic part of the cortex.

(*h*) Near the region of sperm attachment, a small pseudopodium is formed which engulfs the sperm head. This is fertilization cone. It was formerly thought that it is only the property of echinoderm eggs, but recent electron microscopic studies have shown that it happens in most of the eggs including mammals.

Metabolic changes of penetrated egg

(*a*) Within an unpenetrated egg, there is a cytochrome oxidase inhibitor, which is inactivated after fertilization, this results into the increase of oxygen consumption.

(*b*) A mature unpenetrated egg has little ability to incorporate amino acids to protein. It has been found that in spite of this disability, the enzymes isolated from such egg do synthesize protein *in vitro*. This *in vivo* incapaciation of protein synthesis is believed to be due to the suppression of functional state of messenger RNA or ribosomes.

(*c*) DNA synthesis resumes after penetration.

Resumption or beginning of meiosis

During oogenesis, for necessary organisation of ovum, meiosis remains suspended (excepting sea-urchin and coelenterates, where meiosis is independent of fertilization). In different organism the phase of arrest may differ:

(*a*) No meiosis before fertilization, e.g., *Ascaris*, Molluscs and Crustaceans.

(*b*) Arrested at the metaphase of first meiotic division, e.g., Nemertines, Insects.

(*c*) Arrested at the end of first meiotic division, e.g., amphibians. Whatever may be the cause of arrest, it continues and meiosis ends during the period of activation.

Block to polyspermy

In most eggs considerable changes take place during activation by which entry of more than one sperm is prevented. In the eggs of annelid, sea-urchin, starfish, Amphioxus and frog, it is done by the formation of a fertilization membrane.

Formation of fertilization membrane

In unpenetrated sea-urchin eggs, the vitelline membrane is thin and closely set with plasma membrane. Within the cortex the cortical granules remain enclosed in vesicles. Immediately after penetration, the cortical granules are released from the vesicles and the contents adhere to the inner side of the vitelline membrane. The vitelline membrane lifts up from the plasma membrane and this lifting is accompanied by the accumulation of fluid and a hyaline layer between vitelline membrane and plasma membrane. The vitelline membrane becomes thicker and is known as fertilization membrane.

Runnstrom (1952) has shown that fertilization membrane is formed by the separation of cortical and pigment granules.

The eggs of teleostean fishes have hard covering called chorion. An opening called micropyle serves as the passage through which only one sperm can enter at a time. Immediately after the entry of one sperm the cytoplasm of egg invades to close the micropyle.

In mammalian eggs, the prevention of the entry of more than one sperm is done by progressive changes in the zona pellucida, which is known as zona reaction.

In the eggs of many organisms, i.e., insects, elasmobranchs, urodeles, reptiles and birds, more than one sperm may enter within the egg but only one male pronucleus unites with the female pronucleus. The other sperm pronuclei help in the breakdown of yolk.

Fusion of the pronuclei. The process of fusion varies in different organisms—in some there is complete fusion, in others the chromosomal elements arrange themselves along the spindle fibres. The process which happens in frog will be described here:

(1) Sperm head and middle piece enter within the egg cytoplasm. Immediately after the entry, the egg completes its meiosis and loses its centriole.

(2) After the entrance, the sperm rotates 180° and thus the middle part comes in the front end. At this stage the sperm moves within the egg at random. This path is called *penetration path*.

(3) The egg pronucleus moves to the periphery and the direction of the sperm becomes oriented to it. The second path is called *copulation path*. Both the paths are trailed by marks of pigments.

(4) Considerable changes occur within the sperm pronucleus which swells up and the chromosomes become reticular in appearance. In the midpiece, the centrosome and centriole form the asters around them.

(5) The centriole of the sperm divides and moves to the poles forming fibres in between them, where the chromosomes of male and female pronuclei arrange themselves in the equator after the disappearance of nuclear membranes. It resembles metaphase plates of mitosis and soon results into cleavage.

Significance of fertilization

The process of fertilization has three important significances:

Ensures propagation. The union of haploid chromosomes and the attainment of diploid state helps the organism to propagate its own species. In lower organisms, the entire individual is involved, but in higher forms this is achieved through sexual reproduction and by the fusion of two specialized cells. Restoration of normal chromosome number occurs in a variety of ways. In some earthworms and planarians the chromosome number becomes doubled before meiosis. The meiotic division causes the adjustment of normal chromosome number.

Rejuvenates. The union of two cells provides some stimuli which are necessary for the vitality of the individual.

Produces variations. The process of gametogenesis results into the production of gametes which are dissimilar among themselves and to their parents. The recombination of these gametes, therefore, produces large number of variations which face the ordeal of natural selection. Fertilization thus ensures the proudction of variations.

Parthenogenesis

Large number of evidences are available which show that egg may start to develop without being impregnated by sperm. Such development is called *parthenogenesis.* Several agents (heat, cold, ultraviolet-ray and X-ray, hyper- and hypotonic salt solutions, mechanical agitation, electric currents) can stimulate development. The development of an unfertilized egg initiated by such artificial agencies is called *artificial parthenogenesis*. But in many invertebrates, viz., rotifers, aphids, bees, wasps, ants and some vertebrate species especially the domestic turkey, parthenogenetic development is a normal process. This phenomenon is called *natural parthenogenesis*. The parthenogenetic eggs develop mostly into males while the fertilized eggs develop into females. Parthenogenetic individuals are haploid in the initial stage but in number of cases it has been recorded that these haploid individuals become diploid by duplication of their chromosomes. Thus a diploid homozygous individual is formed.

Parthenogenesis reveals the fact that all the factors in development are present in the egg but it requires some external stimuli which are provided by the penetration of the spermatozoon during fertilization.

ANALYSIS OF DEVELOPMENTAL EVENTS—SEGMENTATION

Fertilization results into the formation of zygote. The process of segmentation (cleavage) immediately follows fertilization or any other process which activates the egg. Cleavage consists of division of the zygote into a large number of cellular entities. The cells which are produced during segmentation are called *blastomeres.* At first, the cells remain closely associated, but later on they form the lining of a hollow sphere called *blastula.* The blastula contains a cavity named *blastocoel* and its outer covering is designated as *blastoderm.* The formation of blastula culminates the cleavage period. The process of segmentation prepares the groundwork for the future design of the embryo by producing adequate number of cells. The cleavage also establishes the fundamental conditions for the initiation of next developmental stage —Gastrulation.

Different planes of cleavage

During early cleavage, distinct geometrical relationships exist between the blastomeres, i.e., each plane of cell-division bears a definite relationship with each other. The planes of division are:

Meridional plane of cleavage. When a furrow bisects both the poles of the egg passing through the median axis or centre of egg it is called meridional plane of cleavage. The median axis runs between the centre of animal pole and vegetal pole.

Vertical plane of cleavage. When a furrow passes in any direction (does not pass through the median axis) from the animal pole towards the opposite pole.

Equatorial plane of cleavage. This type of cleavage plane divides the egg halfway between the animal and vegetal poles and the line of division runs at right angle to the median axis.

Latitudinal plane of cleavage. This is almost similar to the equatorial plane of cleavage, but the furrow runs through the cytoplasm on either side of the equatorial plane.

Different types of cleavage

Considerable amount of reorganisation occurs during the period of cleavage and the types of cleavage depend largely upon the cytoplasmic contents. Different types of cleavage encountered in different eggs are catalogued below:

Holoblastic or total cleavage. When the cleavage furrows divide the entire egg. It may be:

Equal. When the cleavage furrow cuts the egg into two equal cells. It may be radially symmetrical, bilaterally, symmetrical, spirally symmetrical or irregular.

Unequal. When the resultant blastomeres become unequal in size.

Meroblastic cleavage. When segmentation takes place only in a small portion of the egg resulting in the formation of blastoderm, it is called meroblastic cleavage. Usually the blastoderm is present in the animal pole and the vegetal pole becomes laden with yolk which remains in an uncleaved state, i.e., the plane of division does not reach the periphery of blastoderm or blastodisc.

Transitional cleavage. In many eggs, the cleavage is atypical which is neither typically holoblastic nor meroblastic, but assumes a transitional stage between the two.

Effects of yolk in segmentation

The fertilized egg in most cases contains yolk, which are inert bodies. During division these bodies exert mechanical influences. In the egg of Amphioxus, the yolk is thin and remains uniformly distributed. Therefore the division is complete and early divisions occur at a very quicker rate. The amphibian egg contains yolk which is localised at the vegetal pole. Here division initiates from the animal pole and extends towards the vegetal pole, where the progress of cleavage slows down considerably. Consequently, the animal pole divides faster than the vegetal pole. The eggs of reptiles and birds are fully laden with large masses of yolk, thus restricting the cytoplasm and nucleus on the periphery as a circular disc on the animal pole. Here the lines of cleavage divide only the small animal pole region. Such effects of yolk on cleavage pattern influence the pattern of further development.

Mechanism of cleavage

The incidence of cleavage provides unique opportunity to study the mechanism of cell division and specially the role of different cell organelles during division. Opinions differ regarding the accumulation of force for the initiation of cleavage and following factors are believed to be responsible for controlling the cleavage: (1) Localised expansion of cortex. (2) Increased stiffness of the cortical cytoplasm. (3) Increase of tangential force activity in the cortex. (4) Contractile nature of the regions near the cortex and (5) Formation of new cell membrane from the subcortical cytoplasm.

Though the abovementioned factors are not clearly understood, it is evident that three structures present within the cell: Cortical layer, Spindle structures and Chromosomes play the important part. The energy which is required during the process is supplied by the metabolic activity of the developing egg. Besides the factors involved in segmentation, there are *cleavage laws* which govern the behaviour of the cells during cleavage.

Sach's Rules. The blastomeres tend to divide into identical daughter cells and a cleavage furrow tends to cut the previous cell at right angles.

Hartwig's Laws. The position of nucleus is vital and it tends to lie at the centre of the protoplasmic content of the cell. The nucleus exerts influence on cleavage. The long axis of mitotic spindle usually coincides with the long axis of the protoplasmic content. During cleavage the long axis of the protoplasm has the tendency to cut transversely.

Balfour's Law. The rate of cleavage is inversely proportional to the amount of yolk material present in the egg.

Chemical changes during cleavage

Significant chemical changes go on in the fertilized egg during cleavage. They are:

Increase of nuclear material: During cleavage a steady increase in nuclear material (predominantly DNA) is observed. Cytoplasm of the egg is the source of such nuclear material. Cytoplasmic DNA contained in mitochondria and yolk platelets are available.

RNA synthesis: During cleavage messenger RNA (mRNA) and transfer RNA (tRNA) are synthesised during cleavage, especially in late stages.

Synthesis of proteins: Throughout the period of cleavage there is steady and spectacular increase in protein synthesis.

Segmentation in different chordates

The pattern of cleavage differs in different animals. The following account will give an idea of the process of cleavage in different chordates.

Amphioxus. The cleavage in Amphioxus is typically holoblastic (Fig. 5.10). The first cleavage is *meridional*. The second cleavage is also *meridional* but at right angle to the first one. Four equal blastomeres are produced. The third cleavage is *latitudinal* and occurs slightly above the equatorial plane resulting in the production of eight blastomeres—four are smaller called the *micromeres* and four are larger known as the *macromeres*. The micromeres are situated towards the animal pole and the macromeres towards the vegetal pole. The fourth cleavage is *meridional* which involves all the eight cells resulting in the formation of eight micromeres and eight macromeres. The fifth cleavage planes are *latitudinal*. Each micromere is divided into an upper and lower micromere and each macromere likewise divides to form an upper and lower macromere. The fifth cleavage planes produce thirty-two blastomeres. The sixth cleavage planes are nearly *meridional* involving all the hirty-two cells resulting in sixty-four cells. At the 64-cell stage a conspicuous space is produced at the centre and this space becomes filled with a fluid. When the eighth cleavage planes take place, the blastula becomes pear-shaped and the blastocoel becomes large.

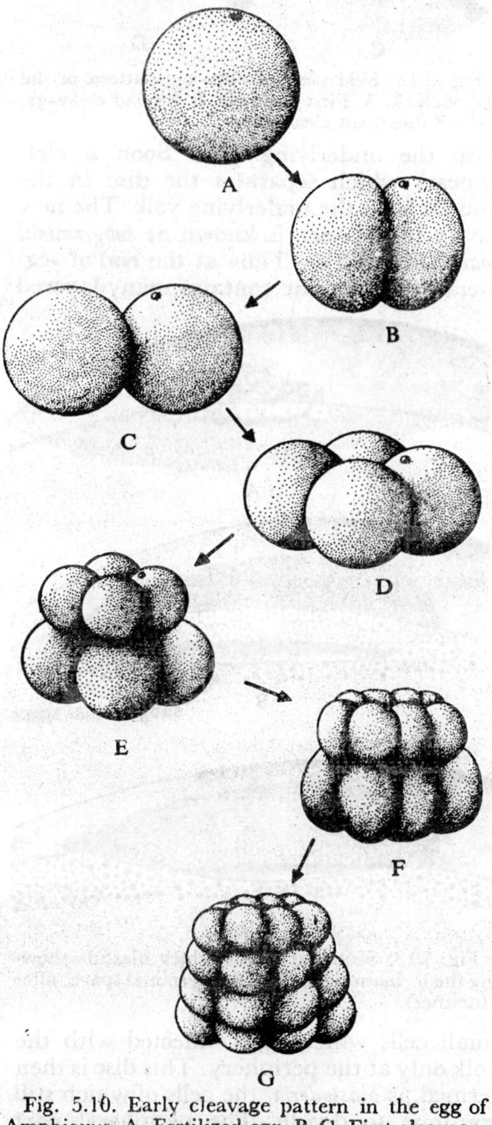

Fig. 5.10. Early cleavage pattern in the egg of Amphioxus. A. Fertilized egg. B–C. First cleavage. D. Second cleavage. E. Third cleavage. F. Fourth cleavage. G. Morula stage.

Frog. The egg of frog is telolecithal with a considerable amount of yolk localized towards the vegetal pole. The cleavage is holoblastic in nature, but differs considerably from that of Amphioxus because of larger quantity of yolk.

The first cleavage plane is *meridional* which occurs at about 3–3½ hours after fertilization. But the time depends largely on extrinsic factors. The first cleavage starts at the animal pole and gradually travels towards the vegetal pole. Thus the egg is bisected along the poles. Two blastomeres of equal size are produced. The second cleavage is almost *meridional* but oriented at right angles to the first cleavage plane (Fig. 5.11). The four blastomeres thus produced are not qualitatively identical, because the grey crescent material is present in two of the four blastomeres. Each blastomere contains dark pigment at the animal pole and yellowish yolk towards the vegetal pole. The third cleavage is *latitudinal* and occurs at right angles to previous cleavage planes but passes slightly above the equator. The furrow produces eight unequal blastomeres, four micromeres in the animal hemisphere and four macromeres in the vegetal part. The fourth cleavage planes are *meridional* which involve the micromeres first and pass on slowly towards the yolk-laden macromeres of the vegetal pole. In Amphioxus, the cleavages occur in a synchronous fashion, while in frog considerable degree of irregularities (asynchronism) appear in later stages. But it is certain that the micromeres always continue to divide at a faster rate than do the macromeres.

At the eight-celled stage, a small space makes its appearance between the four

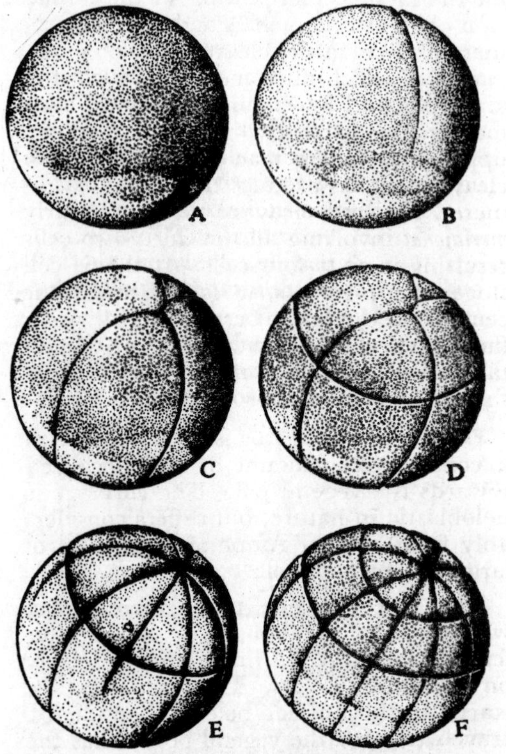

Fig. 5.11. Semidiagrammatic representation of the cleavage pattern in the egg of frog. A. Fertilized egg. B. First cleavage. C. Second cleavage. D. Third cleavage. E. Fourth cleavage. F. Morula stage.

micromeres. As development goes on, this space becomes conspicuous and forms the blastocoel. The floor of the blastocoel is formed of macromeres. The blastocoel (or segmentation cavity) is eccentrically located and becomes displaced towards the animal pole as development proceeds.

Chick. Typical meroblastic cleavage occurs in chick, where the segmentation activity is restricted only at the *blastodisc* or *germinal disc* (Fig. 5.12). Thus the cleavage is incomplete.

The first cleavage starts as a *meridional* furrow near the centre of the blastodisc at about $4\frac{1}{2}$ hours after fertilization when the egg reaches the isthmus of oviduct. This furrow cuts across the blastodisc and passes towards the vegetal pole but does not reach the pole. The second cleavage is also *meridional*, but approximately at right angles to the first one. The third cleavage is *vertical*. The fourth cleavage is also *vertical* but the division is not synchronous. As a consequence eight central cells encircled by twelve marginal cells are produced. From this point onward the cleavage becomes irregular and a disc containing smaller cells appears. This disc remains firmly connected

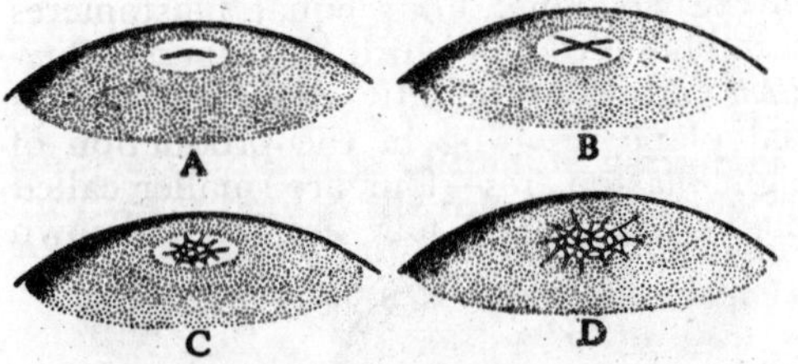

Fig. 5.12. Side view of cleavage pattern of the egg of chick. A. First cleavage. B. Second cleavage. C–D. Subsequent cleavages.

with the underlying yolk. Soon a cleft appears which separates the disc in the middle from the underlying yolk. The new cavity in between is known as *subgerminal space* (Fig. 5.13). Thus at the end of segmentation, the disc contains many-layered

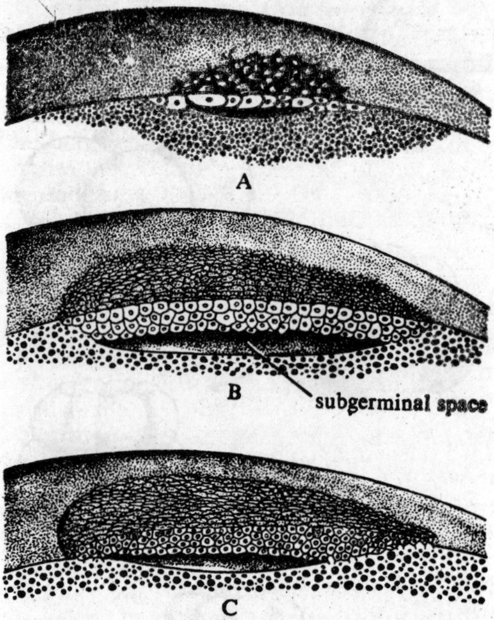

Fig. 5.13. Sectional view of chick blastula showing the formation (A–C) of subgerminal space (after Huettner).

small cells which are connected with the yolk only at the periphery. This disc is then termed as *blastoderm*, the cells of which still continue to divide. The peripheral part which lies in contact with yolk possesses granular cells called *area apaca* and the inner layer having clear portion is called *area pellucida*. At one end of area opaca, aggregation of cells takes place. This denotes the formation of future posterior side.

Rabbit. The egg of rabbit is small and does not contain any yolk (i.e. alecithal type of egg). The cleavage is holoblastic and nearly equal. Irregularities and asynchronism become the rule in the cleavage of rabbit like all other eutherian mammals.

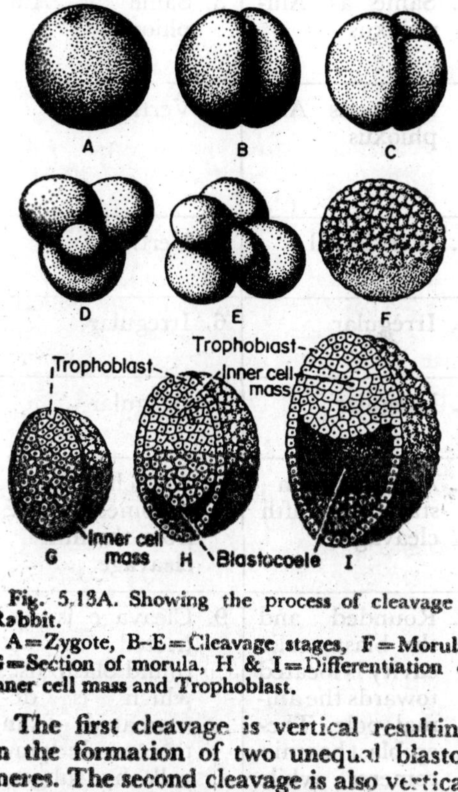

Fig. 5.13A. Showing the process of cleavage in Rabbit.
A=Zygote, B–E=Cleavage stages, F=Morula, G=Section of morula, H & I=Differentiation of inner cell mass and Trophoblast.

The first cleavage is vertical resulting in the formation of two unequal blastomeres. The second cleavage is also vertical but runs at right angle to the first. The third cleavage is horizontal but slightly above the equator. Subsequent divisions are rapid and irregular. The blastomeres thus produced become clustered together to form a solid cellular ball called *morula*. Two types of cells (small and large) are recognised in the morula. The large cells lie at the centre. Soon a cavity appears inside the cell mass on one side. The cavity gradually increases which shifts the central cell mass to one side. The stage is called blastocyst stage. The *inner cell mass* in the centre is attached with the outer cell layer (*trophoblast*) of the blastocyst. The cavity is called the *blastocoel* or *subgerminal cavity* (Fig. 5.13A) which is filled with a fluid. The inner cell mass remains attached at the *embryonic knob* towards the animal pole. From this embryonic knob, the embryo arises. The

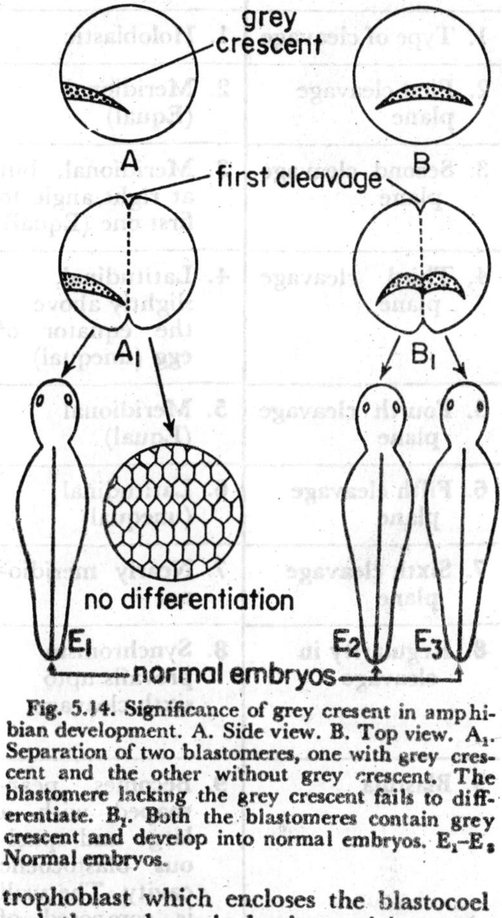

Fig. 5.14. Significance of grey cresent in amphibian development. A. Side view. B. Top view. A_1. Separation of two blastomeres, one with grey crescent and the other without grey crescent. The blastomere lacking the grey crescent fails to differentiate. B_1. Both the blastomeres contain grey crescent and develop into normal embryos. E_1–E_3 Normal embryos.

trophoblast which encloses the blastocoel and the embryonic knob participates in the formation of placenta. The trophoblastic cells overlying the embryonic knob is called *cells of Rauber*.

Importance of cleavage in embryonic pattern

The cleavage phase of development and blastulation are extremely significant, because the blastoderm is morphologically elaborated in such a way that the important presumptive organ forming areas of the future embryo are segregated into definite districts of the blastoderm. Such orientation of the organ forming areas in the blastoderm permits an ordered movement of these areas during gastrulation to take up their fateful position. So the period of cleavage and blastulation is regarded as

TABLE EMBRYOLOGY–2

Comparative account of cleavage in Amphioxus, Frog and Chick

Points	Amphioxus	Frog	Chick
1. Type of cleavage	**1. Holoblastic**	1. Holoblastic	1. Meroblastic
2. First cleavage plane	**2. Meridional (Equal)**	2. Meridional (Equal)	2. Meridional (Equal)
3. Second cleavage plane	**3. Meridional, but at right angle to first one (Equal)**	3. Same as Amphioxus	3. Same as Amphioxus
4. Third cleavage plane	**4. Latitudinal, slightly above the equator of egg (unequal)**	4. Same as Amphioxus	4. Vertical
5. Fourth cleavage plane	**5. Meridional (Equal)**	5. Meridional (Equal)	5. Vertical
6. Fifth cleavage plane	**6. Latitudinal (unequal)**	6. Irregular	6. Irregular
7. Sixth cleavage plane	**7. Nearly meridional**	7. Irregular	7. Irregular
8. Regularity in cleavage	**8. Synchronism prevails upto sixth cleavage.**	8. Asynchronism starts from fifth cleavage.	8. Asynchronism becomes the rule from the fifth cleavage
9. Blastula	**9. Becomes pear-shaped with a large and spacious blastocoelic cavity. The wall is composed of one layer of cells.**	9. Rounded and the blastocoelic cavity is located towards the animal pole. The epiblast has micromeres while the hypoblast has macromeres.	9. Cleavage is restricted to a round blastodisc, which is demarcated from the underlying yolk by subgerminal space. The blastodisc is divided into two zones—(*i*) area pellucida and (*ii*) area opaca.

the phase of preparation for future differentiation.

The cells which are produced at the end of segmentation resemble the zygote—but do they possess the same potentiality as the zygote itself. Driesch (1891), in order to get an answer, separated the two blastomeres at the two-celled stage and found that both the blastomeres developed into complete embryos. His conclusion was that each blastomere has the full potentiality to be an entire embryo. But in 1900, Roux showed that if one of the blastomeres of the two-celled stage is killed, the remaining one produces 'half embryo'. He claimed that each cleavage results into the segregation of specialization in the blastomeres and this is irreversible. This experiment demonstrates that an organising or controlling centre is elaborated to control the development process. The experiment of Spemann and others have shown that it is the grey crescent region which plays the vital role in the process of determination and the blastomeres which are formed due to segmentation are neither completely regulative nor irreversibly determined. Fig. 5.14 shows the importance of grey crescent in the development of amphibian embryo. It has been experimentally established that the grey crescent in the amphibian blastula transforms into the dorsal lip of the blastopore which acts as an instigator and controller of the gastrulation process.

ANALYSIS OF DEVELOPMENTAL EVENTS—GASTRULATION

The blastula passes into the stage called gastrula by the process—Gastrulation. This process is extremely important in the ontogenetic process of an animal, because the blue-print of the future organisation is laid down during this phase. During this crucial and dynamic process major presumptive organ-forming areas of the blastula become reorganised in a fashion that allows their ready transformation into the fundamental body plan of a species. Gastrulation is essentially a process of migration of cells from one place to the other in the embryo. Besides movement of cells, considerable nuclear differentiation also takes place. In almost all animals it results in (*i*) the establishment and differentiation of three primary germinal layers—ectoderm, mesoderm and endoderm, (*ii*) the establishment of nuclear differentiation and (*iii*) the beginning of the control of genetic factors over development.

Basic mechanism in gastrulation

The process of gastrulation involves following three cellular activities, *cell-movement*, *cell-contact* and *cell-division*. All these mechanisms are carried in a nicely co-ordinated and integrated way. Number of factors are believed to be responsible for this coordination, but it has not been possible to pin point the final answer. It is undeniable that this process is controlled largely by intrinsic factors which are correlated with the external as well as internal conditions.

Methods used to study gastrulation

The correct observation of incidences during gastrulation was started from the findings of W. Vogt in 1923. Vogt used vital dyes (Janus green and Neutral red) to mark the cells in an early gastrula and noted that cells during gastrulation actually migrate from one place to the other. The vital dye technique of Vogt resulted into the application of several other methods: (1) Visible differences in the cytoplasmic particles were used as natural marker, (2) tagging of the cells with carbon particles and (3) tagging of the cells with radioactive substances.

Morphogenetic movement of cells

During gastrulation, cells from one region of embryo move to another to take up their future fateful position. Two terms, *emboly* and *epiboly* which are quite opposite in their meanings, are generally applied to explain the process of movement. Emboly means the throwing in or insertion of cells and epiboly signifies the extending upon. The movement of cells establishes a particular form and involved in organ formation in embryo—so this movement is designated as the *morphogenetic movement*. Fig. 5.15 shows the movement of cells in gastrulation.

Fundamentally, the morphogenetic movement is similar but the details of the process vary greatly. Following types of cells movement occur:

Epiboly. It involves the extension along the anteroposterior axis and peripheral divergence.

Emboly. The inward movement of cells is classified into different types depending on the behaviour of migrating cells. These are: (1) *Invagination*. It denotes the infolding of a layer of cells to form a cavity encircled by infolded cells. Generally in the gastrulation of Amphioxus and frog, the wall of the blastoderm is pushed inside the blastocoel. This creates a new cavity called the archenteron which communicates with the exterior by a blastopore. This process of inpushing goes on and the inpushed layer forms the walls of the cavity. The archenteron (or primitive gut) completely obliterates the blastocoel. (2) *Involution*. It implies the inward rotation of cells as seen in the gastrulation of amphibian and avian eggs. From one end near the edge of the blastoderm, the cells begin to move inwards to form the inner lining of the blastoderm. (3) *Convergence*. It means the movement of cells to a particular region of the gastrula. In amphibian egg, the migration of cells to the external edge of the blastoporal lip is designated as convergence. The same phenomenon of convergence of cells is seen in the formation of primitive streak in chick embryo. (4) *Divergence*. This phenomenon is opposite to convergence, when involuted cells diverge to take up their future positions inside the gastrula. (5) *Infiltration*. During this process, cells of the blastoderm infiltrate near the bottom of the blastocoel to form a second layer as seen in the gastrulation of chick. (6) *Delamination*. This is a process of separation of a

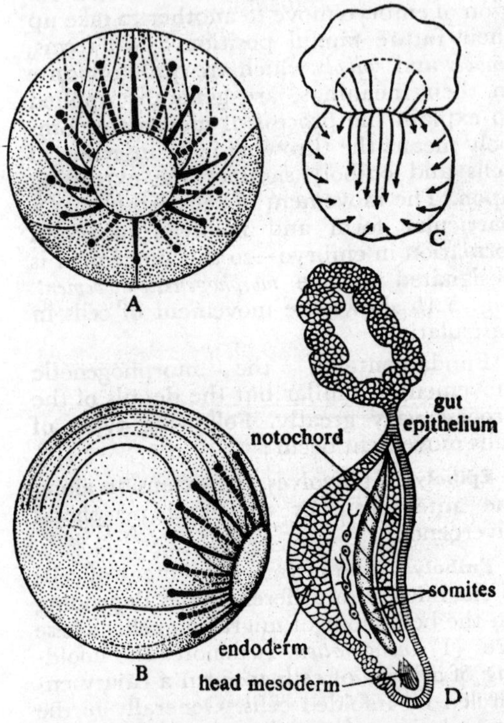

Fig. 5.15. Showing morphogenetic movement in amphibian gastrulation. A–B. Normal gastrulation. C–D. Exogastrulation. Note the failure of formation of nervous structures in the sectional view (D) of the embryo. Heavy lines indicate the movement of cells on the surface of the embryo while the thin lines show the movement inside the gastrula in A and B.

group of cells from others to form discrete cellular masses. (7) *Extension*. The elongation of presumptive areas after they have moved inside the embryo is called the extension. (8) *Cell proliferation*. It means the increase in the number of cells during gastrulation. (9) *Concrescence*. It is similar to convergence. The cells from two sides migrate anteriorly along one axis, but in convergence the cells from two sides unite together and then move anteriorly.

The above terms are coined for the convenience of analysing the events in gastrulation. Recent observations have established that it is essentially a phenomenon of integration. It was, therefore, felt necessary to understand the whole process for a meaningful comprehension of individual event.

GASTRULATION IN DIFFERENT CHORDATES

AMPHIOXUS: The blastula of Amphioxus contains the potential endodermal cells at the vegetal pole, i.e., *hypoblast* which forms the floor of the blastula. The presumptive organ forming cells (i.e., notochordal, mesodermal, epidermal, etc.), form the *epiblast*. The epiblast constitutes the roof of blastula. The blastocoel is large. The *dorsal crescent* (presumptive neural and notochordal cells) lies in the future dorsal lip region of the blastopore while the *ventral crescent* (mesodermal area) occupies the ventral lip.

With the onset of gastrulation, an increase of mitotic activity is observed in the dorsal and ventral crescent regions. With the activity of the different cells, the endodermal plate invaginates into the blastocoel. During this process of invagination, the dorsal portion moves at a faster rate to touch a point which marks the anterior end of the developing embryo. The notochordal cells, occupying the mid-dorsal region of the blastopore, involute and occupy a mid-dorsal position in the developing archenteron. Then the ventral crescents gradually converge on either side of the notochordal cells. Thus the roof of the archenteron is composed of mesodermal and notochordal cells. This process of emboly is accompanied by epiboly when the ectodermal and neural cells extend along the antero-posterior direction. The extension of ectodermal cells and the proliferation, involution and infolding of presumptive endodermal, notochordal and mesodermal cells result in the formation of a double-layered embryo (Fig. 5.17). The external layer forms the ectoderm. The internal layer has a dorsomedian area of notochordal cells with two bands of mesodermal cells. The rest of the inner layer is formed of endodermal cells. Rapid cell proliferation, accompanied by emboly and epiboly causes an anteroposterior elongation of the gastrula.

As the developing gastrula elongates in the anteroposterior direction, the ventral crescent is gradually shifted dorsalward

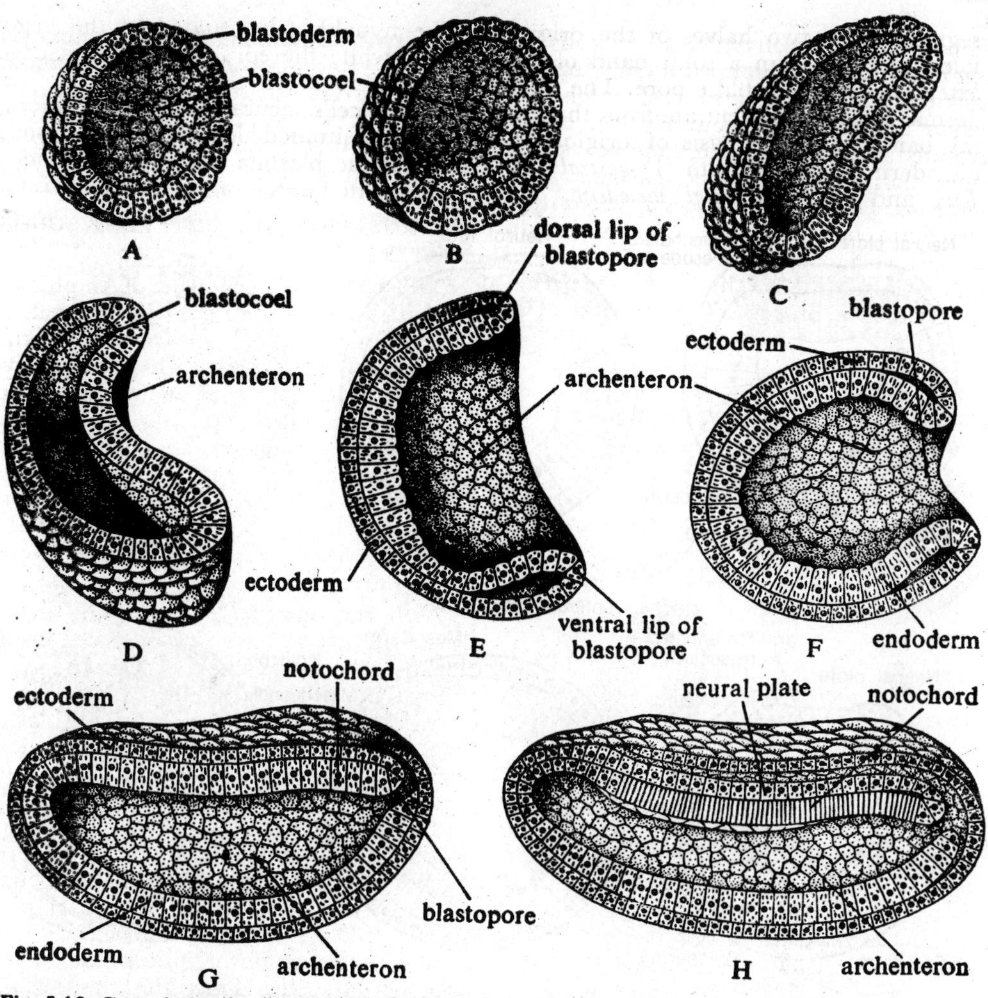

Fig. 5.16. Gastrulation in *Amphioxus*. A. Blastula. B–D. Beginning of invagination. E. The embryo becomes a double-walled cup with broad blastoporal opening. F. Constriction of blastopore. G–H. Completion of gastrulation. A–H. Sectional views.

along the inner lateral side of the blastoporal lip. As a result of convergence, the mesoderm comes to lie on the two sides of the notochordal material at the dorsal blastoporal lip. At the end of gastrulation the blastopore becomes smaller and is closed by ectodermal overgrowth. A neuroenteric canal is formed between the archenteron and developing neural tube.

Mesoderm differentiation. The transformation of the neural plate to form the neural tube is associated with the formative of a shallow groove on either dorsolateral walls of archenteron. The cells forming these two grooves are smaller than other cells. The grooves become deeper and their edges come together. Such fusion results in the separation of a solid notochordal rod along the mid-dorsal line. These two lateral grooves become divided by transverse partitions into *enterocoelic pouches* which grow between the endoderm and ectoderm (Fig. 5.17). The cavities of these pouches retain their connection with the archenteron at the beginning which become subsequently lost. As a consequence paired hollow blocks of mesodermal cells are formed. Formation of hollow mesodermal blocks is observed only in the first two pairs of somites. The posterior enterocoelic pouches are pinched off as solid blocks of mesodermal cells within which coelomic cavities are formed anew. This process is observed upto fourteenth pairs of somites. In the rest of the posterior-segments, the two halves of the original folds meet to form a solid band of cells extending up to the blastopore. The mesodermal somites differentiate from the lateral bands. On the basis of origion the mesoderm is divided into (1) *gastral mesoderm* and (2) *peristomial mesoderm*. The gastral mesoderm develops from the enterocoelic pouches, while the peristomial mesoderm differentiates from the lateral bands.

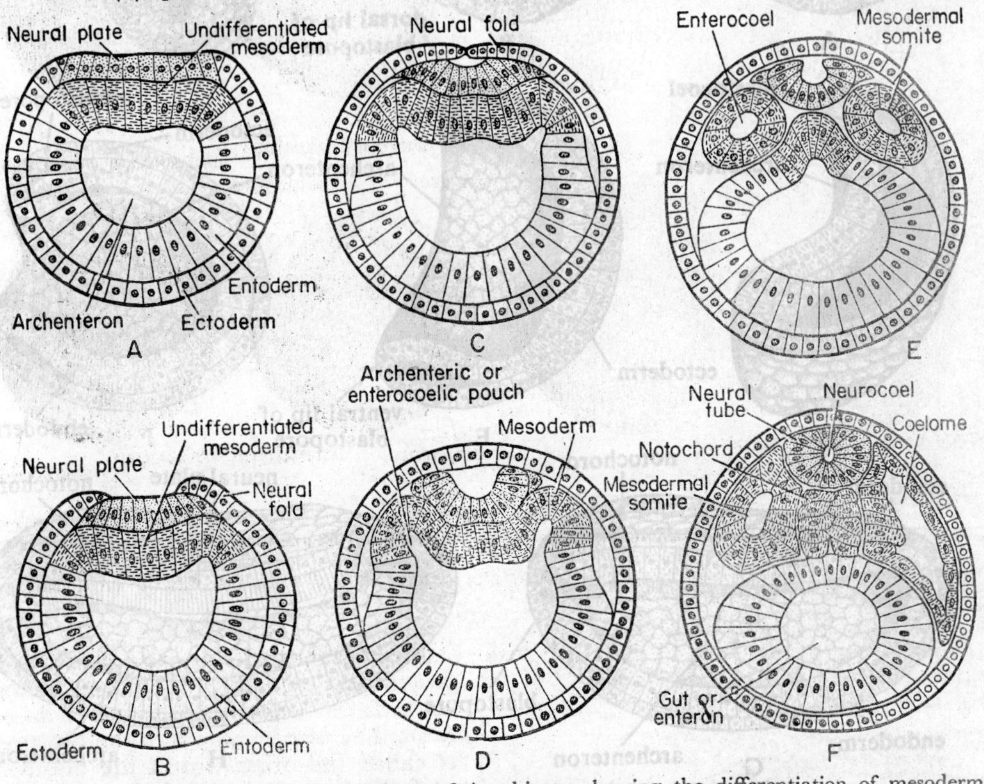

Fig. 5.17. Transverse sections of embryo of Amphioxus showing the differentiation of mesoderm. A–B. Gastrula stage. C–F. Post-gastrula stages. Note the early phase of formation of mesodermal somite on the left and coelome on the right side (after Huettner).

The somites or segmental mesoderms gradually grow ventrally on either side until they meet in the midventral line below the alimentary canal. The mesodermal sheet becomes double-walled enclosing coelome within themselves. The lateral plate mesoderm becomes thus splitted into (*i*) *somatic mesoderm* in association with ectoderm and (*ii*) *splanchnic mesoderm* in association with the endoderm.

FROG. In late amphibian blastula, the presumptive organ forming areas are oriented around the blastocoelic cavity. The hypoblast is situated at the vegetal pole, while the epiblast is located at the animal pole. In the epiblast the notochordal cells, neural plate and epidermal areas are situated along the anteroposterior axis of the blastula with the notochordal cells located at the most posterior position.

At the end of the cleavage all the blastomeres remain stationary and none of them have shifted from its original position. But at the onset of gastrulation a great mass migration started to occupy their definite position in the developing embryo. Gastrulation begins with the appearance of a small cleft-like invagination at one side and just above the grey crescent (Fig. 5.18). This cleft-like invagination is crescent-shaped and represents the dorsal lip of the blastopore. As gastrulation progresses the crescent-shaped cleft continues to expand to assume a semicircular appearance, then becomes horse-shoe-shaped and finally forms a ring. This ring represents the blastopore. The blastopore becomes the focal point for gastrulation activities.

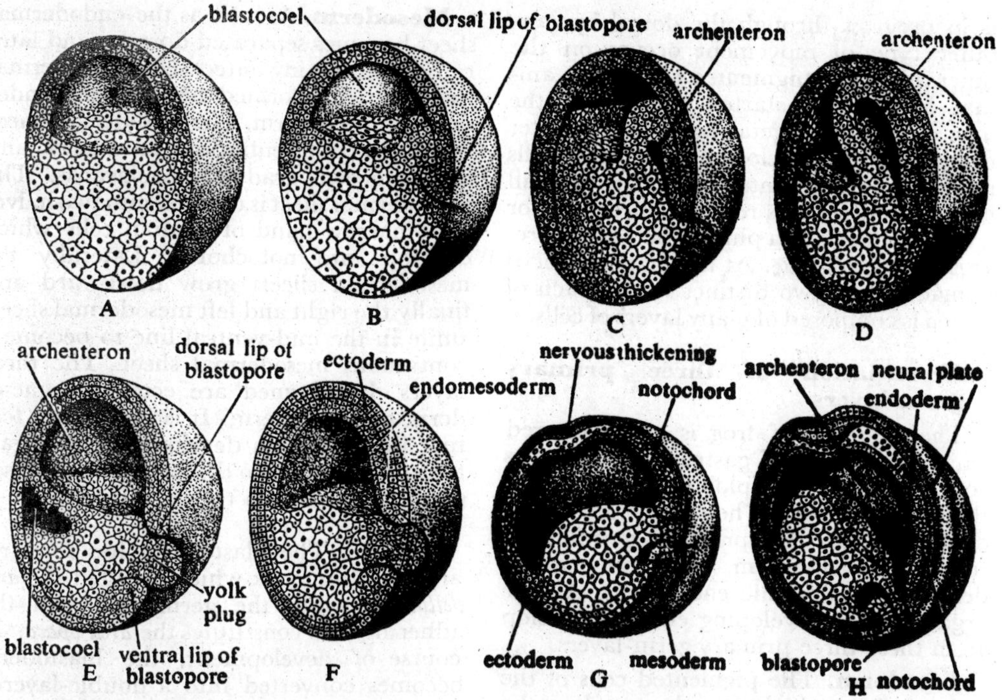

Fig. 5.18. Semidiagrammatic sectional views of the gastrulating embryo of Frog. (after Huettner). A. Blastula. B. Formation of blastopore. C–E. Enlargement of archenteron and gradual obliteration of blastocoel. F. Total obliteration of blastocoel and formation of endomesoderm. G–H. Completion of gastrulation showing the initiation of neural plate, notochord and differentiation of mesoderm as a middle layer.

Migration of cells inside the gastrula starts along the newly-formed dorsal lip of blastopore and this inward pushing is caused by the endodermal cells which are folded inward (Fig. 5.19) and forward towards the future anterior end of the embryo. The upper margin of the blastopore is called the dorsal lip of the blastopore and the lower edge is designated as the ventral lip of the blastopore. As invagination expands within the blastocoel, the prechordal plate cells from the upper part of the dorsal side move inward. The new cavity thus produced is called the archenteron which communicates to the exterior by the blastopore. With the further advancement of invagination, the archenteron continues to expand by obliterating the blastocoel. The inward moving cells form a new border beneath the outer cells. The roof of the archenteron consists of the involuted layer which includes the endoderm and mesoderm. Beyond this layer lies the ectodermal layer. The floor of the archenteron is made up of a layer of endodermal cells, the derivatives of the large yolk cells which were located in vegetal hemisphere of blastula.

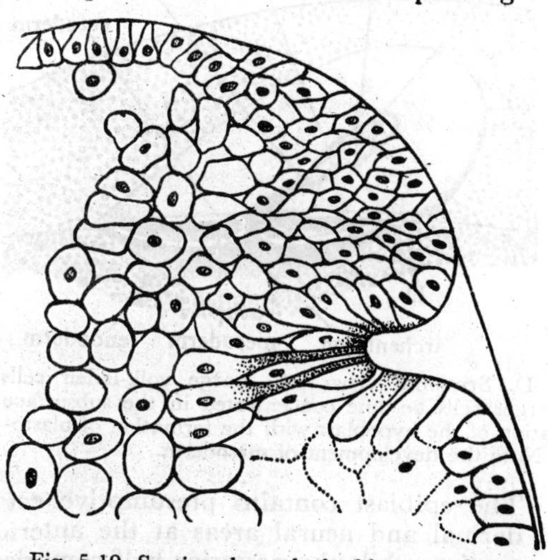

Fig. 5.19. Section through the blastoporal region of an early amphibian gastrula showing the streaming of blastoporal cells into the interior. As invagination begins, certain cells retain their attachment with the surface of the egg but become bottle-like in appearance.

When the inward movement of the cells

is in progress through the dorsal lip, another type of movement occurs on the outer side. The pigmented cells of the animal hemisphere started to enclose the macromeres of vegetal hemisphere. After completing the enclosure, the outer cells reach up to the ventral lip of it. A small mass of macromeres remains uncovered for a while and acts as a plug of the blastopore. It is called *yolk plug*. At this stage, embryo is made up of two distinct strata, each of which is composed of many layers of cells.

Differentiation of three primary germ-layers

The blastula of frog is mono-layered which in course of gastrulation becomes converted into a triploblastic stage, i.e., three cell-layered. These three layers are designated as the primary germ-layers (embryonic ectoderm, embryonic mesoderm and embryonic endoderm). All the organs of the developing embryo develop from these three primary germ-layers.

Ectoderm. The pigmented cells of the animal pole, which spread to enclose the macromeres of the vegetal hemisphere become differentiated into ectoderm.

Endoderm. The dorsal and lateral sheets of cells which form the roof of the archenteron represent the endoderm as well as mesodermal material. Upon completion of gastrulation, the roof and sides of the archenteron become lined by a single layer of endodermal cells which have differentiated from the involuted several celled thick archenteron roof.

Mesoderm. As soon as the endodermal sheet becomes separated dorsally and laterally from the involuted cells, mesodermal sheet is being formed between the endoderm and ectoderm. The mesodermal sheet starts its differentiation anteriorly and then proceeds gradually backwards. The mesodermal sheet is divided into two halves by a narrow band of median cells which develop into notochord. Laterally the mesodermal sheets grow downward and finally the right and left mesodermal sheets unite in the mid-ventral line to become a continuous mesodermal sheet. The three layers thus formed are ectoderm, mesoderm and endoderm. It is the special feature in amphibian development that gastrulation results into the formation of mesoderm first and then the endoderm.

CHICK. The blastoderm has a central area free from yolk which is called the *area pellucida*, while the germ-wall with the adhering yolk constitutes the *area opaca*. In course of development, the blastoderm becomes converted into a double-layered structure—the upper one is the *epiblast* and the lower layer is called the *hypoblast*. The space between these two layers is called the blastocoel while the space below the hypoblast is the primordial archenteron.

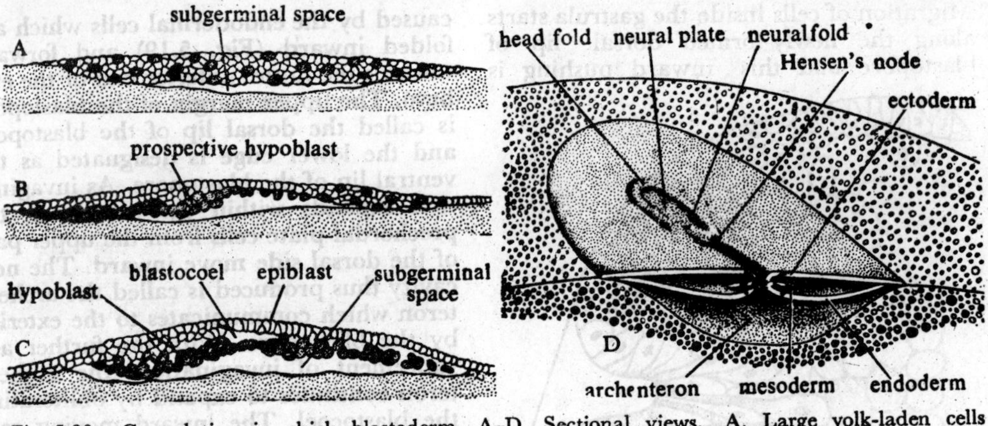

Fig. 5.20. Gastrulation in chick blastoderm. A–D. Sectional views. A. Large yolk-laden cells remain intermingled with smaller cells. B. The larger cells become concentrated in the subsurface position to form the prospective hypoblast. C. Separation of the hypoblast with the formation of blastocoel. D. Transverse section of primitive streak stage. Note the development of mesoderm.

The epiblast contains presumptive ectodermal and neural areas at the anterior portion while the posterior half comprises of presumptive notochordal and mesodermal cells. The hypoblast transforms into the endoderm and the epiblast is converted into ectoderm and mesoderm (Fig. 5.20).

At the initiation of gastrulation, the hypoblast cells from the posterior end start migrating towards the anterior end of the embryo along the median line. Immediately after the inauguration of the movement in the hypoblast, the cells of the epiblast overlying the migrating hypoblast move downward towards the hypoblast. These involuted cells occupy a position between the epiblast and hypoblast and migrate to the lateral and anterior ends between the epiblast and hypoblast. Movement of cells in the blastoderm of chick during gastrulation has been studied by Spratt (1946) by carbon particle technique.

With the activities of the epiblast and hypoblast, the presumptive mesoderm cells from the posterior half of the epiblast move posteriorly and converge from the lateral sides towards the median line. These converging cells begin to accumulate at the posteromedian border of the area pellucida as a raphe-like thickened structure. This marks the appearance of the primitive streak (Fig. 5.21). The migratory cells after coming to this region move inwards and migrate anteriorly and laterally. The migratory cells of the epiblast move downward and occupy the position between the epiblast and hypoblast. These cells then diverge anteriorly and laterally as a broad middle layer of mesodermal cells along the primitive streak. During its forward movement, it approaches the presumptive notochordal areas. With the anterior movement the streak also starts to move backward. The primitive streak is fully formed at about 18–19 hours after incubation. Gradually the area pellucida changes from a round to pear-shaped appearance. The primitive streak represents the posterior region of the developing embryo and the embryo proper develops anterior to it. It is also an area of cell proliferation and rapid growth. The primitive streak becomes very conspicuous in early embryonic life. It consists of a groove (*Primitive groove*) which is flanked on both the sides by two ridges (*Primitive ridges*). It terminates anteriorly in a *primitive pit*, and posteriorly in a *primitive plate*. Immediately anterior to the primitive pit (which represents the defunct neuroenteric canal) lies an elevation, *Hensen's node* or *Head process*. In this area the mesoderm becomes thickened and projects from the primitive streak. With the formation of the Hensen's node, the primitive streak regresses posteriorly and the major organ forming areas become well established. A groove appears on the outer surface of the head process and the two folds unite to form a tube. The formation and closure of the groove continue posteriorly. The entire process may be compared with the action of a zipper. As the closure of the groove comes to the posterior end, the backward movement ceases leaving an opening at the posterior end.

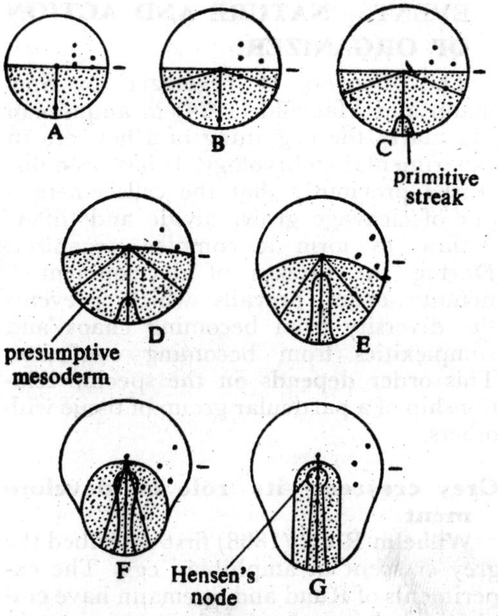

Fig. 5.21. Migration of cells (A–G) of the blastoderm to form the primitive streak in chick (Bodemer).

In the gastrulation of chick, the mesoderm differentiates lastly from the epiblast by the process of involution, elongation, expansion and extension. The hypoblast gives rise to endoderm and the epiblast differentiates into ectoderm and mesoderm.

MAMMAL. In the mammalian blastula (Blastocyst) the formative area (germ disc) is restricted at one end. The germ disc is composed of epiblast and hypoblast. In the embryo of pig, the gastrulation activities are observed at two centres—the posterior end forms the primitive streak while the anterior end forms the Hensen's node. The behaviour of these portions is almost similar to that observed in the gastrulation of chick. The mesodermal cells from the primitive streak move between the epiblast and hypoblast and form two winglike areas. The mesodermal cells are divided into (*i*) embryonic mesoderm confined to the germ disc and (*ii*) extra embryonic mesoderm.

Casual analysis of gastrulation mechanism

Lillie (1913) established that the surface layer of the egg at first remains plastic but in course of development it loses its plasticity and becomes rigid. Spemann (1918) found that up to gastrulation, when the eggs are cut into two halves, each will form a complete embryo. But after gastrulation, each half gives rise to half embryo. He also noticed that the half containing blastopore forms a complete embryo. He came to the conclusion that blastopore plays an important role in gastrulation. Later he and Mangold (1924) grafted blastopore of one to the gastrula of another and demonstrated that the grafted blastopore influences the host tissue to form embryonic axis. They termed the blastopore as "*organiser*" and the influence of organiser as "*induction*". Spemann's lead was soon followed by different workers and considerable information became available regarding the nature of organiser. It may be summarised that induction involves three distinct events: (*a*) *Evocation*, (*b*) *Individuation* and (*c*) *Competence*. The first two, evocation and individuation are the properties of organiser and the competence is the feature of the tissues on which the organiser acts.

In 1943, Holtfreter, working on the gastrulation mechanism of amphibian eggs, demonstrated that superficial cells are united by an extracellular surface coat and the beginning of invagination is due to the expansion of certain cells. This expansion according to him is caused by the change of surface tension due to the high pH of blastocoelic fluid. Though many workers have questioned the findings of Holtfreter, it remained true that initiation of invagination is the property of localized cells. It may be due to local difference of pH or due to differential adhesiveness of the cells.

The work done to explore the nature of involution and epiboly also explained that the entire process is due to the nature of participating cells. It was demonstrated that the cells which are more adhesive are less mobile and on the contrary more mobile cells are less adhesive. Once this was understood attempts were made to explain the mechanism of gastrulation in terms of cellular adhesibility and cellular mobility.

In 1955, Townes and Holtfreter examined the interaction of different cell layers in amphibian gastrulae and demonstrated (*a*) that endoderm cells are less adhesive than mesoderm, (*b*) that outer ectoderm is less adhesive than inner ectoderm and (*c*) mesoderm is less adhesive than inner ectoderm but more adhesive than endoderm.

Basing on this contention Stainberg (1964) proposed that disposition of different layers in a gastrula depends upon the adhesive nature of the cells. Outer ectoderm being less adhesive stays outermost. Mesoderm being less adhesive than inner ectoderm but more adhesive than endoderm remains in between the two.

ANALYSIS OF DEVELOPMENTAL EVENTS—NATURE AND ACTION OF ORGANIZER

The discovery of 'organizer' by Spemann and Mangold (1924) in amphibian egg marks the beginning of a new era in experimental embryology. It has been discussed previously that the cells emerged out of cleavage grow, divide and differentiate to form a complex organism. During the process of development a definite order prevails which prevents the diversity from becoming chaos and complexities from becoming confusion. This order depends on the specific relationship of a particular group of tissue with others.

Grey crescent—its role in development

Wilhelm Roux (1888) first described the grey crescent in amphibian egg. The experiments of Rund and Spemann have emphasised the vital role of grey crescent in amphibian development (see Fig. 5.14). If the egg is surgically divided into two halves, each having one half of the grey crescent, each half of the egg will develop into an entire embryo if cultured in isolation. The same operation if performed on an egg in which one half gets the entire grey crescent, while the other blastomeres lack the grey crescent material and the two blastomeres when cultured in isolation, the blastomere which is devoid of grey crescent will develop a simple sac of ectoderm containing endoderm, while the blastomere having the grey crescent will form an entire embryo. Similar operation of removal of grey crescent from the egg prior to cleavage has shown that the process of cleavage remains unaltered, but gastrulation does not occur.

The vital dye method and other marking devices, such as carbon particle technique and radioactive labelling method have been applied to ascertain the fate of grey crescent during development. These experimental studies have revealed that the grey crescent material of the egg gives rise to the dorsal lip of the blastopore in early gastrula. But in the late stage, i.e., when the gastrula transforms into an embryo, the grey crescent materials become localised in the head endoderm of the primitive gut.

Dorsal lip of Blastopore and its significance

The ectoderm in early gastrula remains in an undifferentiated stage, i.e., it has manifold potencies at that stage. During gastrulation, the potencies of ectoderm become limited from general to specific one. In normal amphibian development the dorsal ectoderm just above the dorsal lip of the blastopore becomes the neural structure. The dorsal lip region is called the chorda-mesoderm because it develops normally into mesoderm and notochord. Spemann and his student, Mangold, by using a transplantation technique, have shown that the dorsal lip of the blastopore, when implanted to the ectodermal area of the same or another embryo at that stage of development, the dorsal lip is able to initiate the formation of a secondary embryonic axis. Fig. 5.22 demonstrates the formation of a secondary head as a result of grafting the dorsal lip of the blastopore. They used the dorsal lip of the blastopore of *Triton cristatus* (Nonpigmented) as the graft to an embryo of *Triton taeniatus* (Pigmented) of the same age. This experiment demonstrates that the neural tissue of the secondary embryo is entirely formed from the tissues of the host—thus suggesting the fact that the dorsal lip of the blastopore is not only the controller of development but

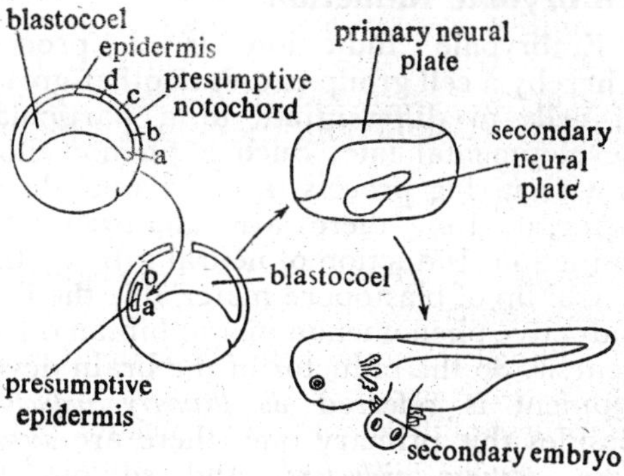

Fig. 5.22. Formation of a secondary head after transplantation of the dorsal lip (a-b region) of the blastopore.

also acts as an instigator to induce the host tissue to differentiate.

Concept of embryonic induction

The foregoing discussion suggests that the commitment of a particular tissue to a particular developmental fate depends generally on the relationships established with surrounding tissues. The development of amphibian neural plate has shown that the dorsal lip of the blastopore invaginates inside the embryo and induces the dorsal ectodermal layer to form the neural structure. If the invagination is prevented by treatment of the amphibian blastula with a solution of 0·35% sodium chloride solution as shown by Holtfreter in Exogastrulation Experiment (see Fig. 5.15 C.D.), the dorsal ectoderm that normally becomes neural tube forms the ectodermal vesicle. So the determination of neural tube is dependent upon the association with the adjoining tissue, such a phenomenon is often referred as *embryonic induction*.

The transplantation experiment of Spemann and Mangold (1924) with the dorsal lip of the blastopore and the exogastrulation experiment of Holtfreter (1933) with amphibian embryo also suggest another phenomenon in embryonic differentiation. When the differentiation of a group of cells depends upon another group of cells —such a process is called the *dependent differentiation*. The development of neural tube is a case of dependent differentiation. When a cell group is capable of differentiation without the intervention of other cells—the differentiation is called the *independent differentiation.*

Embryonic induction

Embryonic induction is a process whereby a cell group causes another group of cells to differentiate to a particular developmental fate. Such a phenomenon is a recurring process in embryonic development, i.e., there are hierarchies of induction. Induction of neural tube by the dorsal lip of blastopore material is the first inductive phenomenon in amphibian ontogenesis. So the inductor of the brain development is referred as *primary inductor*. Besides this primary one, there are *secondary*, *tertiary inductors*, and so on, in hierarchical array. The brain having been induced, develops regional specificity under the influence of head endoderm and mesoderm. The hindbrain exerts inductive influence over the overlying ectoderm to form ear placode which becomes invaginated to form auditory vesicle. The auditory vesicles, in turn, induce the surrounding mesenchyme to develop into the auditory capsule. The development of eye relates a similar story. The diencephalon forms paired optic vesicles which make contact with the overlying head ectoderm. The ectoderm forms a lens which invaginates to form lens vesicle. The eye-cup-lens complex induces the formation of cornea and associated mesodermal derivatives.

Nature of induction

The neural differentiation in amphibian development has shown that the archenteron roof induces the overlying ectoderm to develop into neural structures. This fact suggests a direct influence upon the ectodermal cells by the archenteron roof. Two alternate possibilities can be suggested: (1) *Surface interaction* or (2) *Chemical intervention.* Most of the experimental evidences on this line suggest the existence of chemical mediation between the inductor and induced structures. Many chemical substances suspended in egg albumen (an inert substance) can induce neural tissue when introduced into the blastocoel of early gastrula. As the implanted material contains no cells—the question of surface interaction of cells at the inductive interface becomes impossible. Besides, if a portion of presumptive embryonic ectoderm and a piece of dorsal lip of blastopore are kept separated in a culture medium by a millipore filter paper of 20 micra thick with an average pore size of 8 micra, the ectoderm develops into a neural structure. The filter paper prevents the transfer of cells but permits diffusible chemical substances of low and high-molecular weight. Similar experiments with epithelio-mesenchymal interactions in the differentiation of kidney, pancreas, salivary gland, thyroid and thymus glands, etc., also suggest this 'transfilter phenomenon'. The inductive phenomena that occur in all the cases suggest beyond doubt that induction can occur without cell contact and inductive agent is chemical. It involves the transfer of chemicals from the inducing to reacting tissues.

Chemical nature of inducing substances

The mode of action of inductors and the actual chemical nature of inducing substances are incompletely known. Many chemical substances, like sterols, fatty acids, proteins, glycogen, synthetic carcinogenic agents and even silica particles and dead tissue can induce neural tissue formation. But these substances cannot fully imitate the action of normal inductor, because regional organisation of the brain fails to occur by these artificial chemical stimulants. Experiment with the dorsal lip of the blastopore of an amphibian gastrula gives some idea about the chemical nature of organiser.

A portion of the dorsal lip of blastopore from an amphibian gastrula is explanted and cultured *in vitro* in suitable culture medium. After about 7–10 days, all the cells of the explant are eliminated leaving only the medium in which the explant was cultured. The medium is filtered by millipore filter paper. The medium is called the "*conditioned medium*". A small fragment of embryonic ectoderm, excised from early gastrula, was put in the conditioned medium. After about one to three weeks, the ectoderm cells formed neural and pigment cells. Thus induction is caused in the absence of inducing cells. This fact indicates that the cells of the dorsal lip of blastopore have certainly liberated the inducing substances directly into the culture medium which have caused neural induction. Spectrophotometric and electrophoretic analyses indicate that the inducing substance in the medium as the macromolecules—predominantly nucleoproteins. Now the question arises whether the protein moiety or RNA part of nucleoproteins is more effective as inducing agent. Digestion of RNA part by ribonuclease (an enzyme which digests the RNA) does not reduce the inductive capacity to a considerable extent, but similar treatment with proteolytic enzymes (pepsin or trypsin) which digests the protein component reduces the occurrence of neural inductions in about 80% cases. This experiment suggests to the contention that the protein part of the nucleoprotein is more effective inducing agent.

Biological Organization of Organizer

Extensive experimentations have been performed to gain an insight into the organization of the organizer, particularly in amphibian development. The dorsal lip of blastopore (chordamesoderm) acts as the primary organizer. Fig. 5.23 indicates

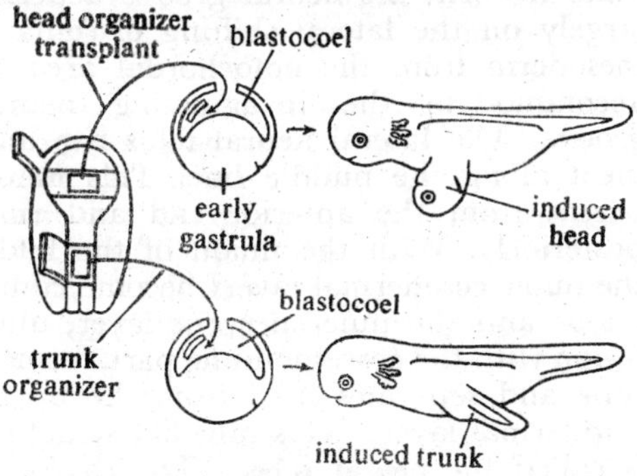

Fig. 5.23. Showing the existence of head and trunk organizer in amphibian development.

that the organizer is not a homogeneous system. The chordamesoderm of a neurula stage is used as the implanted material. The anterior portion used as graft into the blastocoel induces a head, while the posterior part as a graft produces a secondary trunk and tail. Because of this, the anterior portion is called the *head organizer* and the posterior part is designated as the *trunk organizer*. This differential action of the chordamesoderm indicates that the organizer is heterogeneous, but the origin and nature of differences between the different regions are difficult to interpret. There are diverse opinions regarding this aspect. According to one, there are different inductors present in the chordamesoderm, each one is responsible for the expression of a specific part of the nervous system. The other theory advocates that there is a single inducing substance in the chordamesoderm and the differential effect is produced by the different concentration of the same substance. But the existence of both qualitative and quantitative differences in the chordamesoderm is the recent view on neural induction. The chordamesoderm is regarded as a two-dimensional sheet having a pair of gradients. These two gradients are co-ordinated in such a way in the chordamesoderm that a perfect pattern is established. Neural induction is resulted as the outcome of co-ordinated interaction of two different inducing agents—one is the *forebrain inductor* and the other is *mesoderm inductor*. The forebrain inductor can induce forebrain, while the forebrain inductor in conjunction with the mesoderm inductor develops both neural and mesodermal structures. Fig. 5.24

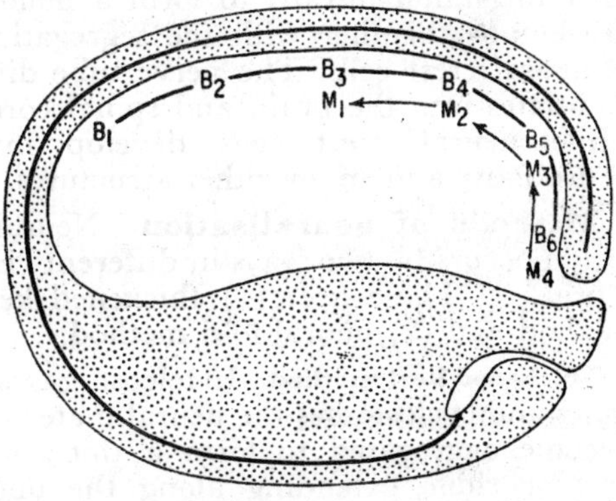

Fig. 5.24. Schematic representation of the quantitative distribution of two types of inductors in amphibian gastrula. B=Forebrain inductor. M=Mesoderm inductor.

shows the quantitative distribution of the two inducing substances. The forebrain inductor is present in same concentration along the length of the archenteron roof while the mesoderm inductor is present at its highest concentration at the posterior end and gradually diminishes to zero at the anterior end. Combinations of these two inductors are believed to be responsible for specific development in the induced structures.

Successive inductions occur in the development of embryonic structures. The embryo, as such, represents a dynamic system and the positional relation of cells always remains in flux. Occurrence of successive inductions indicates the presence of diverse quality of embryonic induction. The theoretical interpretations on embryonic induction discussed herein, are not the final answer to the question of the nature and action of the induction, but it provides valuable information on this line.

ANALYSIS OF DEVELOPMENTAL EVENTS—ORGANOGENESIS

DEVELOPMENT OF BRAIN

The first important morphogenetic change following gastrulation is the development of the central nervous system. The central nervous system starts as a simple tubular nerve tube which, in course of development, transforms into brain, spinal cord and their associated structures. The morphogenetic processes involved in this process are designated as *neuralisation.* It includes the separation of neural materials from the embryonic ectoderm, their migration inward to form a hollow nerve tube together with the segregation of neural crest cells. The nerve tube differentiates into the brain and spinal cord, while neural crest cells develop into neuroblasts and many other structures.

Methods of neuralisation. Neuralisation occurs by two ways in different vertebrates. These are: (1) thickened keel method and (2) neural fold method.

Thickened keel method. In teleost, ganoid fishes and cyclostomes the neural materials become aggregated to form a thickened keel or ridge extending along the mid-dorsal axis of the body. This ridge separates itself from the overlying ectoderm and develops a lumen within to form a tube.

Neural fold method. It occurs in most of the vertebrates where neural cells become aggregated to form a neural plate. This plate folds inward to form neural groove. The neural groove transforms into a neural tube which sinks from the overlying ectoderm.

Events in neural morphogenesis. After the completion of gastrulation, the ectoderm of the future dorsal side of the developing embryo tends to condense to form a thick and compact neural plate with elevated margins. This thickened part is called the **neural (or medullary) plate or neural placode.** The plate is formed by two simultaneous processes: (1) elongation of prospective neural cells in the direction perpendicular to the surface of the developing embryo and (2) shrinking the exposed surfaces both dorsally and ventrally. The neural plate is pear-shaped, i.e. it is broader at the anterior part but gradually narrows towards the posterior end. This particular shape of the neural plate is crucial for shaping the future structures. The shaping of the neural plate is resulted as the consequence of regional differences in the cell contration. Fig. 5.25 relates the stages of neural morphogenesis in frog.

A depression appears along the entire length of the neural plate which folds downward to form a **neural or medullary groove.** The formation of the neural groove is associated with the median and dorsal movement of the ectodermal layer attached to the lateral edges of the neural plate. Thus the raised or folded margin of the neural groove is called the neural fold. The downward movement of the neural plate to form the neural groove depends largely on the lateral shifting of somatic mesoderm from the notochordal area to accommodate the invaginating neural groove. The lateral neural folds rise and meet along the middle line. This union begins from the anterior end and runs posteriorly. With the union of the folds, the outer ectodermal layers become continuous and the inner nervous layer, after fusion with the corresponding part, forms a tube and separates itself from the upper ectodermal layer. This tube-like structure is called the **neural tube.** The cavity of the neural tube is called the neurocoel which is broader at the anterior end and opens to the exterior through an opening called the *neuropore.* The neuropore ultimately closes at the later stage of development.

Associated with the formation of neural tube **neural crest cells** become segregated on the two sides of the neural tube. These neural crest cells lie as two longitudinal strips of cells, one on each dorsal side of the neural tube.

Neural crest and its fate. At the corners of the fusing neural fold during brain formation, groups of neural crest cells become detached to occupy a position over the neural tube. In course of development these cells leave their position and migrate to other parts of the embryo. These cells are versatile in their developmental fate and develop neuroblasts of the spinal and sympathetic ganglia, Schwann sheath cells producing the myelin sheath and

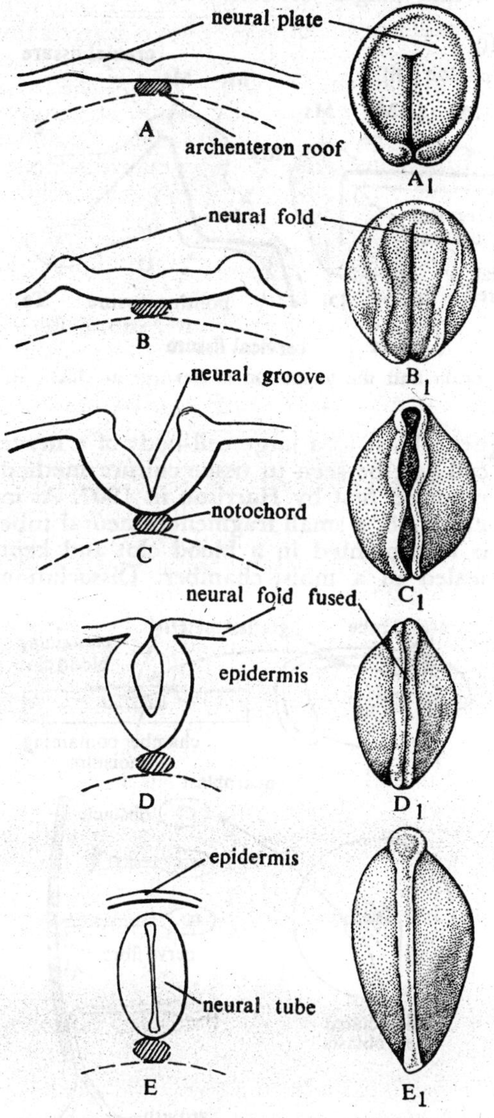

Fig. 5.25. Early stages of morphogenesis of brain. A–E. Sectional (Transverse) views. A_1–E_1. Dorsal views of entire embryos.

neurilemma of the nerve fibres, melanoblasts, chromaffin tissue of adrenal medulla, meninges, cartilages of the jaw, etc. Weston (1963) has shown the migration of neural crest cells. The neural tube and neural crest cells labelled with radioactive isotopes are excised from the trunk of a developing chick embryo and transplanted to a normal (non-labelled) host in place of its counterparts. It has been shown that the neural crest cells migrate along two ways: (*i*) dorsolaterally along the skin and (*ii*) ventrolaterally in relation to the neural tube.

Structural differentiation of the neural tube. The differentiation of the neural tube into the brain and spinal cord depends upon many intrinsic and extrinsic factors. The anterior part of the neural tube transforms into the brain while the posterior narr w part becomes elongated to form the spinal cord. The broad anterior pait is demarcated from the narrow posterior part by *isthmus*. Remarkable changes occur in the anterior part during its conversion into the brain. This is caused by (1) unequal thickening of the neural tube wall, (2) invaginations or evaginations of the wall and (3) various types of bending or folding (flexure formation).

Immediately after the formation of the neural tube, the anterior part swells up and two constrictions develop to divide the anterior part into three general regions: *Prosencephalon*, *Mesencephalon* and *Rhombencephalon*. In course of development, prosencephalon and rhombencephalon become further subdivided thus giving rise to five parts: *Telencephalon*, *Diencephalon*, *Mesencephalon*, *Metencephalon* and *Myelencephalon* (Fig. 5.26). Many factors are responsible in brain morphogenesis. Differential growth and intraventricular pressure are regarded to be the important morphogenetic factors in brain development, especially in flexure formation. Fig. 5.26 relates the development of flexures and different regions of the brain.

Histogenesis in brain development. The early neural tube is fairly uniform in structure. The walls are composed of neural epithelial cells which eventually differentiate into: (1) neuroblasts and (2) spongioblasts. The neuroblasts develop into nerve cells and fibres while the spongioblasts give origin to ependymal and neuroglial cells.

The neural epithelium is composed of pseudostratified columnar epithelial cells which form the *primitive ependymal layer* or *matrix layer*. Gradually the cells of the matrix layer migrate to each lateral side to form a cellular layer called the *mantle layer*. And lateral to the mantle layer lies a cell-free *marginal layer*. The cells of primitive ependymal layer are usually called the *germinal cells*, some of which after a day or two following the closure of neural groove, develop neuroblasts and migrate first to the mantle layer (Fig. 5.27). In the mantle layer, the cells differentiate into (1) neuroblasts and neurons and (2) spongioblasts and neuroglial cells. The neuroglial cells give rise to *astrocytes* and *oligodendrocytes*. The neuroblasts do not remain evenly distributed but are aggregated into clusters. From mature neuroblasts, nerve cells and fibres grow out in a distinct pattern and turn the brain into a 'working unit'.

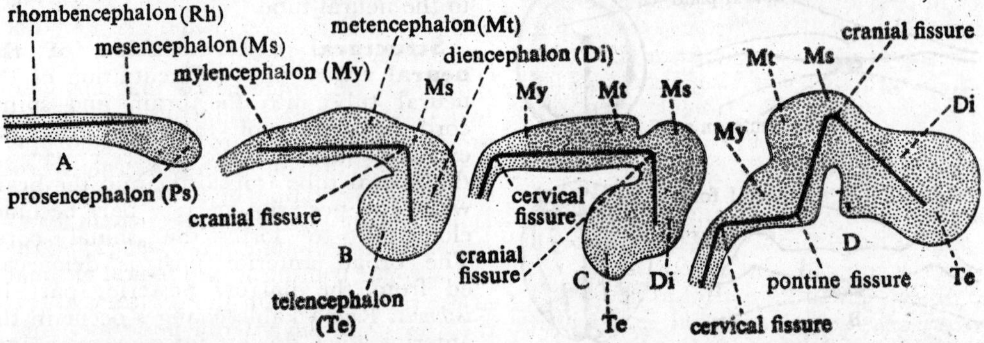

Fig. 5.26. Showing the regional differentiation of brain and the formation of flexures in chick. A. 48 hours. B. 72 hours. C. 4 days. D. 8 days.

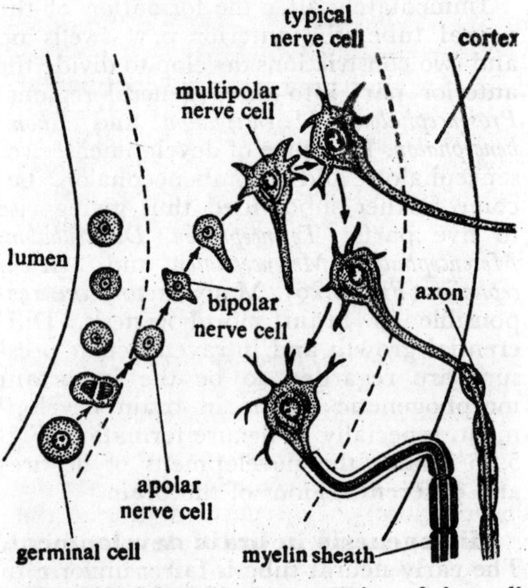

Fig. 5.27. Schematic representation of the wall of neural tube showing the developmental fate of neuroblast—the germinal cell.

Development of nerve cells and fibres. Nerve cells originate from neuroblasts which develop from the neural tube, neural crests and cranial placodes. The actual stages of conversion of a small neuroblast to a large cell-body of a nerve cell can be seen in tissue culture method reported first by Harrison in 1907. As in Fig. 5.28, a small fragment of neural tube is transplanted in a blood clot and kept sealed in a moist chamber. Dissociation and dispersion of cells are the first observable events in tissue culture. The origin of nerve fibres is the most notable event in this process of conversion. Three theories are

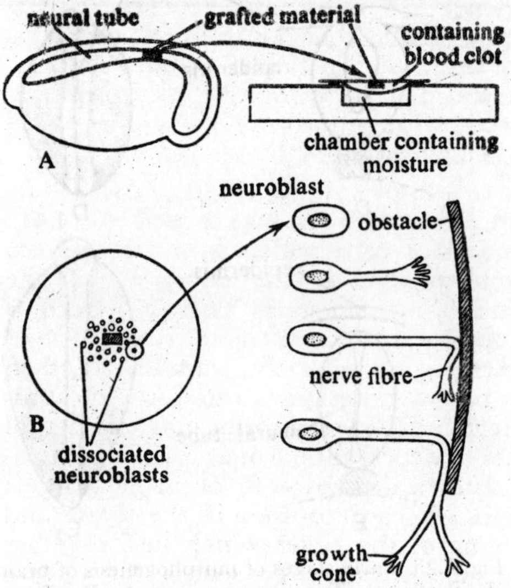

Fig. 5.28. Development of nerve fibre from a single neuroblast. A. A small portion of nerve tube is cultured within a blood clot in a specially designed culture chamber. B. Shows migration of neuroblasts and each transforms into a full-fledged nerve cell with fibres.

extant on this particular issue. They are: (1) *Cell-chain theory*. This theory relates that the fibre is laid down by chains of cells which surround the nerve fibre. (2) *Plasmodism theory*. According to this theory the nerve fibre is laid down on preformed protoplasmic bridge. (3) *Outgrowth theory*. The theory advocates that the fibre is formed as an outgrowth of a single neuroblast. The tissue culture experiment gives support to the last concept and settles the long standing controversy regarding the issue. At the beginning, a thin strand of protoplasm emerges as outgrowth from one side of the neuroblast. This outgrowth becomes amoeboid and creeps along the solid object. The outgrowth has developed a *growth cone* at the terminal end which may branch to form two or more growth cones. The growth of nerve fibre exhibits *streotropism*, i.e. it moves along solid object.

Causal analyses in brain morphogenesis. In the entire process of nervous system formation, a number of inductive events occur. In the amphibian eggs, the dorsal lip of blastopore acts as primary organizer to induce the inward moving cells to form chordamesoderm which in turn induces the dorsal ectoderm to be neuralised. The formation of the neural tube is also guided by the influence of regionally specific inductions.

The neural plate at the beginning is an oval, flattened plate and is formed by the ectodermal cells which have come from lateral regions to the dorsal side. The neural plate elongates rapidly, which is caused by the movement of cells. The cells first move towards the middle and then run in two directions: anteriorly and pos-teriorly. The transformation of neural plate to neural tube which is called neurulation is also known to occur *in vitro*. It begins with a depression in the centre and curving of the edges which fuse together to form the tube. To search the motive force behind the formation of tube, the behaviour of cells in the centre and periphery is intimately studied. Certain suggestions, like differential water uptake, differential cell divisions have been negativated. It is now claimed that elongation of the plate is due to migration of cells but curvature is caused by changes in the cell adhesion.

In further development, the anterior part of the tube swells up considerably to form brain vesicles. Considerable amounts of cell division and cell movement occur during the process. The different parts of the brain in course of its development induce the formation of structures like optic, auditory and nasal placodes on the outer ectodermal covering.

It must be remembered that mesodermal cells which immediately remain around neural tube are believed to play most important role in the epigenetic process. The formation of brain establishes: (1) many histological features remain determined at neural plate stage and (2) all the cells do not transform into neural element at the same time. On the contrary a gradient exists in the anterior-posterior plane.

DEVELOPMENT OF EYE

The early stages of eye development follow a generalised pattern in all vertebrates and the details of eye morphogenesis in chick will give an idea of the process in general. The eye is a very complicated structural unit. The development of eye reveals the incorporation of different **tissues which follow an orderly fashion to give the geometry of pattern. Beacuse of the fact, the incidences involved in eye morphogenesis are regarded as a perfect model to explore the general problems of embryology.**

The development of eye is discussed under three steps:

(1) *Development of sensory areas.*
(2) *Development of lens.*
(3) *Development of associated structures.*

Development of sensory areas.

Formation of optic placode and optic vesicle. The primordial eye rudiment lies at the very anterior end of neural plate in the form of two closely placed oval areas, one on either side of the middle line. They are lined below by mesoderm. In course of the formation of brain, these two lateral sides of the forebrain, which are destined to be the future diencephalon, become thickened. These parts are known as *optic placodes*. These two placodes extend laterally as small blunt bulgings, which become known as *optic vesicles* (Fig. 5.29). The vesicles elongate through the loose mesenchyme towards the epidermal covering and remain connected with the brain by a narrow stalk called *optic stalk*.

Formation of optic cup. As the optic vesicle touches the ectoderm, the ectoderm cells elongate perpendicularly to the region of contact to form the *lens placode* which invaginates to form lens vesicle. With the invagination of lens placode to form *lens vesicle*, the optic vesicle reverses its outward bulging and turns inwards to form the *optic cup* to accommodate the lens. The optic cup is double-walled. Such

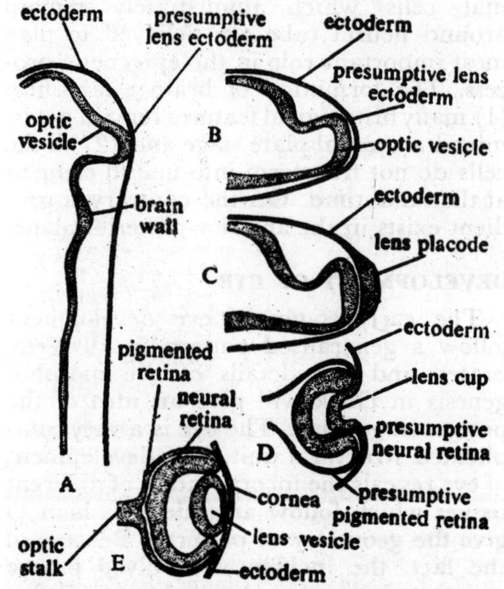

Fig. 5.29. Showing the developmental stages of optic cup, lens and cornea. A. Diagrammatic longitudinal section of the embryo. Note that the optic vesicle originates from the diencephalic region of the brain and makes contact with the overlying ectoderm. B–E. Transverse sectional views. B. Transverse sectional view of A-stage. C. Showing the induction of lens placode by the optic vesicle. D. Invagination of optic vesicle to form optic cup and the lens vesicle. E. Separation of lens vesicle and differentiation of neural and pigmented retina.

inpushing takes place asymetrically and continues obliquely into the *optic stalk*. Near the optic stalk, a slit is left in the ventral side which is called *choroid fissure*. This fissure acts as an outlet for optic nerve (Fig. 5.30). The blood vessels also find an impass into the optic cup. The outer wall of the optic cup remains thin and gives rise to *pigmented retina* while the inner wall (*neural retina*) becomes greatly thickened and elaborated to transform into light sensitive retina.

Formation of retina and optic nerves. The inner lining of the optic cup transforms into light sensitive retina, which in turn differentiates into seven layers of nervous elements. The neural retina gives rise to rods and cones and some other types of cells with which the visual cells synapse. The histogenesis of retina is divided into three phases: (*i*) a phase of cell multiplication, (*ii*) a phase of cellular readjustment and (*iii*) a phase of final differentiation.

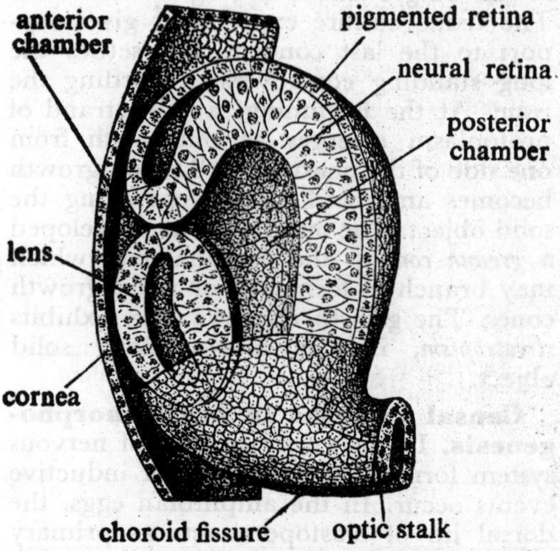

Fig. 5.30. An enlarged schematic model of eye showing the relative position of the structural components.

Fig. 5.31 gives the details of development of retinal cells. Rods and cones are arranged in the outermost part, i.e. towards the pigmented layer of the optic cup. The position of visual cells is due to the migration and stratification of neural layer of the retina. The fibres of rods and cones unite with the fibres of the ganglionated layer of the retina. These fibres converge towards the optic stalk to form the optic nerve.

Development of lens. The region of the outer ectoderm which comes in contact with the optic vesicle thickens and is known as *lens placode*. The placode invaginates to form a *lens pit* or *lens cup*. The two ends of the lens cup unite and remain within the space between optic cup and outer ectoderm. It is then called *lens vesicle*. The inner cells of the vesicle transform into *lens fibres* and the cell layer next to ectoderm forms the epithelium of the lens. At about 96 hours in chick embryo, the lens cavity becomes reduced and the cells of the median wall of the lens become elongated to obliterate the lens cavity. The cytoplasm of the cells becomes clear and these cells transform into lens fibres. The extreme elongation of the cells is evidenced by the placement of their nuclei in the equitorial region of the lens and the cells

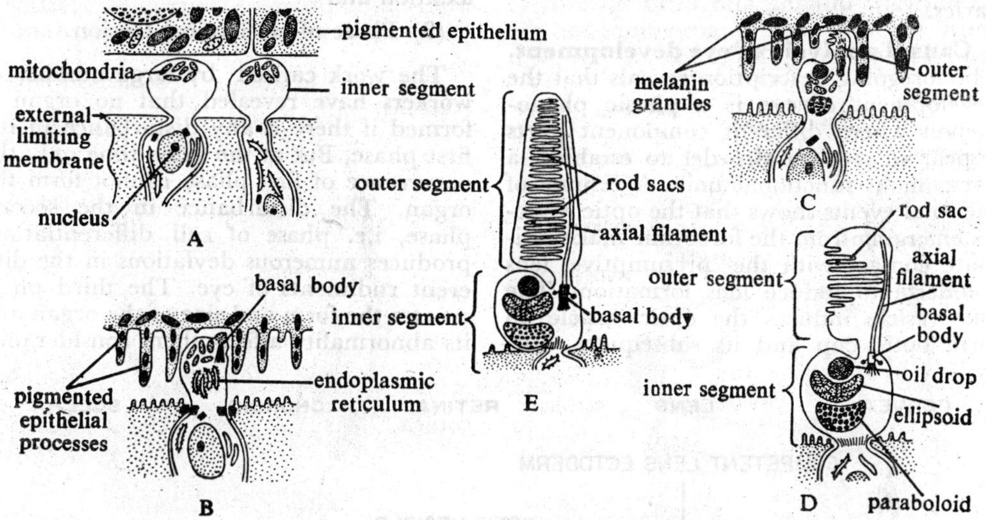

Fig. 5.31. Schematic diagrams showing the maturation of rod. Each visual cell produces a minute cytoplasmic bud which projects through a pore in the external limiting membrane (A). This bud develops into the rod. Inside this bud, several specialised cell organelles develop (B). The bud forms the inner segment of the rod and from this segment the outer segment arises (C). The basal body of the inner segment produces cilia which participate in the differentiation of the outer segment (D). In the outer segment, rod sacs are produced by the folding of the cell surface. These sacs are stacked one above the other and the outer segment elongates with the growth of the sacs (E).

are stretched extending from one surface of the lens to the other. A lens fibre may reach 10 mm as seen in man.

Development of associated structures

Choroid and sclera. The optic cup and optic stalk become invested with a layer of mesenchyme which later forms an outer densely fibrous layer called *sclera* and an inner pigmented and richly vascularised layer called *choroid*.

Conjunctiva, cornea and aquous humor. After the detachment of the lens the ectodermal epithelium forms a cell-free lining. Along the inner border of the ectodermal epithelium lies the mesenchyme, which forms the sclera. The inter epidermal layer is called *conjunctiva* and inner mesenchymal continuation of sclera is known as *cornea.* The transparency of cornea is vital for the admission of light into the cavity of eye. The cornea is composed of an outer epithelial layer (derived from ectoderm) and a postepithelial stroma (derived from immigrating mesenchymal cells). In the stroma, layers of collagen fibres accumulate parallel to the surface. The stroma undergoes significant biochemical changes and undergoes dehydration. Due to loss of water and suppression of pigmentation, the cornea attains perfect transparency. A space develops between lens and cornea, which is called *anterior chamber*. A watery fluid, called aqueous humor accumulates within the space.

Iris. The choroid layer extends in front of the lens to form a circular curtain known as *iris,* which has a hole in the centre called *pupil*. The iris is pigmented and possesses *papillary muscles* which regulate the diameter of the pupil. The iris develops from the pigmented retina.

Ciliary muscles. The pigmented outer layer of the optic cup together with mesenchymal elements extends in front to form ciliary muscles or ciliary bodies.

Vitreous humor. The cavity between lens and retina is known as *posterior chamber*. It becomes filled with a gelatinous matrix called vitreous humor.

Muscles of the eye. Several muscles are formed by the condensation of outer head mesenchyme. These muscles are involved in rotating the eye-ball within the orbit.

Eyelid. The outermost epidermal layer in front of the eye becomes skin. It splits into two halves to form the *eyelids*. In different vertebrates the shape of eyelids varies.

Causal analysis of eye development. The foregoing description reveals that the development of eye is a phasic phenomenon where different component parts appear in sequential order to establish a harmonious functional unit. A resume of the total events shows that the optic vesicles emerging from the forebrain make intimate contact with the presumptive lens ectoderm to induce lens formation. The lens vesicle induces the optic vesicle to form optic cup and its subsequent differentiation. The optic cup-lens complex induces the overlying ectoderm with some mesenchyme to form the cornea. Thus a reciprocity of induction occurs in eye morphogenesis—and this type of induction is called the *synergistic induction*. Fig. 5.32 gives the schematic representation of the participation of different tissues and the inductive phenomena in the development of eye.

Events of eye development. Events of eye development may be divided into three phases:

(1) Phase of induction,

(2) Phase of cell differentiation and axiation and

(3) Phase of mechanical tension.

The work carried by large number of workers have revealed that no organ is formed if there is any disturbance in the first phase. But at the same time only the occurrence of first phase cannot form the organ. The disturbance in the second phase, i.e. phase of cell differentiation, produces numerous deviations in the different rudiments of eye. The third phase creates the form and size of the organ and its abnormality affects them considerably.

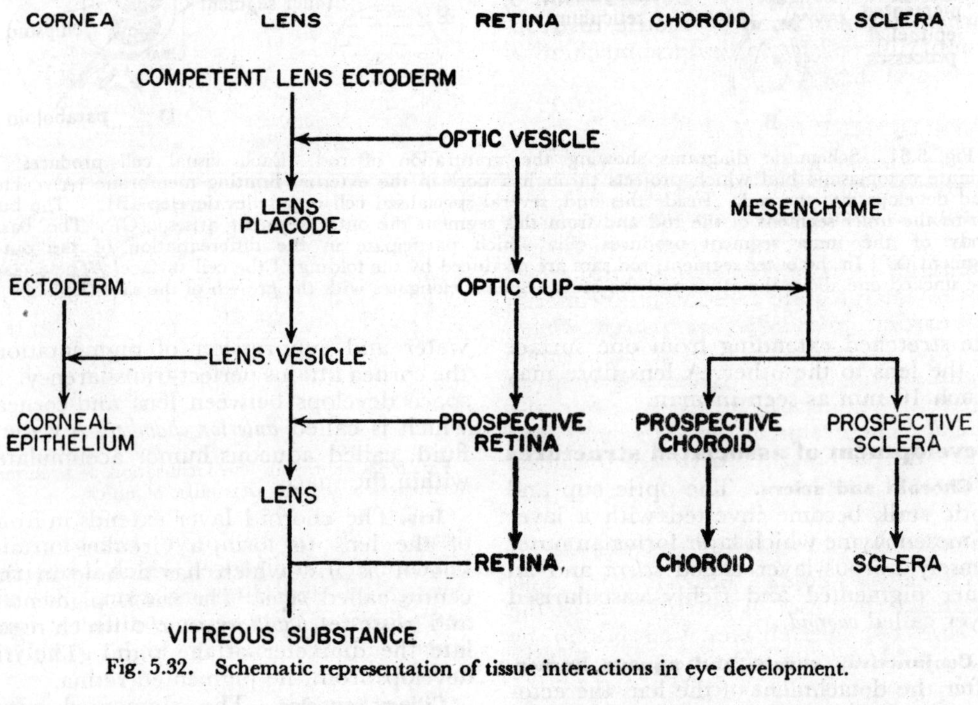

Fig. 5.32. Schematic representation of tissue interactions in eye development.

Some of the important findings to explain the mechanism of eye formation are discussed below:

(*i*) The lateral extension of the optic vesicles to reach the ectodermal layer is caused by the pressure exerted by the intraventricular fluid of the brain. (*ii*) The loose surrounding mesenchyme of the primordial eye rudiment plays important part. Experimental evidences suggest that the mesoderm first contributes most actively to the development of the eye and then exhibits its formation in the middle of the brain. (*iii*) In the primary eye rudiment at the beginning, the different layers have the capacity to undergo mutual transformation. The inner wall of the optic cup transforms into the neural retina while the outer wall develops into the pigmented retina. If the position of these layers is reserved, the original inner layer may form neural retina and *vice versa*. The transformation of the inner wall of the optic cup to form neural retina is caused by the inductive influence of the lens vesicle. The outer layer is converted into pigmented retina by the influence of mesenchyme. (*iv*) The proportion of pigmented retina and neural retina depends on the extent of tension produced by the fluid which accumulated inside. (*v*) The formation of cornea depends upon geometric distribution of different layers and the mechanical tension of the eye.

DEVELOPMENT OF HEART

The formation of heart (cardiogenesis) in vertebrates is one of the most dynamic events in embryonic development. The heart is a mesodermal organ, which differentiates initially from the ventral edges of the lateral plate mesoderm. Primarily the cardiac primordia are paired which, however, become fused to form a single organ. The process of cardiogenesis in different vertebrate forms is essentially similar. The events of the development of heart in the chick embryo are discussed below.

Localisation of cardiac primordia

The localisation of heart-forming cells occurs at the onset of gastrulation in the embryos of all vertebrates. Vogt (1929), by using vital staining technique, has localised the prospective heart cells in amphibian embryo. Butler (1935) and Spratt (1942) have shown that heart-forming cells are widespread in the blastodisc of chick (Fig. 5.33A). With the movement of cells of the epiblast to form the primitive streak, heart-forming cells become restricted to the epiblastic region anterior to the developing primitive streak. With the migration of epiblastic cells to form the mesodermal layer, the heart-forming cells become concentrated about Hensen's node. The heart-forming cells, then, migrate to join the mesoderm and move laterally. When the definitive primitive streak is formed, the heart-forming cells take lateral position as paired cardiac primordia. Each primordium is capable of developing a whole heart. If the paired primordia are prevented from fusion, two independently beating hearts ('cardiac bifida') will result in an embryo.

Before the formation of heart, the presumptive heart-forming cells acquire specific biochemical characteristics from their neighbours and have an inherent capacity of undergoing self-differentiation. This is attested by the fact that these bilaterally located cardiac primordia, when transplanted into an indifferent location, are capable of differentiating into cardiac

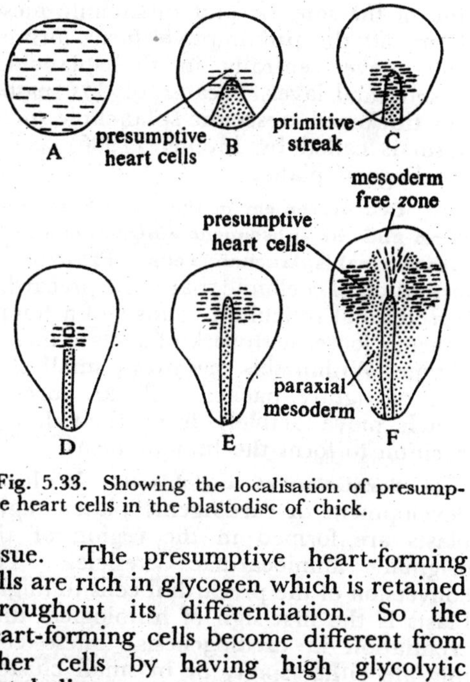

Fig. 5.33. Showing the localisation of presumptive heart cells in the blastodisc of chick.

tissue. The presumptive heart-forming cells are rich in glycogen which is retained throughout its differentiation. So the heart-forming cells become different from other cells by having high glycolytic metabolism.

Stages of heart formation

In course of development of heart, the paired cardiac primordia come together in the midventral line. This is brought about by the action of four types of morphogenetic movements. These are:

(1) **Folding movements of ectoderm and endoderm.** This movement of the endoderm to develop into crescentic pouch of the anterior intestinal portal and early

foregut is of great importance. The presumptive heart-forming cells use the endodermal layer as the substratum for their migratory activity. The folding movements bring the paired cardiac primordia together in the midline to develop into an unpaired median tube.

(2) **Formation of amniocardiac vesicles.** This process of development of embryonic coelom or amniocardiac vesicle is also important in heart formation. With the formation of head fold and initiation of foregut, the lateral plate mesoderm in the region of cardiac primordia splits to form a dorsal layer (somatic mesoderm) and a ventral layer (splanchnic mesoderm). The coelomic space thus enclosed by these two layers is called the early pericardial or amniocardiac vesicles. With the separation of the somatic and splanchnic mesoderm, all the presumptive heart-forming cells move ventrally in the splanchnic mesodermal layer. Because of this reason, this thickened crescentic splanchnic mesoderm is called by Mollier (1906) as the 'cardiogenic plate'.

(3) **Cell movement in the splanchnic mesoderm and the subsequent emigration of the mesodermal (splanchnic) cells.** Prior to the formation of coelomic space, the precardial mesodermal reticulum consists of a homogeneous loose meshwork of stellate mesenchyme. Within this meshwork small clusters of tightly packed cells are present which move actively from their lateral position to form the tubular heart.

(4) **Formation of angioblasts.** With the development of cardiogenic plate, angioblasts are formed in the region of the original amniocardiac vesicle. The conversion of the precardial cells to angioblasts is the first sign of histological differentiation in cardiogenesis. These cells migrate either singly or in small clusters out of the mesoderm and form a loose layer (*vascular layer of Pander*) in the meso-endodermal space. This layer forms the endocardium of the heart and in the posterior region it produces the blood islands. These islands produce the endothelium of the remaining vasculature, erythroblasts and blood plasma.

Formation of primitive tubular heart

As stated earlier, the primordial endocardial cells begin to differentiate independently as a pair of delicate tubular hearts. These paired tubular hearts are arranged on either side of the anterior intestinal portal (the opening from the yolk into the foregut). The folding movement of the ectoderm and endoderm to form the head fold and foregut, causes the migration of the paired tubular hearts together when they fuse to form an unpaired tubular heart tube. The paired rudiments meet and fuse when the foregut is separated from the yolk sac. This process begins at 7 to 8 somite stages in chick and is completed when the embryo comes to 20-somite stage. Each of the paired rudiments has an inner endothelial lining (*endocardium*) and the outer is the *epimyocardium*. Simultaneously with the migration of splanchnic mesoderm anteromedially and separation from the folding endoderm, it becomes thickened to form paired epimyocardia. This layer develops later into the thick *myocardium* and a thin nonmuscular *epicardium* (or *visceral pericardium*). Fig. 5.34 shows the origin and subsequent fusion of paired cardiac primordia during cardiogenesis. The epimyocardium remains attached ventrally by *ventral mesocardium* and dorsally by *dorsal mesocardium*. Both these mesocardia disappear subsequently.

The fusion of the heart tubes begins at the anterior end and extends gradually to the posterior sides. Fusion starts in the region of the future ventricle and the auricle is still represented by double tubes. Then gradual union occurs in anteroposterior direction and the process of fusion is completed when the embryo becomes 20-somite stage in chick.

Inductive relationship during cardiogenesis

Many embryologists claim that the development of heart is intimately related to the developing endoderm. In amphibian embryo, the removal of endoderm causes the failure of heart formation. But in the development of chick, the removal of endoderm does not prevent normal cardiogenesis (Waddington, 1932). Many embryologists have claimed that the migration of cardiac cells and the folding movement of the endoderm is independent processes, normal cardiogenesis is not hampered if such relationship is disturbed. But experimental evidences on this line are unsatisfactory to ascertain the actual role of endoderm on the precardial mesodermal cells.

Histological differentiation in cardiogenesis

The differentiation of angioblasts from the splanchnic mesoderm and the transformation of the splanchnic mesoderm itself to form the myocardium and epimyocardial mantle are the first indication of histogenesis in heart development.

When the first tubular heart tube is produced, the space between the endocardium and myocardium becomes filled by 'cardiac jelly'. It is a thick gelatinous mass containing aldehyde, acid mucopolysaccharides. Many cells from the endocardial and myocardial layers migrate into this gelatinous layer to form a loose meshwork of stellate cells which characterise the early heart tube.

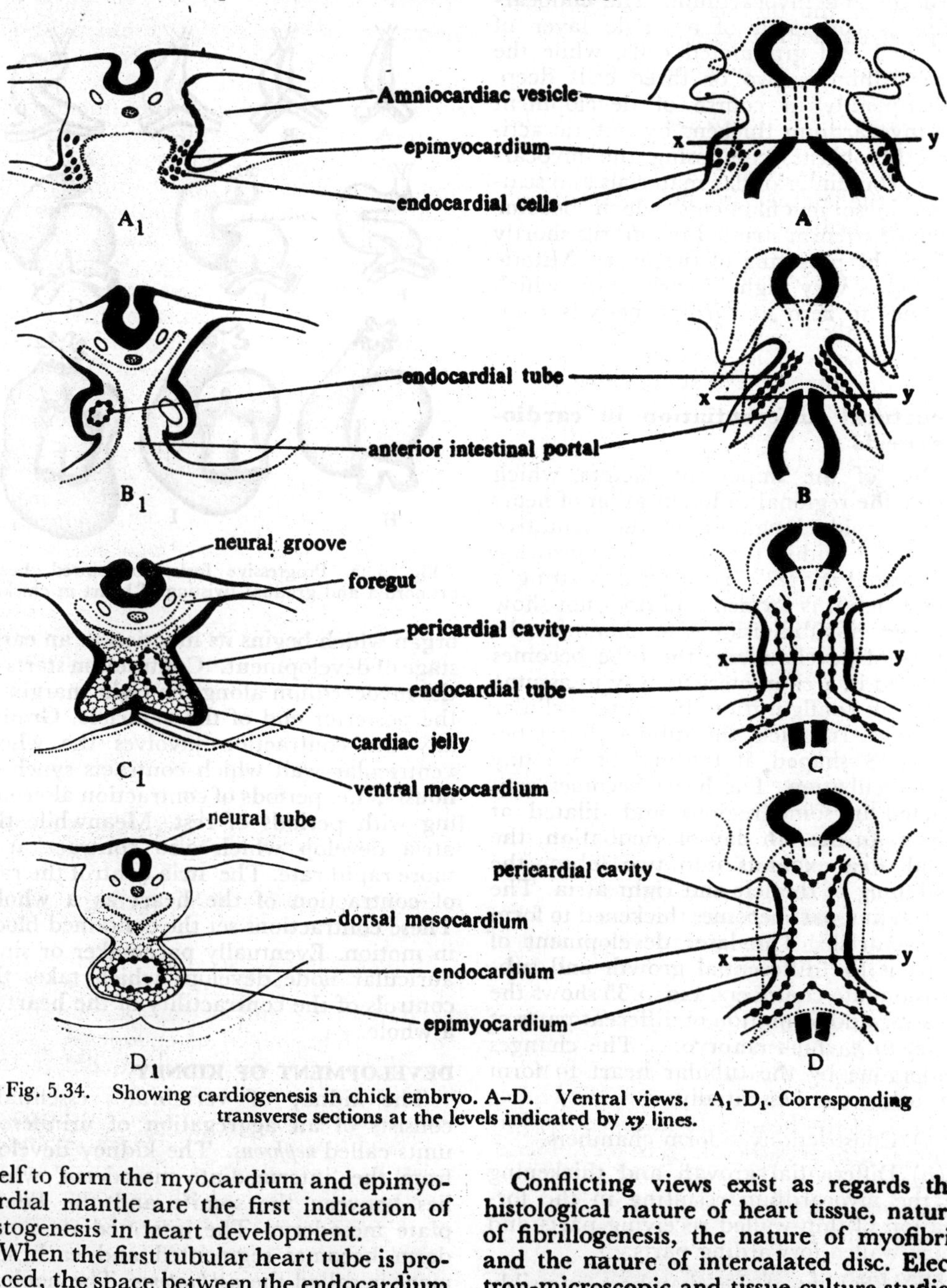

Fig. 5.34. Showing cardiogenesis in chick embryo. A–D. Ventral views. A_1–D_1. Corresponding transverse sections at the levels indicated by *xy* lines.

Conflicting views exist as regards the histological nature of heart tissue, nature of fibrillogenesis, the nature of myofibril and the nature of intercalated disc. Electron-microscopic and tissue culture studies have revealed that heart tissue is syncytial in nature and the intercalated disc consists of a pair of apposed cell membranes. The region is c. ered with electron-dense granules. The myofibrils do not cross the intercalated discs and there is no protoplasmic continuity across the apposed membranes.

The early tubular heart consists of endocardium and myocardium. The endocardium is composed of a single layer of flattened and granulated cells, while the myocardium is two or three cells deep. Subsequently in course of development the myocardium thickens by mitotic activity. Myoblasts, composing the myocardium, contain granular materials and scattered loose myofilaments which become grouped to form striated myofibrils shortly before the pulsating of the heart. Mitotic activity is very high in early stage which declines to zero as cardiogenesis is completed.

Structural differentiation in cardiogenesis

One of the important factors, which causes the regional differentiation of heart is the rapid elongation of the primitive heart tube within the less-rapidly growing pericardial space. The heart is a straight tube when it is formed and does not show any sign of subdivision into chambers. In course of development, the tube becomes inflected in a characteristic way to assume the adult configuration due to the cellular activities. In chick, the tubular heart becomes 'S'-shaped at the end of 3rd day after incubation. The heart becomes constricted in some regions and dilated at others. In the 4th day of incubation, the atrial area expands into two lobes—the beginning of the left and right atria. The descending part becomes thickened to form the ventricle. The later development of heart is the differential growth and subdivision into chambers. Fig. 5.35 shows the twisting and formation of different parts of heart in a chick embryo. The changes undergone by the tubular heart to form adult heart are essentially:

(*i*) Constrictions to form chambers.

(*ii*) Differential growth and thickening of the myocardium resulting in the formation of thin-walled receiving parts and thick-walled forwarding parts.

(*iii*) Kinking of the chambers—possibly due to rapid growth within crowded quarters.

(*iv*) Formation of septa, valves, etc.

Functional changes in cardiogenesis

The definitive function of the heart starts as the paired cardiac tubes fuse and the contraction begins as soon as the primitive ventricle is formed. So the heart is the

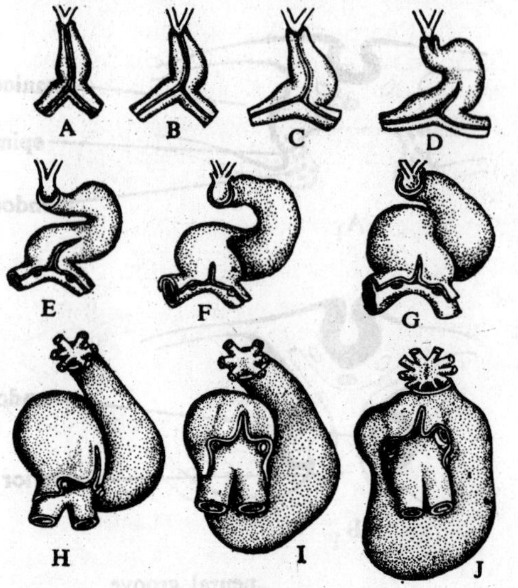

Fig. 5.35. Progressive fusion of paired heart primordia and gradual twisting of heart in chick.

organ which begins its function at an early stage of development. Contraction starts in the myocardium along the right margin of the posterior end of the ventricle. Gradually the contraction involves the whole ventricular wall which contracts synchronously, i.e. periods of contraction alternating with periods of rest. Meanwhile the atria develop which also contract at a more rapid rate. The atria control the rate of contraction of the heart as a whole. These contractions set the contained blood in motion. Eventually pacemaker or sino-auricular node develops which takes the controls of the contractility of the heart as a whole.

DEVELOPMENT OF KIDNEY

The kidney of vertebrates essentially consists of an aggregation of uriniferous units called *nephrons*. The kidney develops from the 'intermediate mesoderm' which lies between the somite and the lateral plate mesoderm. The intermediate mesoderm becomes segmented and each segment is called the *nephrotome*. The nephrotome is transformed into the nephrons which involves significant cellular events.

Mode of origin

A nephrotome contains a coelomic space, called the *nephrocoel* which communicates into the adjacent splanchnocoel by the *peritoneal funnel* (Fig. 5.36). The nephrotome is converted into a nephron in the following ways:

(*i*) The nephrotome, prior to its transformation, is a strand of cells between the somite and lateral plate mesoderm.

(*ii*) A tubular outgrowth develops from the dorsolateral wall of nephrotome.

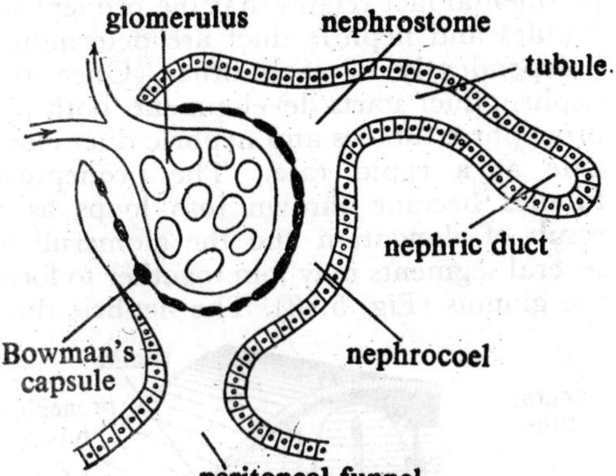

Fig. 5.36. Schematic diagram of a typical nephron.

(*iii*) The principle tubule originates from the tubular outgrowth which communicates with the nephrocoel through *nephrostome*, i.e. the cavity of tubule is actually an extension of the nephrocoel.

(*iv*) The median wall of the nephrotome invests a tuft of blood vessels (arterial capillaries) to form the renal corpuscle.

(*v*) The actual mode of origin of renal corpuscle is controversial. It was believed that Bowman's capsule is formed by a process of invagination of the glomerular mass into the wall of the nephrotome. But the electron-microscopic studies of Kurz (1958) have established that the double-walled Bowman's capsule is not formed as a result of invagination, but due to a cleft within a compact cellular mass. The inner layer becomes reflected over the glomerulus while the outer one forms the capsular wall.

The basic pattern of the development of nephron becomes greatly modified in different vertebrates. The deviation is due to: (*i*) Typically hollow nephrotomes are not found in embryos of higher vertebrate forms, instead the tubules develop within a continuous nephrogenic cord without exhibiting segmental disposition. (*ii*) All the nephrons do not differentiate at a time, rather, the nephrons appear in a sequential order from the anterior to the posterior end. (*iii*) The structural organisation of the nephrons also shows gradual complexity progressively from the anterior to the posterior end.

Developmental events of nephrons in vertebrates

In primitive vertebrates, the distinction between the anterior and posterior nephrons is not well marked but in amniotes (reptiles, birds and mammals) the development of nephric system shows the manifestation of three distinct entities which succeed each other during ontogenic development. The entities are: Pronephros, Mesonephros and Metanephros. The fishes and amphibians possess first pronephros which gives way to the mesonephros—the final kidney of an adult. In amniotes, besides these two units, a third entity, the metanephros arises as the definitive adult kidney. All the types of nephrons exhibit a striking similarity in their cellular transport mechanism and physiological performances.

Pronephros. The segmental origin of the pronephric tubules is the characteristic feature in nephric development. In amphibians, the pronephric tubules are developed from the nephrotomes beneath third and fourth somites in salamanders and second, third and fourth somites in frogs. It is to be noted that the number of the pronephric tubules corresponds directly to the number of segments involved. In chick, the pronephros develops from nephrotomes between fifth to sixteen somites. The pronephric tubules begin to form when the embryo attains 12 to 13 pairs of somites (40–45 hours of incubation). The tubules become well developed in 16–21 somites stage. At 35-somite stage and at about 65–70 hours of incubation, the pronephros undergoes degeneration. The sequence of events of transformation of the nephrotomes into the pronephric tubules is clear in lower vertebrates, but in higher tetrapods the stages are not so clear. When several pairs of pronephric tubules are developed, they open into the coelomic cavity proximally while the distal ends join the pronephric duct (Fig. 5.37). The nephric duct is called the pronephric duct which not only serves as the drainage channel for the pronephros, but becomes involved with the development of mesonephros. Typical pronephros is a functional kidney in the larval stages of fishes and

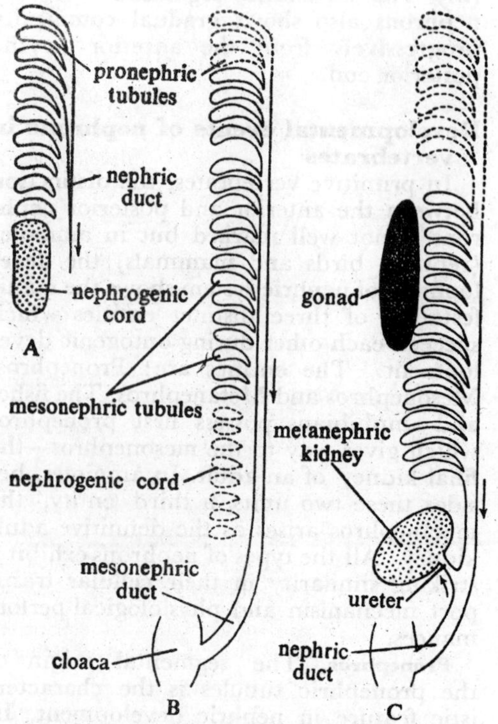

Fig. 5.37. Diagrammatic representation of the development of nephric system (after Torrey). A. Mode of origin of nephric duct. B and C. Showing the relationship between nephric duct and mesonephros and metanephros.

amphibians. But in the embryos of reptiles, birds and mammals, the pronephros develops in the anterior nephrotomes and is not functional at any stage. In human embryo, about seven pairs of pronephric tubules develop which start degeneration immediately after the initiation of the nephric duct.

Independency in the differentiation of pronephric tubules and nephric duct. The nephric duct (pronephric duct) starts development from the mesodermal blocks situated more posterior to the segment from which pronephric tubules begin to form. In amphibians, the nephric duct originates from the nephrogenous mesoderm behind that which provides the pronephric tubules. The somite 5 usually marks the level of the nephric duct primordium in amphibians. O'Connor (1938) has applied vital stain to pronephric swellings below the third and fourth somites in *Ambystoma*. It was observed that the stain appeared only in the pronephric tubules. When the stain was applied below the fifth and seventh somites, the stain became confined to the nephric duct. Holtfreter (1943) has bisected the embryo between the levels of fourth and fifth somites and has observed that in the hind piece, though devoid of pronephric tubules, the nephric duct still develops perfectly. The above experimental fact relates that the pronephric tubules and nephric duct are determined independently of each other. Once the nephric duct starts development, both the pronephric tubules and nephric duct elongate at a rapid rate. The pronephric tubules become thrown into loops as a result of elongation and the glomeruli of several segments may join together to form the glomus (Fig. 5.38). The nephric duct

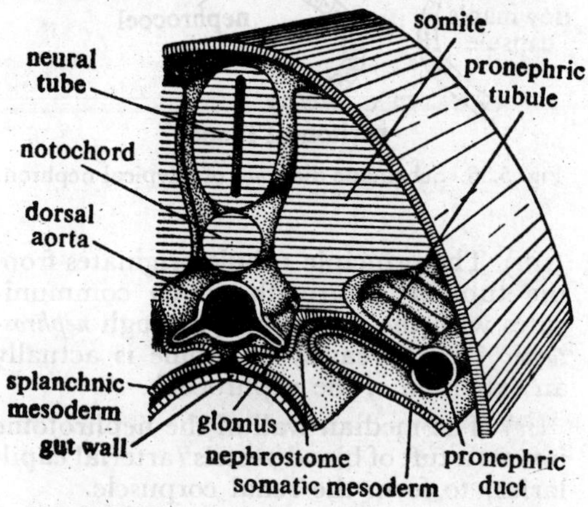

Fig. 5.38. Showing the disposition of pronephros (after Huettner).

(now designated as pronephric duct) after inauguration, pushes itself backward along the lower ends of the somites and the posterior movement is stopped as it reaches the cloaca. The duct fuses with the wall of the cloaca and its lumen opens into the cloacal cavity.

The backward elongation of the pronephric duct towards the cloaca is possibly due to either by (*i*) progressive addition of new material or (*ii*) due to free terminal growth. Extensive literature exists on this particular issue. Overton (1959) advocated that the duct increases by independent caudal growth. Holtfreter opined that the growth of pronephric duct towards the cloaca is due to selective cell-adhesions rather than chemotaxis as advanced by many. Holtfreter also suggested the role of blood vessel during the process, but this issue remains open for further investigation.

Mesonephros. The mesonephros is derived from the nephrotomes posterior to the pronephros. In majority of amphibians and amniotes, the component mesodermal cells of the nephrotomes dissolve into an aggregation of mesenchyme. These aggregated cells stretch on each side of the body along the dorsal margin of the lateral plates. This mass of mesenchymal cells is called the *nephrogenic cord* or *nephrogenous tissue*. The mesonephric tubules develop from the nephrogenic cord extending between 17 to 30 somites in chick embryo.

The proliferation of mesenchymal cells leads to the formation of elongated solid cords. Each such cord elongates and assumes 'S'-shaped appearance. It becomes hollow to form a cavity. One end of such a tubule connects itself to the existing pronephric duct (the mesonephric tubules do not form a duct of their own), while the proximal end forms a double-walled Bowman's capsule. The pronephric duct is now designated as the *mesonephric* or *Wolffian duct* because it serves as a drainage duct for the mesonephros. The Bowman's capsule is supplied by small blood vessels from the dorsal aorta (Fig. 5.39). Several

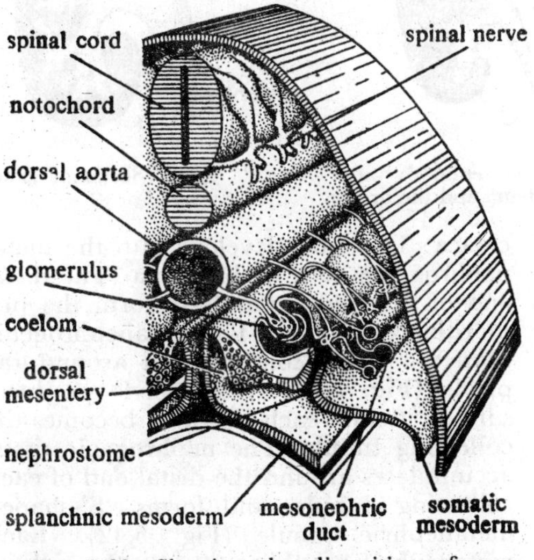

Fig. 5.39. Showing the disposition of mesonephros (after Huettner).

mesonephric tubules are developed in a segment, i.e. the number of mesonephric tubules do not correspond to the number of somites involved in nephrogenesis. When first formed, one mesonephric tubule develops in a segment, but subsequently each tubule gives origin to secondary and tertiary tubules by budding (Fig. 5.40). In

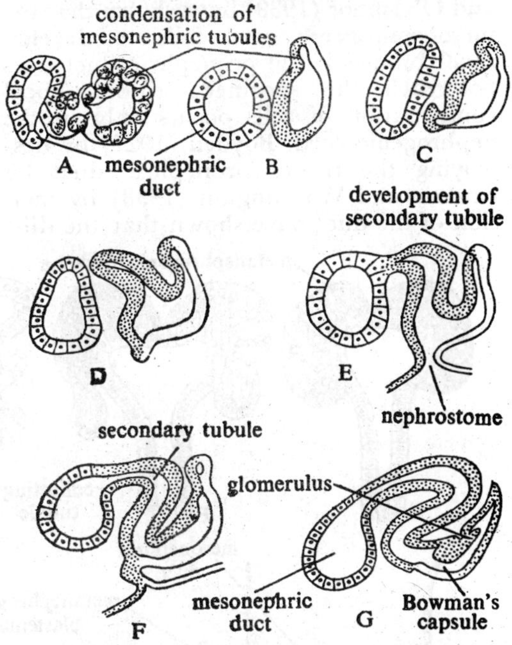

Fig. 5.40. Stages of formation of mesonephric renal unit in frog. Note the origin of secondary tubule.

case of chick embryo, the mesonephros becomes functional from 5th to 11th days. The tubular system becomes extensively coiled in 8th to 10th days of incubation. After this period the mesonephric tubules start degeneration along the anteroposterior direction and their function is taken over by metanephros which differentiates subsequently in the region posterior to that of mesonephros.

Role of pronephric duct in mesonephros differentiation. The mesonephrogenic cord will start differentiation into mesonephric tubules as soon as the pronephric duct is in touch with it. The mesonephrogenic cord develops mesonephric tubules only if stimulated by the pronephric duct. From this observation it is natural to think that the pronephric duct serves as an *inductor* for the differentiation of the mesonephros. Extensive experimentations have been done on this issue to ascertain the inductive role of pronephric duct in mesonephros differentiation.

Humphrey (1928), Burns (1938) and Holtfreter (1944) have experimentally obstructed the backward extension of the primordial nephric duct and have found the formation of mere clump of cells in the mesonephrogenic cord. Waddington (1938) and O'Connor (1939) have shown that the mesonephrogenic cord fails to develop renal tubules if the pronephric duct does not reach the specified region. Local condensation of cells occurs only in the nephrogenic cord. Boyden (1927) by destroying the tip of pronephric duct by cautery and Waddington (1938) by incision of the duct have shown that the differentiation of mesonephrogenic tissue into the mesonephric tubules occurs only when the pronephric duct makes contact with the tissue. But Gruenwald (1942) and Calame (1962) have cast doubt on such induction and reported that the mesonephrogenic cord is capable of a considerable degree of self-differentiation. Gruenwald (1942) and van Geertruyden (1946) have shown that the nervous tissue, when transplanted into the competent mesonephrogenic cells, can induce mesonephros differentiation. So it is not unreasonable to think that other tissues (possibly somites) are also involved in this process. So the generalisation that differentiation of mesonephric tubules depends solely upon induction by the pronephric duct appears premature to accept.

Metanephros. The metanephros is the functional kidney in the postembryonic life of amniotes. It develops from the nephrogenic cord which is derived from the nephrotomes posterior to the mesonephros adjacent to the cloaca. In chick embryo, the metanephros begins its development at the end of the 4th day of incubation and between the 31–33 somites.

In the embryos of amniotes, the ureteric bud emerges as a diverticulum from the mesonephric duct near its junction with cloaca. This bud develops into the metanephrogenic cord or blastema. The distal end of the bud expands to form the primordial renal pelvis. The metanephrogenic blastema starts condensation around the pelvis (Fig. 5.41). The pelvis produces subdiverticula, each of which becomes the collecting tubule. The nephrogenic tissue accumulates around the distal end of each collecting tubule and forms 'S'-shaped metanephric tubule (Fig. 5.42). Each metanephric tubule opens into the collecting tubule at one end and the other end forms a double-walled Bowman's capsule. The metanephros uses the mesonephric duct for the elimination of urine, but the

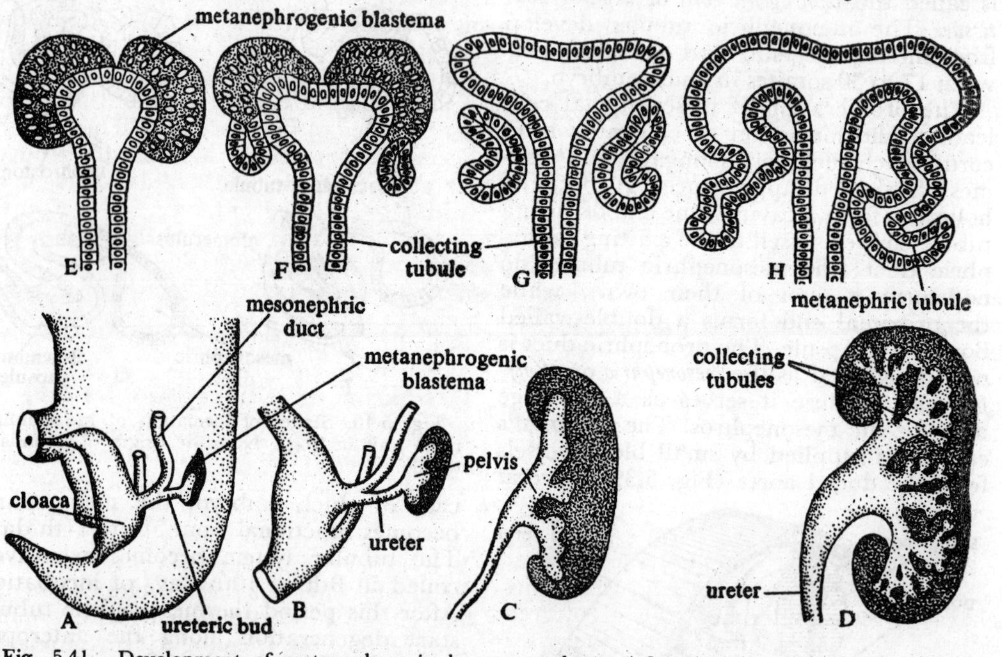

Fig. 5.41. Development of metanephros in human embryos (after Arey). A–F. Sectional views showing the relationship of the metanephrogenic blastema and the ureter.

connection between them is not direct, and is established by means of a special outgrowth, the ureteric bud which transforms into the renal pelvis and the ureter.

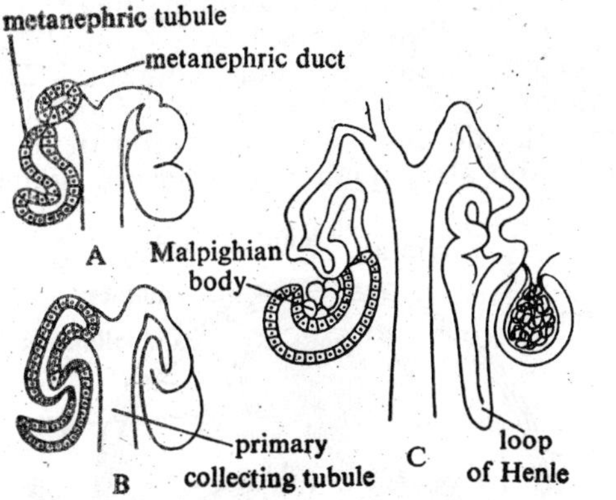

Fig. 5.42. Development of metanephric kidney tubule. A. Condensation of rudiment of renal tubule. B. Renal tubule unites with collecting tubule. C. Subsequent differentiation of renal unit.

Role of ureteric bud in metanephros differentiation. The conversion of the metanephrogenic tissue into metanephric tubules is dependent on an induction from the ureteric bud developing from the mesonephric duct. Because the extirpation of either mesonephric duct or ureteric bud causes the failure of the formation of metanephros. This phenomenon suggests the inductive phenomenon (Boyden, 1927 and Gruenwald, 1937).

It has been experimentally tested that the metanephric primordium undergoes characteristic development when cultured *in vitro*. The pelvis component develops a system of collecting tubules while the blastema forms coiled tubules. When these two components, after separation with trypsin, are cultured independently, neither of them is able to carry through characteristic morphogenesis. The subdivision of the renal pelvis is dependent upon the metanephrogenic blastema, while the tubule differentiation in the blastema rests upon an inductive stimulus from the ureteric bud. So the existence of inductive role played by ureteric bud in metanephric differentiation seems to be positive.

DEVELOPMENT OF EXTRAEMBRYONIC STRUCTURES IN CHICK EMBRYO

The embryo of chick possesses four extraembryonic or foetal membranes: namely, the *yolk sac*, the *allantois*, the *amnion* and the *serosa* or *chorion*. In amphibian embryo, the yolk sac and the allantois are present in rudimentary condition. The amnion and the serosa are developed in the reptilian embryo for the first time in evolutionary history of the vertebrates. In birds these two structures are retained while in mammals these are also present in a modified form. All the extraembryonic membranes are discarded at hatching while the yolk sac is incorporated into the small intestine.

As development goes on, the closely set ectoderm and somatopleure (somatic mesoderm) as well as the endoderm and splanchnopleure (splanchnic mesoderm) extend into the extraembryonic area. The developing embryo becomes located at the central area of the blastodisc. The embryo becomes separated by undercutting grooves formed by limiting body folds. These folds initiate with the formation of crescentic head fold which extends backward as the body folds. The body folds subsequently merge with the tail fold. By this way the embryo undercuts and separates itself from the underlying yolk mass.

Formation of yolk sac and its fate

The yolk, though supplies nourishment to the embryo, is not considered as part of the embryo. But in cases where the quantity of yolk is very large, special yolk sac develops to store the yolk. As the embryo raises itself, the yolk is enclosed by the yolk sac. The yolk sac develops from the margin of the blastoderm which advances around the yolk mass until it surrounds the mass completely.

In chick embryo, the gut is represented by a round cavity beneath the primitive streak at about sixteen hours of incubation. At about 24 hours, a small enteric pocket develops into the developing head as the foregut. The foregut is surrounded by mesoderm. At about 48-hour stage, with the formation of tail bud, a sac-like hindgut develops. The undifferentiated portion of the gut between the fore- and hindgut is called the midgut which remains as a small portion of the small intestine as the embryo grows older. The midgut leads into the reduced yolk sac by yolk stalk (Fig. 5.43 A_1). The yolk sac becomes smaller due to consumption of the yolk and before hatching the yolk sac exists as a projection from the small intestine. The yolk sac is covered by splanchnopleure. The yolk sac is finally incorporated with the small intestine.

The yolk is used as food by the embryo. The endoderm of the yolk sac secretes enzymes which break the yolk into diffusible substances. These substances are carried by the vitelline veins and finally to the heart. From the heart these are conveyed to the actually a membrane covering the embryo and thus housing it like a bag. It separates the embryo from the immediate contact with the environment. Between the embryo and the amnion, there is a space called the amniotic cavity containing amniotic fluid, a saline solution. The amnion as well as the amniotic fluid protect the developing embryo from desiccation and also equalise the pressure against

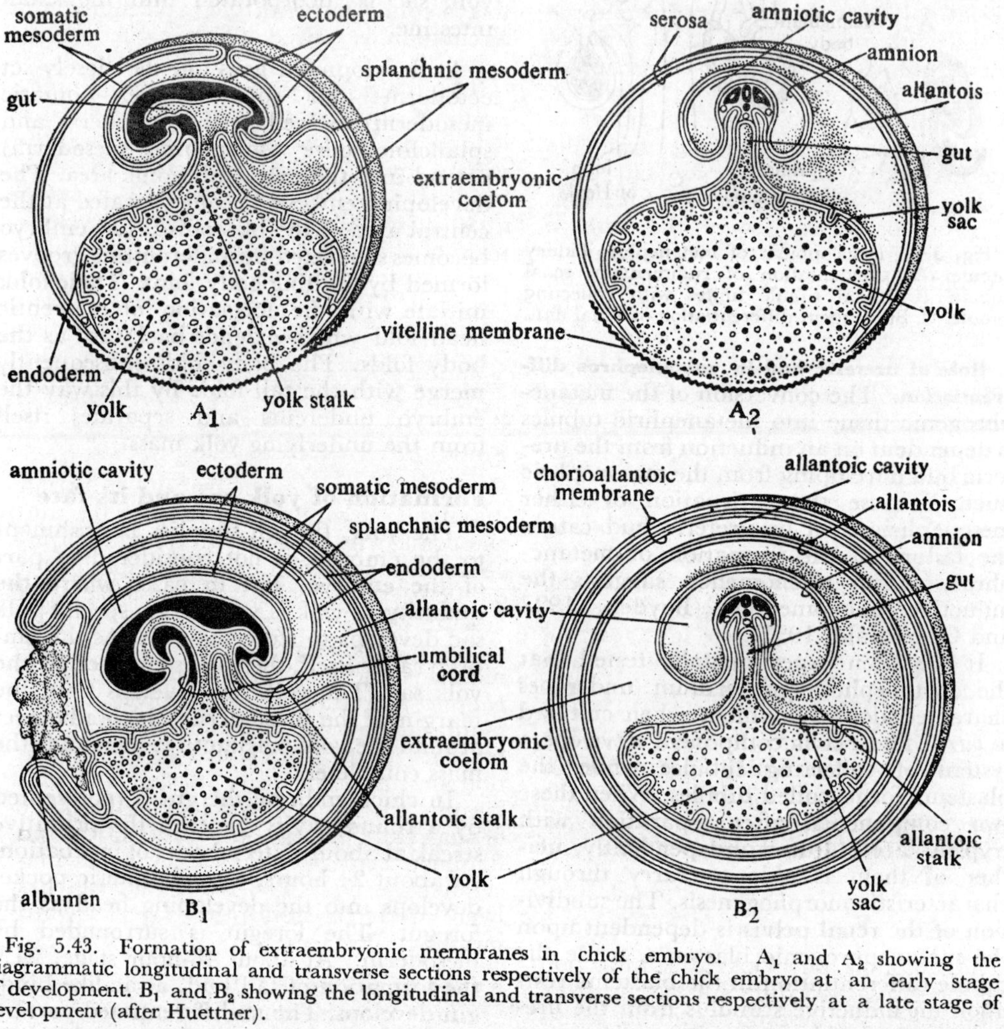

Fig. 5.43. Formation of extraembryonic membranes in chick embryo. A_1 and A_2 showing the diagrammatic longitudinal and transverse sections respectively of the chick embryo at an early stage of development. B_1 and B_2 showing the longitudinal and transverse sections respectively at a late stage of development (after Huettner).

different parts of the embryo and extraembryonic structures.

Formation of amnion and serosa

The amnion and the serosa (or chorion) are the two extraembryonic membranes which develop together. The amnion is the embryo by physical forces. The fluid acts as a buffer and gives protection from any kind of shock. Adhesion of the embryo to the amniotic wall is prevented by the rhythmic movement of the amnion.

As the embryo grows, the weight of the body is enhanced and as a result the embryo sinks into the soft yolk. Due to this, the blastodisc remains elevated over the embryo. At 30 hours of incubation, a well-formed head of the embryo of chick pushes forward and sinks into the underlying yolk. This causes the elevation of a blastodermic fold called the *amniotic head fold*. This fold bends backward over the head and covers it as a hood. The lateral ends of the head fold prolong backward along the two sides of the embryo as *lateral amniotic folds*. The lateral folds come over the embryo and fuse with each other along the antero-posterior direction. As a consequence of fusion, the embryo becomes more or less covered by folds. After 48 hours of incubation another fold appears from the posterior end of the embryo (tail bud) which is called the *amniotic tail fold*. This fold proceeds forward over the embryo and finally meets and fuses with amniotic head fold during the 4th day of incubation (Fig. 5.43 A_1). The region of union of the amniotic folds is marked by a scar called the *seroamniotic connection*.

The amniotic fold is composed of two layers—an inner and an outer. So the fusion of such folds will result in the formation of an inner and an outer layer. The inner layer is named as the *amnion* and the outer one is called the *serosa* (the name *chorion* is reserved from the same layer in mammals by many embryologists). The amnion is composed of somatic mesoderm on the outer side and ectoderm on the inner side. The serosa has the somatic mesoderm on the inner side and the ectoderm on the outerside (Fig. 5.43 A_2). In between the serosa and the amnion lies the *extraembryonic coelom* which is continuous with the *intraembryonic coelom* of the embryo proper. The serosa grows around the yolk sac and encloses it completely at the end of second week.

Formation of allantois

The allantois develops in conjunction with the serosa and differs in nature from the other extraembryonic membranes. It is not discarded in mammals but its intra-embryonic portion is retained as the urinary bladder.

The allantois starts its development as a ventral diverticulum in the 72-hour old chick embryo from the hindgut. This outgrowth consists of an inner endoderm with a layer of splanchnic mesoderm on the outer side. The diverticulum grows rapidly to invade the extraembryonic coelom (Fig. 5.43 B_1, B_2) and into the space between the yolk sac, the amnion and the serosa. Although the distal part of the allantois expands considerably, it remains connected with the hindgut by a narrow *allantoic stalk*. As the embryo separates itself from the extraembryonic parts, the allantoic stalk is enclosed together with the yolk stalk to form the *umbilical cord*. The allantois is connected with the extraembryonic circulation by paired allantoic arteries and veins.

The allantois is partly excretory in function and also acts as a reservoir by storing uric acid. It also serves as a respiratory surface by supplying oxygen to the embryo. It also helps to absorb a large quantity of albumen. With the development of allantoic circulation a considerable portion of calcium is absorbed from the egg shell. This absorbed calcium is utilised by the embryo for the formation of bones. The egg shell by losing calcium becomes thin and delicate and thus facilitates hatching.

FORMATION OF PLACENTA

The embryo, specially in eutherian mammals, becomes implanted to the uterine wall. The process of implantation involves tissue interaction and establishment of connection between the uterine wall and the extraembryonic membranes. The region of attachment between the embryonic tissue and the uterine wall is called the *placenta* and the process involved in implantation is called the *placentation*. The placenta is usually defined as an apposition or fusion between uterine and embryonic tissues for physiological exchange of materials. Human placenta is a round flattened mass from which the name placenta is derived. The name placenta has been derived from the Greek word meaning a flat cake.

Types of placenta

The placenta is divided into several types:

A. Depending on the involvement of embryonic tissue:

(*a*) *Yolk-sac placenta*. When the midgut extension of the splanchnopleure enclosing the yolk fuses with the extraembryonic somatopleure to make embryonic contact with the uterine wall. Examples: *Mustelus*.

(*b*) *Chorio-allantoic placenta*. The allantoic evagination of the hindgut unites with the extraembryonic somatopleure to make contact with the uterine tissue. *Examples*: Eutherian mammals and a lizard, *Chalcides*.

B. Depending on the distribution of villi:

(*a*) *Diffused placenta*. The villi are numerous and distributed uniformly over the whole of chorion. *Examples*: Ungulates, Cetacea.

(*b*) *Cotyledonary placenta*. The villi become aggregated in special regions to form small tufts. *Examples*: Ruminants.

(*c*) *Zonary placenta*. The villi are confined to an annular zone on the chorion. *Examples*: Carnivora (Pinnipedia).

(*d*) *Discoidal placenta*. The villi become restricted to a discoidal area as seen in rodents and insectivores. In apes and man, the placenta is of metadiscoidal type.

C. Based on the relationship of villi with the uterine wall:

(*a*) *Deciduate placenta*. The villi become intimately connected with the mucous membrane of the uterine wall which comes out with the embryo at the time of birth.

(*b*) *Indeciduate or Adeciduate placenta*. The villi are loosely united with the uterine walls which separate from the uterus at birth.

D. Based on the degree of involvement of foetal and maternal tissues:

(*a*) *Epitheliochorial placenta*. The epithelium of uterus remains in simple apposition with the chorion of the embryo. *Examples*: Pig and Horse.

(*b*) ***Syndesmochorial placenta*. The epithelium of the uterus disappears and the chorion comes in direct contact either with the glandular epithelium or endometrium of the uterus. *Example*: Sheep.**

(*c*) ***Vasochorial or Endotheliochorial placenta*. Both the glandular epithelium and the endometrium disappear and the chorion comes in close contact with the endothelium of the uterine capillaries. *Examples*: Dogs and Cats.**

(*d*) ***Haemochorial placenta*. The glandular epithelium, endometrium and endothelium of the capillaries disappear and the chorion is bathed with circulating maternal blood. *Example*: Man.**

(*e*) ***Haemoendothelial placenta*. Like that of haemochorial type of placenta, the glandular epithelium, endometrium and the endothelium of the maternal blood capillaries disappear. With the disappearance of these maternal structures, the tropoblastic epithelium (outer layer of the blastocyst) of the foetus also disappears, as a result the foetal endothelium separates the maternal and foetal circulating blood stream. *Examples*: Many rodents.**

Organogenesis of human placenta

In human females, implantation of the developing embryo occurs in the early luteal phase when the endometrium of the uterus remains in optimum condition. The developing egg reaches the uterus in blastocyst condition with a greatly enlarged blastocoelic space. The placental organogenesis is described under two broad aspects:

I. Previllous period (6th–13th day).

II. Villous period (14th day to term).

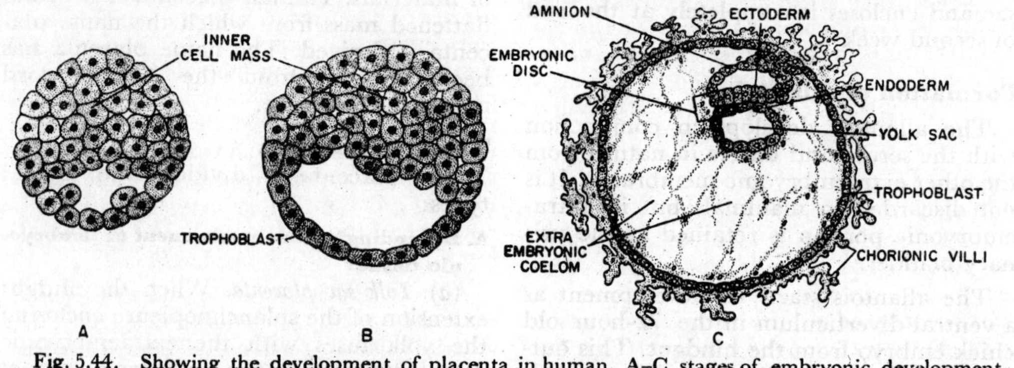

Fig. 5.44. Showing the development of placenta in human. A–C. stages of embryonic development during first week following fertilization (after Patten).

Previllous period:

The implantation of human embryo takes place about 6th to 9th days after fertilization (Figs. 5.44, 5.45). The end of the blastocyst containing the developing germinal disc attaches itself to the uterine wall (FIG. 5.46A). The uterine epithelium is eroded at the region of contact. The trophoblast tissue increases in thickness in this contact area due to the division of epithelial cells of the trophoblast layer. primary villi, when first formed, lack mesodermal core. The mesoderm of the somatopleure invade into them to form the *secondary villi* (Fig. 5.46D). With the formation of the blood islands and appearance of blood vessels, the secondary villi transform into the definitive *tertiary villi*. At this time, some endometrial tissue including blood vessels near to the invading chorionic vesicle break down to produce liquefied areas called the *embryotroph*. The liquefied

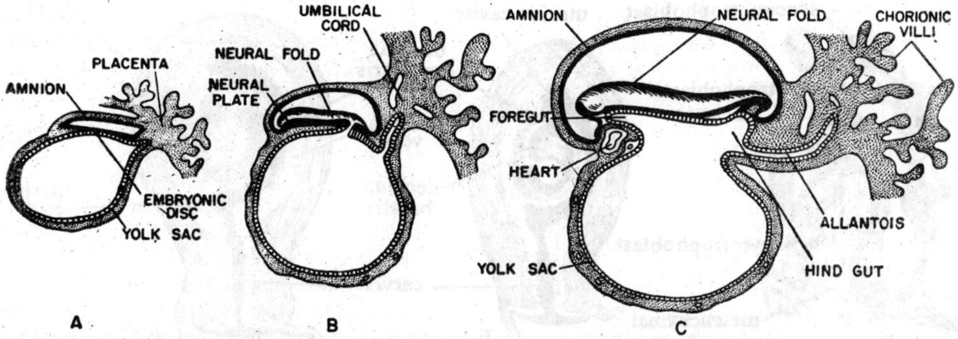

Fig. 5.45. Showing the development of placenta in human (contd.). A–C. Stages of embryonic development during the third week after fertilization (after Arey).

Prelacunar stage. The implanted embryo, consisting of a bilaminar disc, is protruded into a cavity (lecithocoel). This cavity is enclosed by the trophoblast. Profound cytological changes occur in the trophoblast layer. The inner trophoblast cells remain cellular and are designated as the *cytotrophoblast* while the outer cells fuse together to form a syncytium called the *syncytiotrophoblast*. The syncytiotrophoblast serves as the invading tissue of the embryo into the uterine wall. The growth of this syncytiotrophoblast is caused by differentiation of the cytotrophoblast and by amitotic division of the syncytial nuclei.

Lacunar stage (10th to 13th days of placental development). As the syncytiotrophoblast invades and increases in quantity, irregular spaces are produced in the syncytiotrophoblast. These spaces are called *trophoblastic lacunae* (Fig. 5.46B).

Villous period:

During the period between 14 and 18 days, the trophoblastic lacunae merge with one another to form large cavities bordered by syncytiotrophoblast. Such a cavity is named as the intervillous space and the *primary villi* are formed through the proliferation of the cytotrophoblastic elements into the syncytial trabeculae. The

material from the embryotroph is assimilated by the syncytiotrophoblast for the growth of the embryo. This particular type of nutrition is called the *histotrophic nutrition*.

From the physiological point of view the foetal villi developing from chorionic plate can be divided into three categories: (i) *Chorionic villi*, through which physiological exchange of materials taking place between the foetus and the mother. (ii) *Anchoring villi* for mechanical anchorage of foetus and (iii) *Free villi*.

The developing chorionic vesicle grows and invades the endometrium of the uterus. The uterine mucosa extends over the invading vesicle. The endometrial tissue covering the chorionic vesicle is called the *decidua capsularis* while the endometrial portion which is not concerned with the closure of such vesicle is called the *decidua parietalis* or *decidua vera*. The endometrial portion lying between the musculature of the uterine wall and the invading villi is called *decidua basalis*.

Although the chorionic villi are formed over the entire chorionic vesicle, only the villi in relation to decidua basalis are retained and those grown to decidua parietalis are reabsorbed to form a smooth area

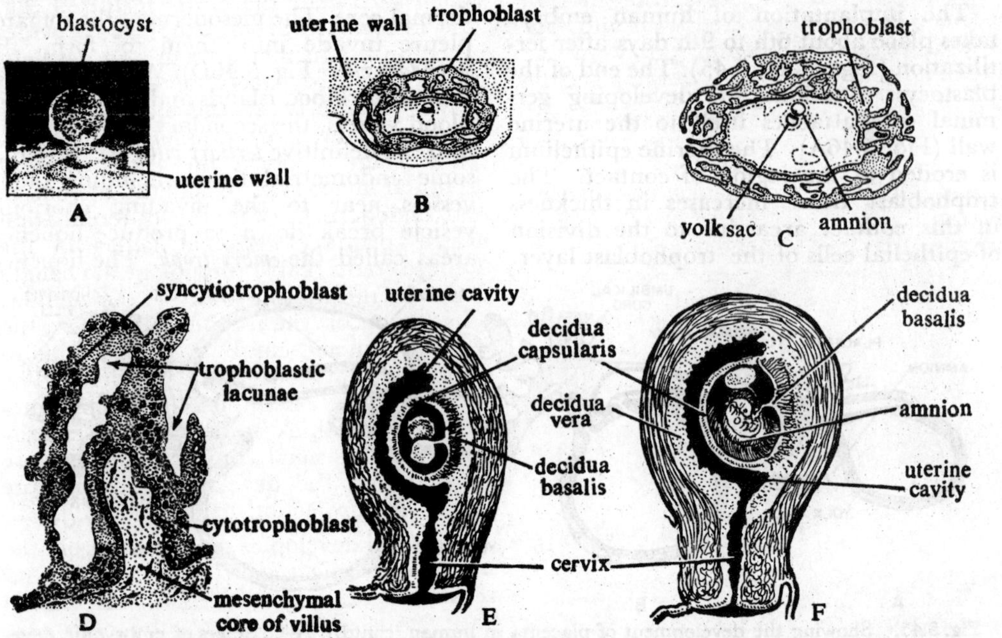

Fig. 5.46. Showing the development of placenta in human (contd.). A. Surface view of the implantation site of a human blastocyst. B. Sectional view of blastocyst of about 12 days which is engulfed completely inside the endometrium. C. Sectional view of embryonic vesicle of about 13–15 days. Note the presence of trophoblastic lacunae. D. Formation of secondary villus. E. Sectional view of uterus along with the developing embryo of 4 weeks old. F. Sectional view of six weeks old embryo. Note the disappearance of villi at one side and elaboration of villi on chorion frondosum (after various sources).

named as the *chorion laeve*. The villi within decidua basalis become greatly enlarged to serve the main role of physiological interchange (Fig. 5.46E). This area of chorionic vesicle with the villi is the *chorion frondosum* which together with the tissue of decidua basalis forms the actual *placenta*. Thus the placenta is composed of decidua basalis (maternal placenta) and chorion frondosum (foetal placenta) (Fig. 5.46F).

The villi, at the early phase of development, consist of blood capillaries within mesodermal core covered over by cytotrophoblast and syncytiotrophoblast on the outer side. As development goes on the blood capillaries grow enormously and the cytotrophoblast layer is extremely reduced to few scattered cells below syncytiotrophoblast.

The villi are aggregated into groups called the cotyledons which are separated by incomplete *placental septa*. The villi in each cotyledon remain surrounded by a pool of maternal blood and by this way a hemochorial type of placenta is established.

Organogenesis of placenta in rabbit

In rabbit, as in other eutherian mammals, placentation involves an interaction and attachment between the uterine wall and the extraembryonic membranes (chorion). The initial contact of the embryonic tissues with the uterine wall forms an *epitheliochorial relationship* in rabbit. Subsequently in course of development, it changes into a hemochorial condition in which the endothelial tissue of the uterine blood vessels is destroyed by the erosive action of the foetal tissue. As a consequence, the chorionic epithelium of the embryonic part of the placenta comes in direct contact with the maternal blood. During the later stages of pregnancy, even the chorionic epithelium also disappears thus leaving the endothelial lining of the foetal blood vessels in contact with the maternal blood. This type of placental relationship is called the *hemoendothelial* type (Fig. 5.47) and is regarded to be the most intimate grade of placental contact in animal kingdom. Depending upon the distribution of villi, the placenta in rabbit is categorized as the *discoidal* type. At the onset, the chorion becomes uniformly covered with villi, but at late stage the villi remain only on one side. The villi continue to develop on the

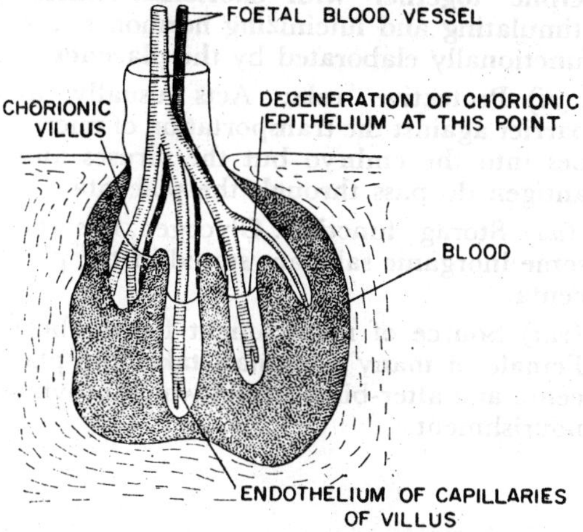

Fig. 5.47. Showing the hemoendothelial type of placenta during late gestation period in rabbit. Note that the chorionic epithelium is eroded and the capillaries of the chorionic villi remain surrounded by maternal blood.

side which turned away from the restricted lumen of the uterus, while the villi on the other parts of villi become atrophied.

Events in placentation in rabbit

The egg of rabbit, after fertilization, comes to lie in one of the crypts of the uterine wall (Fig. 5.48). As soon as the

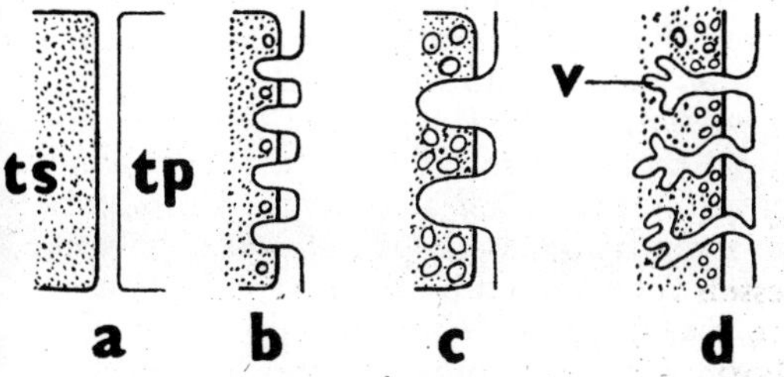

Fig. 5.48. Showing the gradual contact between the maternal and foetal tissues during placenta formation. a=two tissues come in close contact [dotted area=Trophospongia (ts) or uterine tissue and clean area=trophoblast or foetal tissue]. b=rupture of trophospongia (maternal tissue) and emergence of villi from trophoblast (tp). c–d=formation of villi (v) and gradual penetration into the trophospongia. Note vascularisation in trophospongia shown by small round circles in trophospongia.

fertilization is completed the corona radiata (covering of the egg) breaks down to form a frothy substance (*histotroph*) which surrounds the entire fertilized egg. The histotroph, around the egg, provides nutrition for further growth and development. The egg starts development and grows vigorously at the expense of the histotroph and soon completes the gastrulation stage to form the three primary germinal layers (viz., ectoderm, mesoderm and endoderm). With the differentiation of the primary germinal layers in the embryo, remarkable physiological changes occur in the uterine wall of the mother. The region of the uterine wall where the embryo touches, continues to bleed due to some special enzymatic action. The 'maternal site of implantation' becomes subsequently thick, spongy and vascularised. The superficial contact is made more intimate by the development of minute finger-like projections called the villi.

The villi develop as outgrowths from the extraembryonic region of the foetus and penetrate into depressions in the wall of the uterus. These outgrowths (villi) are initially formed by the trophoblast but the connective tissue and blood vessels enter into the outgrowths later on. They are designated as the 'chorionic villi' and the blood vessels are actually the ramifications from allantoic blood vessels. The region of the uterine wall which has passed the preparatory stage (vascularised) is called the *trophospongia*. The trophoblast and trophospongia come to establish intimate physiological relationship during placentation. The villi of the trophoblast penetrate into trophospongia and ramify extensively deep into the maternal tissues. During this action, the villi are constantly bathed by the maternal blood vessels because of excessive vascular state of trophospongia.

After coming into close physiological unition, the trophoblast and trophospongia form a nutritional bridge where the maternal blood will bring nutrient materials for the developing embryo throughout the entire period of gestation. In rabbit, the epithelium of the chorion may disappear in some regions during the later stages of gestation—thus permitting an exposure of foetal blood vessels to the maternal blood. The placenta is represented jointly by the trophoblast (chorion with its villi) and the trophospongia (uterine wall). At the time of parturition the chorionic villi are simply drawn out from the depressions in the uterine wall and the foetal tissues are separated without causing damage to the uterine wall and without causing any bleeding.

Functions of placenta

The functions of the placenta are manyfold. It serves:

(*i*) Adhesion or anchorage of the developing embryo with the uterine wall.

(*ii*) Nutritional role—the food materials from the maternal blood circulation reach the blood stream of the embryo for supplying nutrition.

(*iii*) Excretory role—the waste products from embryonic circulation are eliminated to the maternal blood stream.

(*iv*) Respiratory role—it serves as the external respiratory surface for the developing embryo.

(*v*) Elaborate endocrine functions: Two ovarian hormones—estrogen and progesterone together with chorionic follicle-stimulating and luteinizing hormones are functionally elaborated by the placenta.

(*vi*) Protective role: Acts usually as barrier against the transportation of microbes into the embryo but the viruses and antigen do pass through the placenta.

(*vii*) Storage function: Glycogen, fats and some inorganic salts are stored in the placenta.

(*viii*) Source of nourishment for mother: Female of many mammals takes the placenta and after-birth tissues which provide nourishment.

PART SIX

HISTORY OF ZOOLOGY

The main objective of Life Science is to study the form, the functioning and the diversity of life. But the evolution of Life Science itself is a fascinating story. It is a stupendous task to unravel this evolution. However, attempts have been made in the following pages to give a commentary on the march past of Life Science.

Biology now occupies a commanding position in the domain of science and with the march of time it is becoming a field of compelling interest for all intelligent people. The discoveries and conquests of Biology specially during the nineteenth and twentieth centuries have brought about revolution in the minds of all thinking men. The history of Biology is as interesting as the history and transformation of life itself.

BIOLOGY—ITS MEANING

Biology is one of the fundamental sciences and encompasses all questions relating to life in its different phases and manifestations. The structure, development and evolution of living organisms together with their physiology come under Biology.

Biology, in its true sense, did not or rather could not arise till the nature of the protoplasm was made known by Schultze in 1860. The growth of Biology is due mainly to the progress in the three fields namely, the knowledge about the architecture of living organisms, their development and their physiology. A parallel advance in the systematic classification of plants and animals, the doctrine of evolution and finally the knowledge about the intricacies of heredity have placed Biology on a solid platform.

MILESTONES IN ZOOLOGY

The humble beginning. From the very dawn of his appearance on earth, man has always been interested in animals. These interests were centered, chiefly, round his need to kill animals for food and also to escape from being killed. As a necessary corollary, primitive men had to know something about the habit and distribution of the animals round him. Thus the knowledge about the habit and distribution of animals constitutes the dawn of the Science of Zoology and at the same time it may perhaps be regarded as the oldest of all sciences. Cro-Magnon men had drawn many pictures of animals in the caves they used to live (Fig. 6.1). These pictures were

Fig. 6.1. Drawing of Animals by cave-dwellers.

that of Bison, Reindeer, hairy rhinoceros, mammoths, birds and fishes. The pictures are so vivid that it is possible to identify the animals up to the genus level. These drawings are certainly indicative of the types and distribution of animals of that age.

The next important record that is available to us dates from about 5000 years ago and that from the fertile crescent of the Tigris and Euphrates. That was the age of Barbarism. Records show that men in those ages were engaged in some agriculture and animal husbandry. Both these practices demanded some preliminary knowledge of Biology.

As time passed on, the civilization of Babylon arose in the same geographic area and with it written language came into existence. A written language provided an opportunity in which observations could be recorded and read. In 2000 B.C., the Babylonian ruler, **Hummurabi** made the existing laws, including medical laws codified. He arranged to record the codes on clay tablets and kept the tablets in a special building which undoubtedly formed the first library.

In close parallel to the civilization in the fertile crescent of the Tigris and Euphrates, the Egyptian civilization was emerging in the valley of the Nile. The ability of these people to make paper from the water plant, Papyrus, did a tremendous job to advance the recording of knowledge and, therefore, the advance of science. The Ebers Papyrus is an encyclopedia of medical information of that time and Smith Papyrus (both named after their discoverers) was a text book on surgery. The recent translation of these ancient documents shows that the treatises dated from the fifteenth century B.C. All these indicate to the fact that the science of medicine attained a wonderfully high grade of development amongst those people. As the formulation of a system of medicine in the early hours of mankind needs observation and practice of centuries, it is reasonable to believe that these manuscripts were the attempts at reducing medicine to a system. It is apparent that the system is built upon much scientific knowledge and must have been inspired by earlier writings on medicine and allied sciences.

The civilization of Egypt is divided into three periods. They are: (i) *Pre-dynastic period*, (ii) *First dynasty* and (ii) *New kingdom.* The New kingdom lasted till 600 B.C. Low relief carvings of animal figures of the pre-dynastic period have been discovered. During the first dynasty period step pyramids were erected. This period saw the evolution of a *Polytheistic*[1] *religion* in which the Pharaohs were gods and kings at the same time. At the fag end of this period there was a switch over to *Monotheistic*[2] *religion.* **Ihknaton** was the king then. He encouraged realism in arts and the picture of animals drawn in his regime are so accurate that the animals drawn could be identified up to the species level. All these advancements were, however, temporary. During the period of New-kingdom a reversion to old polytheism came and mythical god animals saturated the work of art. From this time onwards the civilization of Egypt underwent a decline.

Overlapping somewhat with the Egyptian civilization the Aegeans on the island of Crete, shores of Greece and parts of Asia minor adjacent to it arose. Picture of animals represented mostly by fishes and cattle of this age have been discovered. Pain killing drugs were known at this period. The serpent was considered as god. The Aegean civilization fell to invaders by 1100 B.C.

The Greeks. The Greeks were a collection of tribes of mixed origin that invaded the coasts and lands of the eastern part of the Mediterranean. They used to call themselves as Hellenes. They had settled there about the beginning of the second millennium before Christ. Before the coming of Greek-speaking people, the eastern coast of the Mediterranean was inhabited by people who were in a neolithic state of culture. They were either displaced or absorbed by the Greeks. The Myceneans and their colonial offshoot Minoans in the island of Crete were the first Greek tribes to attain civilized state. Frequent warfare between the Myceneans and Minoans resulted in the fall and decay of both the civilizations round about 1400 B.C. From the ruin left by these warfares, a distinct Greek civilization arose gradually.

The first Greek tribes who have left any literary remains were the Ionians and the Dorians. The Ionians had greater inclination towards philosophy, mathematics and astronomy. Theories concerning the structure of the universe and attempts to classify substances as elements were offered by *Thales*. The Dorians on the other hand were less civilized but more practical. In the island of Cos within the Dorian territory sprang up a medical school about 600 B.C. It is the earliest scientific institution that is known to us. Knowledge for its own sake was sought and as a result science started rising to new height.

The germ of evolution ideas had its inception in the minds of the Greek philosophers during this period. The thoughts were speculative and often coloured with mythology. One of these Greek philosophers was **Anaximander** whose adult life spanned the first half of the sixth century B.C. According to him men were first formed as fishes. Later on they cast off their fish skins and took up life on dry land. **Xenophanes** was a contemporary of Anaximander in part and lived on into the fifth century B.C. He was the first person to recognise fossils and to say that fossils represent the remains of animals that once lived. He also realised that the presence of fossils of marine animals on dry lands is suggestive of the fact that these fossil laden lands were once under the sea.

The centre of civilization gradually was shifted to southern Italy and Sicily, and this shifting became complete by 5th century B.C. **Pythagoras** migrated from Greece to Italy at about this time. He was a politician, religious leader and scientist and attempted at realistic interpretations of natural phenomenon. **Alcameon** was another man of letter of the same period who made dissections and performed many crude physiological experiments. Alcameon for the first time saw the nerves of the eye and the eustachian tube. At about the same time in the island of Cos there lived the physician of all times—**Hippocrates** (460–370 B.C.). The modern Hippocratic oath of doctors stems from his name. He learnt medicine on his native island of Cos. He practised and taught it in Cos, various other islands and on the main land of Greece. He was a man of noble character, dignified bearing and humane feeling.

A good number of books bear the name of Hippocrates but all of them were not certainly written by him. Some of the best and most interesting of them are of about his date. Probably these were his work. The books reveal a good idea of what he knew and how he worked.

These early work of medicine contain no information whatsoever of the knowledge that is now considered necessary for

1. Doctrine of many Gods.
2. Doctrine of one God.

a doctor. No anatomical, physiological or chemical knowledge is portrayed in these work. Instruments were unknown.

Hippocrates had to work with his own sense. The real scientific value of his work lies in the careful record of what was seen. His work not only recorded things seen but also tell us how these observations were made. These help us to follow the physician into the sickroom and watch him at work.

Hippocrates advocated the doctrine of four humours. The ancients supposed that all matters were made up of four essential elements. They are *earth*, *air*, *fire* and *water*. It was further supposed that the elements work either in opposition or in alliance with each other. Thus, water was opposed to fire but allied to earth. The oppositions and affinities were associated with the view that each element is compounded of a pair of *primary qualities* in the form of *heat* and *cold*, *moisture* and *dryness*. Hippocrates supposed that all living bodies are made of *four humours*. They are *sanguis* (blood), *cholera* (yellow bile), *melancholia* (black bile) and *Pituita* (Phlegm). These four humours bear special relationship with the four elements. Health depends upon the mixing of the humours in the right proportions. If one or the other is in excess the patient will suffer accordingly.

Though this belief is long abandoned still it finds a place in many expressions in current use.

Besides these, Hippocratic work contains some striking and short sayings called *Apromism*. Some of the Hippocratic apromisms which have passed into common speech and are eternal in use are given below:

(*i*) Art is long and life is short.

(*ii*) Desperate diseases need desperate remedies.

(*iii*) Sleeping too much is as bad as walking too much.

(*iv*) One man's meat is another man's poison.

The fifth century B.C. had also to donate some ideas about evolution. It saw the man **Empedocles** who has been hailed by Osborn (1896) as 'the father of evolution idea'. According to this philosopher, plants were the first to rise out of the earth, as did, subsequently, animals. Unattached organs and parts of animal body were formed first and then these organs and parts started combining with each other in haphazard manner. Most of these combinations became monsters or freaks and could not survive. But occasionally there had been happy combination of organs, i.e., right organ in the right place. Such 'fit' combinations survived and populated the earth. It is interesting to note that the idea of fit combination was reflected in the Darwinian idea of survival of the fittest after twenty-three century.

Another scientist of the Greek age was **Socrates** (469–399 B.C.). Socrates laid a good foundation for the development of future science but unfortunately he was against experimentation and was a believer of supernaturalism. **Plato** was another great philosopher of the Greek age. Plato had a strong bias towards mathematics and according to him the world and nature were mere illusions. The Institution where Plato taught was called *Academy*. The meaning of the word was simply the name of a groove of trees. The word is much in use even today to indicate a place of learning.

The greatest of all scientists amongst the Greeks was **Aristotle** (284–322 B.C.). He was the son of the physician to the royal family of Phillip II of Macedonia and was trained in the Academy of Plato. Plato influenced Aristotle a great deal, but in certain philosophical matters there was much difference between them. Aristotle was essentially a biologist but remained attached to his teacher till Plato's death in 347 B.C. Aristotle felt disappointed when he was not made the head of Plato's academy and established a rival academy which he called *Lyceum*. Here his bias towards biology and natural history had full play. In addition to major contributions to the other sciences, he spent ten years observing marine life and wrote several books on Zoology. He devised the first system on taxonomy. He was the teacher of Alexander the Great.

After the Peloponnesian wars[1] there came a decline of Greek civilization and was replaced by the Macedonian Empire centered in Alexandria on the north-coast of Africa. Two important investigators are worth mentioning of this era. One was **Herophilus** (280 B.C.) who is considered

1. A Great War towards the end of 5th century B.C. between Sparta, Peloponnesian confederacy and Athens.

as the 'father of medical Anatomy'. He established the difference between artery, vein and nerves. He also recognised the brain as being the organ of thinking. Following him came up **Erasistratus** who is considered as 'the father of medical physiology'. He could distinguish between sensory and motor nerves and did formulate the primitive ideas about the function of the blood vessels.

The Romans. The Athenian scientific school virtually came to an end after the death of Alexander the Great in 323 B.C. His empire broke into fragments to be seized by his generals. Meanwhile, the Roman civilization began its rise about 500 B.C. and became the dominant centre of the civilized world within two hundred years. Alexandria at this time became the centre of scientific world. A definite 'Alexandrian period' came up in the history of science. It covers the last three hundred years before the Christian era. The biological work of the Alexandrian school have perished save and except fragmentary writings of two biologists of that age, i.e., Herophilus and Erasistratus. As both of them were Greeks, they have been discussed earlier under the Greeks.

During the early phase of Roman civilization there had been little advance in Zoology. **Varro** during this period expressed the belief that diseases are caused by organisms which cannot be seen—a concept that was proved to be true two thousand years later by Pasteur.

In the later period of Roman civilization there came **Pliny** (Pliny the Elder), the Roman naturalist. He was a man of an entirely different character and outlook. He was a man of immense industry with an enthusiasm for collection. He put together a vast number of extracts known in his time about nature and compiled the book *Historia Naturalis*. The book consisted of tales of wonder, of travellers and sailors yarn and superstition. Thus there was more misinformation than information.

After Pliny the most important biologist investigator in antiquity was **Galen** (A.D. 130–200). He was a physician. He made frequent dissection of animals. He examined the structure of sheep, oxen, pigs, dogs, bears and many other animals. He could recognise the resemblances between man and monkey. He gave good descriptions of muscles, bones and joints. Galen did a good deal of work on the function of organs. In his role as a physiologist he investigated the functions of spinal cord and also the purposes served by air breathing. He believed that air enters into heart directly from the lungs. He believed in one God and developed a personal philosophy with the idea that every organ in the human body was created by God in perfect form. This idea fitted in well with the Moslem and the Christian faith and as a result his ideas were incorporated into religious dogma. He commanded so much respect that no one dared to re-examine or question his work even 1500 years after his death.

The dark ages. After Galen, there was no expansion of scientific knowledge for many centuries. It is believed that the advent of Christianity is the root cause behind this dark phase. The early Christians were intolerant of the ideas of science that had been developed until this time. The scientists who developed them had been the religious enemies of the Christians and little of their culture was acceptable to the Christians. The early Christians believed that the return of Messiah[1] was in the offing and their attention was directed to this event. Science was of no interest to them.

When the knowledge and intelligence of men of the Western Europe were on the decline, the same were rising in Syria, Asia Minor and Constantinople. Greek scientific work was studied there.

In the seventh century the great movement of Islam came. It established Arabic as a literary language and by the ninth century the intellectual leadership went to the people of Arabic speech. It remained with them till the thirteenth century. In the hands of the Moslem physicians great advances were made in medicine, mathematics and other sciences. Their science, however, was based on translation of the Greek work into Arabic. These translations were made in large number and ardently studied throughout the Islamic countries. Islam was a rapidly conquering religion and soon extended at the point of sword from Near East to North Africa, Spain, Portugal, Southern Italy, Sicily and many Mediterranean islands. Thus when the Moslem civilization was dominating

1. A Hebrew title applied to a promised deliverer of the Jewish nation.

the stage, Europe was plunged into dark ages and the progress of European science came virtually to a standstill position. When the European people came to recognise their intellectual inferiority they sprang up to secure retranslation of their past scientific work. These were made from the Arabic to Latin with the help of the Jews. The heritage of science from the Greek and Roman periods would have been lost had it not been translated into Arabic.

The view that the rise of Christianity led to the destruction of science in Europe has been challenged recently. Two main reasons have been advocated against the view. The challenger historians advocate that science in Europe began its downward course long before the rise of Christianity. Moreover at the end of the second century of the Christian era, when ancient science almost decomposed, the Christian formed a too little and insignificant sect to influence the people who would study science. It cannot, however, be denied that too much spiritual interest of Christianity in the early centuries did nothing to promote science and on the contrary, discouraged any revival of science in the later era.

Another important factor, according to this, was the mental make-up of the ruling class in the Roman Empire. The Romans were very much practical people. They were engaged very much in governing their empire. They produced great military leaders, great lawyers alright but never cared or appreciated theoretical investigations. Science can be applied to meet the practical needs of life but mere application is never in itself a science. Science will fail to flourish until theoretical investigations are undertaken for their own sake.

The Roman military organisation had certainly medical services. This service was well and carefully organised. Thus far the Roman rulers showed their practical bend of minds. But these leaders never thought of getting their medical officers trained on scientific lines. They never established medical schools. Effective training in anatomy and physiology was totally lacking in the Roman Empire. Mere reading cannot sustain science. Science can only sustain by systematic observations and experiments. These two essentials of science were unheard of in the Roman regime.

The contempt for and aversion to science of the Roman leaders is significant because it teaches a lesson which is of value today and of value in all the departments of science.

The Renaissance. Dust for centuries over the scientific discourses started to be washed out with the advent of the spring of Renaissance. It began in the twelfth century and with it scientific investigations again began to progress in Europe. **Roger Bacon,** a monk, studied optics, the tides of the ocean and human anatomy. His religious affiliation prevented him from dissection of animals. **Copernicus** and **Galileo,** though non-biologists, were other two scientists of the early Renaissance period. The contributions made by them were indications of freedom of thought and at the same time the rising interest in science at that time.

The early part of Renaissance was followed by a period of unrest. The Magna Charta in 1215, the death of Joan of Arc in 1431 and the auto-da-fe of Gregory IX[1] are all indicative of the unrest.

The fourteenth century began with a period of travel. Many information depicting rare and strange things, as a result, came in from overseas. Trade with the East became escalated and drugs started to be imported. Trade and commerce brought with them travellers tales. Some of the tales were true while majority were false. Those of *Marco Polo* (1254–1324) are true but those associated with the name of *Mandeville* are false.

In the year 1453 Constantinople (Istanbul of modern times) was captured by Moslems. They cut off the trade route between Asia and Europe. The search for alternate route to the Orient was stimulated. Vasco-da-Gama, Diaz, Columbus and others went out in search of the routes. The exploration of the world soon reflected itself in biological literature. Voyages of exploration resulted in the recognition of many foreign plants and animals.

Renaissance period gave the world a number of scientists who contributed to Zoology in Europe. The most important of them was **Leonardo-da-Vinci.** He was a biologist of all times and enjoys the same fame in a number of other branches of science

1. Port. word for faith-Execution of this is by burning the Heratic.

as well as in arts. His knowledge of Anatomy and concept of Physiology was outstanding. Paraffin embedding of tissues and casting of body cavities were the techniques invented by him. These techniques have become extremely productive in the hands of later investigators.

Vesalius (1514–1564) was another important scientist of the Renaissance period. He threw a challange to Galen's anatomy and started dissecting animals. He, by publishing the first true atlas of anatomy has laid the foundation of the subject.

William Harvey in 1628 published a book on the function of heart and circulatory system to become the father of experimental physiology. Harvey's thought has undergone little change in modern physiology.

The advent of 'Encyclopedists' is another important feature of the Renaissance period. During the sixteenth century an acute need was felt for some orderly arrangement to the available knowledge in Biology and accordingly, a group of Encyclopedists arose. They attempted to collect all known facts about living things. **Conrad Gesner** (1516–1565), the Swiss naturalist is the best representative of this school.

The other important events of the Renaissance period (later period) are the establishment of scientific societies and advent of scientific journals. These led to exchange of scientific information and views amongst the scientists of the world.

The story of the science of Biology forms the beginning of a story of continuity. It began in an humble way and was all mingled up with medicine in the initial phase. The character of the progress of Biology shows that—(1) It was a crusade against superstition. (2) It was a battle to overthrow the authority. (3) It was a gradual switch over from descriptive to comparative and then to experimental. (4) Instrumentation has aided the progress to a great extent.

The germ of Biology was united with medicine for a long time. Its separation from medicine and its colossal rise as an independent subject is due mainly to the 'steady growth of that zest for exploration into unknown fields which began with the new birth of science in the sixteenth century'. The modern biology did not arise till the nature of the protoplasm was made known by **Schultze** in 1860. From that year onward the history of Zoology is necessarily the history of its individual disciplines. The growth of knowledge became so rapid and the number of significant contributions are so large that it is better to spell out the chronology of development of zoology in an abbreviated fashion.

CHRONOLOGICAL EVENTS IN THE PROGRESS OF BIOLOGY

A. Tools and Techniques

1590—Dutch spectacles makers Janssen and Janssen are credited with building first successful microscope.

1665—Robert Hooke improvised the microscopes (Fig. 6.2) and drew attention to the world of minute organisms.

Fig. 6.2. Microscope used by Hooke. Portion of feather on the right side.

1878—Brandt demonstrated that certain stains or dyes have the property of being incorporated in the living cells. He called them vital stains.

1895—Rontgen discovered the X-rays. Application of X-rays made advances in Physiology, Anatomy and Genetics.

1911—Carrel developed tissue culture technique by growing live tissue outside the body of the animal.

1912—Kite began microsurgical studies by devising micromanipulators.

1923—Hevesay utilised isotopes of chemical elements for tracing fate of molecules in living system.

1932—Construction of emission type of electron microscope by Knoll and Ruska.

1938—Svedberg developed ultra centrifuge.

B. Systematics

1665—Discovery of protozoa by Leeuwenhoek. It enlarged the concept of animal kingdom.

1693—Concept of genus and species on similarity of form was advocated by John Ray.

1758—Binomial system of nomenclature was introduced by Linnaeus.

1848—Status of protozoa and its unicellular nature was determined by Siebold.

1864—Modern zoological classification and development of basic divisions were introduced by Haeckel.

C. Cells and Tissues

1658—Description of cell (RBC of frog) was given by Swammerdam.

1665—Robert Hooke first used the term cell.

1800—Classification of tissue (21 different types) was made by Bichat.

1824—Cell division was first described by Prevost from cleavage division in frog's egg.

1831—Nucleus of plant cells was named and described by Robert Brown.

1835—Description of Sarcode (protoplasm) as the basis of life was made by Dujardin.

1838–39—Schwann and Schleiden advocated the cell doctrine.

1846—The name protoplasm was given by Hugo von Mohl.

1848—First description of chromosomes and drawings of pollen mother cell was done by Hofmeister.

1858—Doctrine of germinal continuity was expressed by Virchow in his motto—*Omnis cellula e cellula.*

1875—Mitosis was described by Stratsburger.

1882—Counts of chromosomes were done by Flemming.

1887—Meiosis was predicted by Weissmann.

1898—Mitochondria and Golgi apparatus was named by Benda and Golgi respectively.

1934—Isolation and analysis of mitochondria was done by Bensley and Hoerr.

1952—Detailed structure of mitochondria with the aid of electron microscope was done by Palade.

1953—DNA as a double helix was proved by Watson and Crick.

1953—Detailed structure of endoplasmic reticulum was given by Porter.

1956—Revision of the count of human chromosome, i.e., diploid–46 in place of 48 was done by Tijo and Levan.

D. Embryology

1651—Concept of continuity of life was given by Harvey—*Omne vivum ex ovo* (All life from egg).

1672—Description of ovarian follicle was made by Graff.

1745—Parthenogenesis was described by Bonnet.

1759—Theory of epigenesis was advocated by Wolff overthrowing the concept of preformation.

1817—Recognition of three germ layers in chick was made by Pander which later was extended by Von Baer to all vertebrates.

1827—Mammalian ovum was discovered by Von Baer.

1854—Description of entrance of sperm into ovum was given by Newport.

1875—Union of male and female nuclear material in fertilization was shown by O. Hertwig.

1885—Cleavage dividing the zygote of frog into cells was described by Roux.

1885—Germplasm theory was advocated by Weismann.

1897—Influence of the yolk present in an egg was demonstrated by O. Hertwig.

1897—Successful heteroplastic grafting was conducted by Born.

1900—Artificial fertilization was conducted by Loeb.

1911—Axial gradient theory was proposed by Child.

1914—Role of fertilizin in fertilization was established by Lillie.

1921—Organiser concept was introduced by Spemann.

1933—Embryological inductor concept was presented by Holtfreter.

1936—Artificial parthenogenesis in rabbit was done by Pincus.

E. Distribution

1858—Zoogeographical realms were recognised by Sclater.

1860—Line of faunal delimitation between the Oriental and Australian zoogeographical realm was made by Wallace.

1894—Life zone concept was advocated by Merriam.

1935—The concept of ecosystem was proposed by Tansley.

F. Palaeontology

1669—Fossils as organic remains was recognised by Stensen.

1791—Correlation between geological strata and fossils was made by Smith.

1796—Vertebrate palaeontology was started by Cuvier.

1830—Lyell's book 'The principle of Geology' brought relation between Zoology and Geology.

1923—Discovery of fossilized eggs of Dinosaurs from Mongolia was made by Chapman.

G. Genetics

1866—Mendel's laws of genetics came to light.

1866—Nuclear control of inheritance was foreshadowed by Haeckel.

1883—Chromosome control of inheritance was proposed by Roux without any experimental verification.

1900—Rediscovery of Mendel's laws was made by De Vries, Correns and Tshermark.

1902—Discovery of X-chromosome was made by McClung and its significance in sex-determination was hinted at.

1906—Linkage groups in sweet pea was discovered by Bateson and Punnet.

1909—Origin of terms—gene, genotype and phenotype was made by Johannsen.

1909—Crossing over as a genetic process was observed by Janssens. Twenty years later Morgan proposed the theory of crossing over.

1910–1920—Morgan established the gene theory of inheritance.

1910—Discovery of sex-linked inheritance was made by Morgan.

1913—First chromosome map was made by Sturtevant.

1925—Sex-determination by environmental factor was demonstrated by Baltzar in *Bonellia*.

1927—Artificial increase in mutation rate was demonstrated by Muller in *Drosophila*.

1931—Exchange of genes during crossing over was demonstrated by Stern.

1933—Giant salivary gland chromosomes were rediscovered in *Drosophila* by Painter and others.

1935—Target theory of mechanism of gene mutation was established by Ressovsky.

1941—Biochemical mutation was produced in Neurospora by Beadle and Tatum. They were awarded Nobel Prize in 1958.

1943—Plasmagene inheritance in Paramoecium was demonstrated by Sonneborn.

1949—Genetic control of protein structure in animals was shown by Pauling.

H. Evolution

1693—Concept of a species was provided by John Ray.

1809—Theory of use and disuse by Lamarck.

1830—Biogenetic law of Von Baer.

1859—Theory of natural selection of Darwin.

1860—Spontaneous generation was disproved by Pasteur.

1874—Gastrea theory of metazoan ancestry and recapitulation theory of Haeckel.

1901—Mutation theory of De Vries.

1932—Effect of genetic drift on small population was outlined by Swell Wright.

1933—Heterotroph theory of origin of life of Haldane and Oparin.

1953—Laboratory demonstration of formation of organic substances by Urey and Miller.

Some outstanding makers of Life Science

ARISTOTLE (384–322 B.C.). Aristotle was born in 384 B.C. at Stagira. Stagira was a small town lay far away from Athens which was the centre of learning.

Aristotle's father was a physician to the King of Macedon. In all probabilities Aristotle got his taste for biological investigation from his father. At the age of seventeen, Aristotle was sent to Athens to become a student of the great philosopher—Plato. Plato influenced his pupil Aristotle a great deal. But they had profound differences between themselves. Plato had a strong bias for mathematics while Aristotle was essentially a biologist.

Aristotle (Fig. 6.3) expected to become the head of the academy after the death of his teacher, Plato in 347 B.C. He was not made the head and in disappointment he left Athens to reside in a small Greek state on the coast of Asia Minor opposite to the island of Lesbos. Here his bias for natural history was in the full cry. He had ample leisure and he spent these time making investigations on *marine life*.

King Philip of Macedon invited him to become the tutor of Alexander. He accepted the invitation and came to Macedon in the year 342 B.C. and stayed there for seven years teaching Alexander. King

Philip was assassinated in that year and Alexander became the King. Aristotle now left Macedon and came back to Athens.

Fig. 6.3. Aristotle.

Aristotle never returned to Plato's Academy but set up his own institution called *Lyceum*. He soon gathered round him an able band of students. He taught there for thirteen years and died in the year 322 B.C.

Biological work of Aristotle

Of the biological work of Aristotle, four major treatises are surviving in a fairly complete form. They are:

1. On Psyche.
2. Histories about animals.
3. On the generation of animals.
4. On the parts of animals.

Aristotle was interested in watching animals and their ways of life. He used to keep notes of what he saw. From his notes the following remarkable accounts can be cited.

1. *On the habits of fish*—Aristotle gave a good account of the breeding habits of cat-fishes and pointed for the first time the ability of these fishes to emit an articulate sound. Of the marine fishes, *Torpedo* and Angler fishes drew his attention. According to him "In marine creatures one may observe many ingenious devices adapted to the circumstances of their lives." His accounts on Torpedo and Angler fish are perfectly true.

2. *On octopuses and their allies*—Aristotle has given a perfect and delightful description of the development of *Octopus* and *Sepia*.

3. *On whales, porpoises and dolphins*—Aristotle had a clear idea about the important anatomical, physiological and embryological differences between a man and fish. Thus he knew that r in has lungs, it is warm-blooded and brings forth its young alive, i.e., viviparity. These characters are absent in fishes. From these differences he could separate the group of mammals that includes whales, porpoises and dolphins from the fishes.

4. *On placental dog fishes*—Aristotle knew well that the fishes differ from mammals in being *oviparous*. But he could find out a group of fishes in which the youngs are born alive.

Aristotle tells us that a shark like fish called *Galeos* lays eggs within itself. These eggs during development acquire an attachment with the womb much like the placenta of mammal.

5. *As bee-master*—The diversities present in the methods of working and general habits of bees is usually printed in Aristotle's *Historia animalium.* Indications of observation and recording of the habits of animals are present here. But in all probabilities the habits of bees were not written by Aristotle himself. It might so happen that a practical bee-master under the influence of Aristotle has written the part.

6. *Aristotelian classification*—Aristotle never made a formal classification. His work contain many terms. Though scattered, if these terms are arranged sequentially, it certainly will speak of a classification scheme or arrangement which was in Aristotle's mind. The scheme is given below.

I. Enaima = Vertebrates

(With red blood and either viviparous or oviparous)

Viviparous:

(*a*) Man.
(*b*) Cetaceans.
(*c*) Viviparous quadrupeds:
  (*i*) Ruminants with cutting teeth on the lower jaw, hoofs cloven—sheep, oxen, etc.

(*ii*) Solid hoofed animals—Horses, asses, etc.
(*iii*) Other viviparous quadrupeds.

Oviparous though at times externally viviparous:

Eggs perfect:
(*a*) Birds.
(*b*) Amphibians.
(*c*) Reptiles.
(*d*) Serpents.

Eggs imperfect:
Fishes:
(*i*) Cartilagenous except *Galeos*.
(*ii*) Other fishes.

II. Anaima=Invertebrates

(No red blood and viviparous, vermiparous, budding or spontaneous)

With perfect eggs:
(*a*) Cephalopods.
(*b*) Crustaceans.

With eggs of special type:
(*c*) Insects, spiders, scorpions.

With generative slime, buds or spontaneous generation:
(*d*) Molluscs (except cephalopods).

With spontaneous generation:
(*e*) Sponges, coelenterates.

7. *Aristotle on the nature of life*—Aristotle was not only a practical biologist but also a theoretician. The most fundamental division among biologists has been between the schools known as *Vitalists* and *Mechanists*. According to vitalists living things are endowed with some principles of a quite peculiar nature and they are quite different from non-living matters. This principle of the living objects not only exists but also is expressed in some manner. The mechanists on the other hand denied the truth of both these beliefs of the vitalists and claimed that all actions, movements, etc., of living things can be expressed as the outcome of forces and can be investigated in non-living matters. Aristotle belonged to the vitalist group.

Aristotle embarked upon the word *Psyche* to indicate *soul*. According to him there are vegetative soul and animal soul. The first or lowest was the vegetative soul and as such plants were regarded by him as the lowest living form. The qualities of life essential for this lowest form of soul were nutrition, growth and the power of reproduction. The animals had all these qualities plus the power of movements. Animals thus possessed, he thought, a second order of soul that includes sensitiveness and motive powers.

8. *Aristotle on evolution*—He maintained that there is a gradation in nature. The lowest grade is *inorganic*, from which arose the higher grade of *organic* by a process of metamorphosis. He conceived that organic world exists in three states—Plants, Plant-animals and Animals. Plant-animals are transitional groups and include sponge and sea anemone. Aristotle failed altogether to appreciate the true significance of fossils. His towering intellect was a failure here.

Aristotle's role in the development of science

Aristotle lived at a time when science was at its infancy and errors and crudities were the rules. So the measurement of the height of Aristotelian achievements in the yard stick of present day standard will not be justified. He had extensive knowledge of animals and specially the life histories of animals. He was the founder of classificatory schemes though it was incidental. He was extremely interested in the problems of how creatures came into existence and he can very well be called the "father of embryology". His observations of the structure, development and his anticipation of the idea of organic evolution, are the ones on which his fame rests. The work of Aristotle place him as one amongst the greatest biologists. He was centuries ahead of his time. He was an original thinker and had an unrivalled capacity for arranging his own and others materials.

Last but not the least is that his influence guided biologists to the right direction.

PLINY THE ELDER (23–78 A.D.). Pliny was a Roman naturalist of the first century A.D. Pliny was a well-born gentleman and was an able and efficient civil servant. By profession he was a General.

Pliny (Fig. 6.4) was a man of immense industry and had a special aptitude for collection. These collections were not of natural history objects but of information about them. He was a good collector—"collector of antedotes—who forsaking observation indiscriminately, mixed fable, fact and fancy taken from the writing of others". The book he had compiled was known as "Natural History". It was written in thirty seven volumes and has been preserved with greater completeness.

His influences. Pliny was more an illusion and his work did nothing to extend the horizon of knowledge. The misinformation compiled by him retrograded the

Fig. 6.4. Pliny.

advancement of science as his work of low quality was widely read during the ages which followed. He recasts Aristotle's plan of classification and formulated a higher artificial one in its place. His scheme was based on incidental circumstances of abode of animals, i.e., air, water and land.

GALEN (130–200 A.D.). After Pliny there remained only one biological investigator in antiquity. He was Galen.

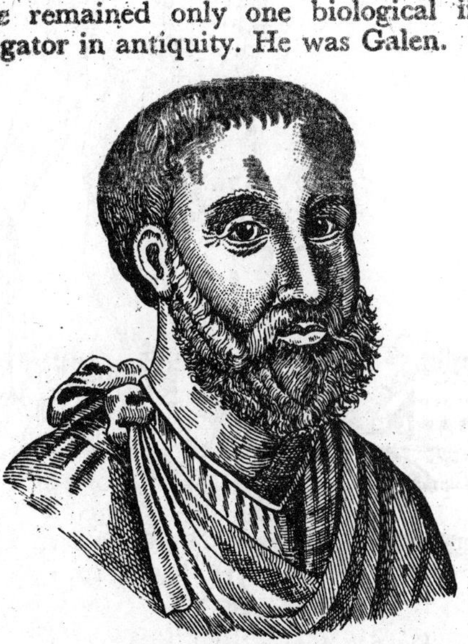

Fig. 6.5. Galen.

Galen (Fig. 6.5) was born in Pergamon in Asia Minor. Pergamon and Alexandria were the two important but rival centres of learning. At the beginning, Galen was a student of philosophy, but at the age of sixteen he switched over to medicine. According to the custom of those days Galen went travelling for studies at the age of twenty. He learned anatomy at Alexandria medical school and came back to his native land at the age of twenty-eight and remained there for four years. He worked as surgeon to the Gladiators during these years. Fig. 6.6 demonstrates the dissection of human body in

Fig. 6.6. The earliest known representation of dissected body of a female. Operator is guided by physician and monk. Kidney, head, stomach are thrown around. Liver is in the hand of the operator.

earlier days. After that he migrated to Rome to seek a career. At Rome he became the physician to the emperor. His fame and eminence as a good physician, anatomist and physiologist soon became established.

His contributions. Galen dissected animals like sheep, oxen, dogs, pigs, etc. He never dissected a human body at least publicly. He could recognise the resemblance between human anatomy and that of monkey.

Galen did a good deal of work on the functioning of organs. In his role as a physiologist he made investigations on the function of spinal cord and tried to find out the purposes behind breathing and heart movement. He, however, could not discover the function of blood and was of the opinion that air enters from lungs directly to heart.

Galen established himself as an author at the age of twenty and remained an industrious writer throughout his life. His work are enormous and cover all departments of medicine.

His influences. Galen had a tremendous influence over the men of his time and even for centuries his works have not been challenged. Thus the errors he made were passed on to the succeeding ages that followed him. The extraordinary respect that Galen commanded was due to his religion. He was a pagan and believed in one God. He was able to develop the idea that every organ in the human body was created by God in the most perfect possible form. This fitted in well with the concept of Christianity of his time. This explains why the name and work of Galen were held for long fifteen hundred years.

VESALIUS (1514–1564). Vesalius was one of the founders of biology who believed in observation and reason. He was born in Brussels on 31st December, 1514. His leaning towards scientific pursuit was

Fig. 6.7. Vesalius.

inherited from an ancestry of physicians and learned men. From his early life he had a passion for anatomy. After his early training at Brussels and at the University of Louvain he went to Paris at the age of eighteen to study medicine. He was there for three years and when wars broke out in Belgium he had to go back to Louvain without the medical degree. After earning some experience as surgeon on the field of battle he went to Padua being lured by the reports that he will have the opportunity of making practical dissections there. His talents were recognised at Padua. After receiving the degree of Doctor of Medicine in 1537 he joined the university as a surgeon.

His contributions. Vesalius (Fig. 6.7), as a great teacher of anatomy used to teach in his own new way. He never allowed unskilled and ignorant barbers to make dissection of human for demonstration before his students. He used to do such dissections by his own hands.

His book on the "Fabrics of Human Body" is a biological masterpiece. The book opens with a description of bones and joints, the general classification of which forms Galen's work. The skull has been considered as the first bone. Skulls have been systematically measured and individuals and races have been classified as broad headed, round headed and long headed. Such classifications are of much importance to Anthropologists. A comparison between the skull of man and dog is also found in his book.

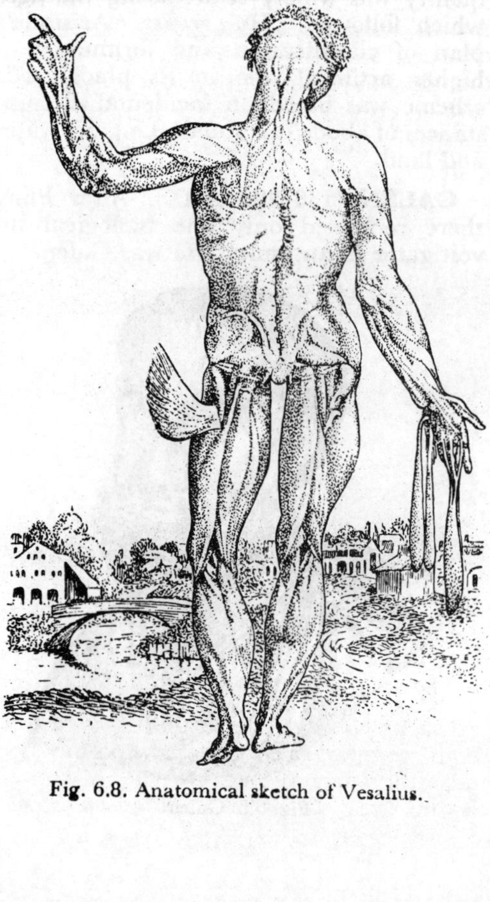

Fig. 6.8. Anatomical sketch of Vesalius.

Vesalius was most successful with his work on muscles. He was successful in describing and drawing these muscles with much accuracy. Fig. 6.8. portrays the anatomical sketch of man.

He gave a fair description of the structure of heart but his account of the septum between the ventricles was not correct.

His influences. Vesalius was the first man to do battle against tradition. He believed in Galen rather than his own eyes but gradually lost his faith in Galen and at the end he casts Galen and his writings into the wind. Thus at the age of thirty the idol of authority was made shattered by him.

He used to teach only what he himself had seen and what he could make his students see. He brought into anatomy the new wave of the time and the men of the time specially the young answered the new wave.

His great book with its illustrations has certainly laid the foundation of modern biological sciences.

The importance of Vesalius lies in the fact that he overthrew the adherence to authority as a method for ascertaining the truth and introduced observation and reason in science. He saw new truth with clearness and had the bravery to force his thoughts on an unsympathetic public.

HARVEY (1578–1657). As the sixteenth century was closing, there came William Harvey (Fig. 6.9) an Englishman

Fig. 6.9. Harvey.

with a new and hitherto unknown vision about comparative anatomy, physiology and embryology. He was born at Folkstone, on the south coast of England in 1578. He belonged to a family of wealthy yeoman. He had his schooling in the King's School at Canterbury and at the age of sixteen he entered Cains College. He was endowed with a power for observation upon organisation of animals but could not cultivate it at this university. Here he studied Latin and Greek to become a graduate with the Arts degree.

In 1598 he came to the great University of Padua in search of best medical instruction. He was fortunate in receiving his instruction in Anatomy and Physiology from Fabricius. Fabricius was the most learned and honoured teacher in Italy during that time. A fast friendship grew between the young medical student and the mature anatomists. Harvey had his first familiarity with the circulatory system under the stimulus from his teacher. The knowledge of circulatory system was further amplified and geared up by Harvey.

Harvey was a student of originality and force and after graduating in Arts, studied for four or five years in scientific and medical lines to obtain the degree of Doctor of Physics. He came back to England in 1602 and obtained the degree of M.D. from Cambridge. In 1615 he was called to lecture on anatomy at the Royal College of Physicians in London. He was also appointed physician to King James I. He continued, however, to lecture regularly at the Royal College.

His contributions. Harvey's book on the circulation of blood or "Anatomical Dissertation concerning the motion of the heart and blood" is still considered as one of the greatest monuments along the high-road of biology. The first edition of the book came out in 1628. It contained 72 pages in miserably printed little Latin quarto. The action of heart in forty species including worms, insects, crustaceans and fishes was recorded in it. In his book he showed for the first time ever in print, that all the blood in the body moves in a circuit and that the heart beat is the propelling force behind circulation. His demonstration of the movement of blood in a circuit was based on careful observations, experiments and cogent reasonings. The central point of Harvey's reasoning was that the quantity of blood which leaves the left cavity of heart in a given span of time comes back again to heart after some lapse (30 minutes or less) of time.

Thus Harvey's idea of movement of

heart was new, his notion about circulation was new and his method of demonstration was new. His lectures at the Royal College embodied the most revolutionary doctrine ever delivered.

It is said that he worked also on comparative anatomy and the development of insects. The manuscripts and drawings of these work were lost due to war.

His influences. Vesalius undoubtedly revealed in a splendid way the construction of human body and Harvey went one step further by introducing experiment to determine the purpose of the structures. The work of Harvey are complemental to that of Vesalius. Their joint work directed and influenced the path of biological investigations in the right way. The versatility of Harvey is beyond any question. He was the first amongst the biologists to give a really reasonable explanation, in physical terms, of any bodily process. His investigations on the anatomy of about sixty animals and embryology of insects and vertebrates had a tremendous influence in promoting future biological investigations.

Harvey initiated a total change in the attitude towards living processes. Biological investigations before his time had been full of vague terms intended to explain living activities. Innate heat, animal spirit, pneumatic force, etc., are some of the vague vocabularies. Harvey's book rang the death knell of a whole vocabulary of such mysterious phrases.

CLASSICAL MICROSCOPISTS

In the last forty years of the seventeenth century there arose a great discovery. The invention of microscope is a crowning event in the history of biology. The introduction of microscope in biological observations diverted the attention of biologists from the plane gross anatomy to that of minute structure and thus the world to the unseen became opened up.

The scientists who made series of observations during this period have been termed classical microscopists. They are Robert Hooke and Grew of England, Leeuwenhoek and Swammerdam of Germany and Malpighi of Italy.

ROBERT HOOKE (1635–1703). Of all the classical microscopists, Robert Hooke was unquestionably the most brilliant. He was very sickly from his childhood and his ill health prevented him from receiving a normal education. He, however, received a good scientific training of the University of Cambridge. After the foundation of the Royal Society he entered its service as a salaried 'curator of instruments'.

His contributions. Hooke was primarily a physical experimenter and most of his best work lie outside the domain of biology. He gave his attention to microscopic studies for some time and published his work in his book '*Micrographia*' published in London in 1665. The book opens with a description and figure of the microscope he used. In his figure of the microscopic structure of the cork he has designated the walls as "cell". The word cell is still alive in modern biology. He has depicted a polyzoon for the first time. The markings on the scales of fishes, the structure of the sting of the bee, the radulae of molluscs, foot of the fly and structure of the feathers are some of the instances of his microscopic observations.

His influences. He lacked fixedness of purpose in the proper employment of his talents. His pioneering work in microscopy gave a powerful stimulus to microscopic studies in England and systematic studies with microscope became escalated.

LEEUWENHOEK (1632–1723). Leeuwenhoek was born in 1632 at Delft. He came of a wealthy brewers' family. His schooling came to an end at the age of sixteen and he never learnt any language but his own Dutch. He was sent to Amsterdam to join a clothing business as

Fig. 6.10. Leeuwenhoek.

book-keeper and cashier. But he never liked the appointment and came back to Delft where he owned a draper's shop and spent all his spare time on microscopy.

His contributions. Leeuwenhoek (Fig. 6.10)

used to send his communications to the Royal Society. These were published by the society in English or in Latin translations from the original Dutch. He turned his microscope in all directions—to the minerals as well as to the animal and vegetable kingdoms.

He observed the minute circulation of blood and demonstrated the presence of capillaries (Fig. 6.11) between arteries and

Fig. 6.11. Capillary circulation after Leeuwenhoek.

veins and thus completed the chain of observations which began with Harvey. He followed this by descriptions of blood corpuscles. His observations that blood corpuscles of fish and frog are oval and that of man is round are correct. He also made descriptions of the blood corpuscles of some invertebrates. He gave exellent account of muscles, the lens of eye, teeth, skin and many others.

Leeuwenhoek's work on the compound eyes of insects are remarkable. He showed that these lenses form numerous inverted images and the insects are endowed with the power of quickness of sight. Other worth mentioning observations of Leeuwenhoek are the development of ant, poison and spinning apparatus of spiders and the life history of Aphids. Parthenogenetical development of Aphids was recorded by him.

Leeuwenhoek focussed his attention to the world of the unseen too. His papers on minute fresh water creatures like Rotifers, Hydra and Volvox were the first of their types. His remarkable achievement is that he could catch a glimpse of bacteria.

His influences. Leeuwenhoek impressed his contemporaries most. He led an easy and prosperous life and showed an avidity amounting to passion towards the microscope. He gave good descriptions and drawings of his instruments. He was much balanced and had a good power for observation. But he was never a solver of problems.

SWAMMERDAM (1637–1680). Swammerdam (Fig. 6.12) was born in 1637. His father had a taste for collection and thus "from the earliest dawn of his understanding young Swammerdam was surrounded by zoological specimens". He thus became passionately devoted to the study of natural history.

Fig. 6.12. Swammerdam.

His father wanted that he should go to the church. But he had no taste for theology. At the age of twenty-six he entered the University of Leyden. He went to Paris and took his degree of Doctor of Medicine in 1667.

His contributions. His work were collected in a book entitled "Biblia Naturae", that was published fifty-seven years after his death. The book contains about a dozen of life histories of insects. Of these, the Mayfly (*Ephimera*) is the most famous and that one on Honey bee the most elaborate. The greater amount of his work deals with structural Entomology (Fig. 6.13). His other work include the fine anatomy of the snail, structure of the clam and squid, development of frog and observations on the contraction of muscles.

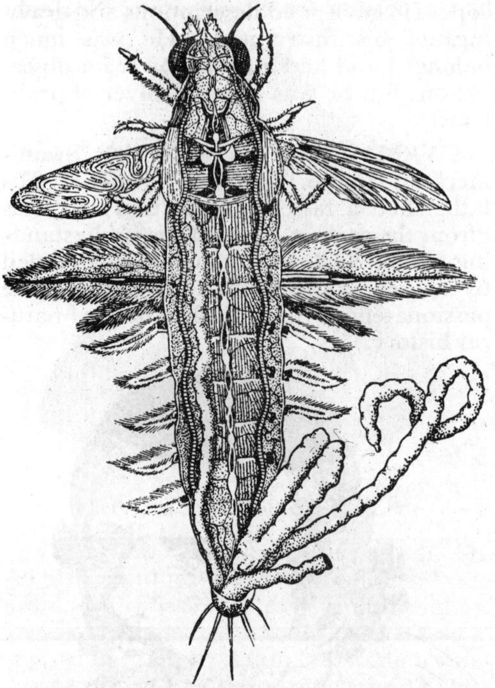

Fig. 6.13. Anatomy of an insect. Drawn by and dissected by Swammerdam.

His influences. After graduating in medicine, he never did practice but devoted his time in examining minute anatomy. He was a skilful and judicious experimenter. His work on respiratory processes and his demonstration that muscles though alter in form but do not alter in size when they contract was a new idea. He improved the technique of injection. His work did a lot to advance physiological knowledge.

MALPIGHI (1628–1694). Malpighi was born in 1628 near Bologona. He came of an affluent peasant family and under the patronage of his parents had a good schooling. He soon developed a taste for belles-letters and philosophy. At the age of twenty-three, for a profession, he started studying medicine and obtained the degree of Doctor of Medicine in 1653 from the University of Bologona. In 1656 he got a post there and started his career as a teacher and investigator. Later he joined the University of Pisa and after three years again came back to Bologona University.

His contributions. Malpighi (Fig. 6.14) remained always busy with experiments

Fig. 6.14. Malpighi.

and research. He made a number of important observations which advanced the knowledge of biology to a great extent.

He first demonstrated the structures of lungs and showed the presence of air cells in it. He presented a tolerable idea of how the air and blood are brought together in lungs.

He extended the work of Harvey and demonstrated the presence of capillary system in the lungs of toad. He also demonstrated the mucous layer or 'pigment layer of the skin'.

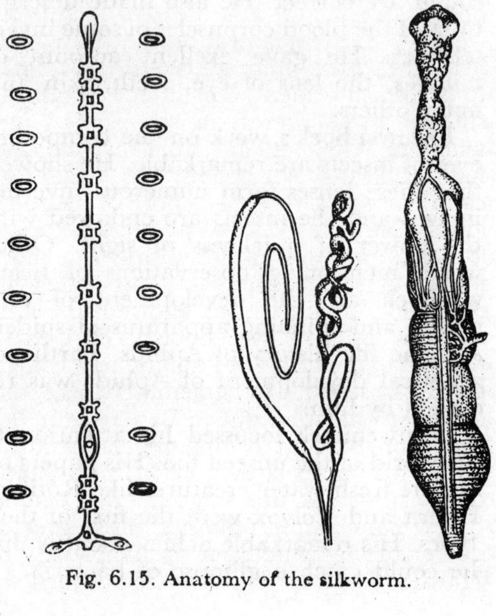

Fig. 6.15. Anatomy of the silkworm.

Malpighi extended the work of Fabricius on the development of chick. With the aid of his microscope he observed that during the early stages of development of the chick a series of vessels arises from the aorta as it leaves the heart. He was, however, unaware of the nature and meaning of these vessels.

The most striking Malpighi's research is the monograph on the anatomy of silk moth (Fig. 6.15). The monograph was the first of its kind in the history of Zoology. Malpighi was the first to observe the tracheae and spiracles of the insect body and could arrive at a correct conclusion about their function.

He also did many work on plant anatomy.

His influences. Malpighi influenced his contemporary world to a great extent. He was honoured at home and abroad. In 1667, the Royal Society requested him to send his communications to them. Most of his work were published in London under the auspices of the society. His Latin style was, however, not good and contained many vague and obscure theoretical explanations. His work show that he was neither apt at designing experiments nor was he a skilful interpreter of his results. As an observer, however, he has seldom been equalled.

LINNAEUS (1707–1778). The great Swedish naturalist, Linnaeus or Linné was born in Rashult in 1707. His father was a poor clergyman. His father's resources were limited and he was persuaded to send his son to Lund and afterwards to Upsala to become a student of medicine. Linnaeus showed little interest in stereotyped instructions. He spent his time in collecting natural history specimens. His father became very upset at the report of his son's low scholarship. He was about to arrange for an apprenticeship for Linné to a shoemaker. He, however, was prevented in doing so and Linné was allowed to continue. In 1732, Royal Society of Upsala wanted to send him to Lapland as a collector and observer. He immediately accepted the offer and went out to Lapland for the excursion without completing his studies. On returning to Upsala he fell into financial distress. At this stage he was helped by his fiancee and obtained the degree of doctor of medicine from the University of Hardewyk in 1735. On his return to Sweden he was appointed Professor of Natural History at Upsala.

His contributions. Linnaeus (Fig. 6.16) had a passion for classification. Not only

Fig. 6.16. Linnaeus.

did he classify plants and animals but also the minerals and even diseases. He introduced the system of *Binomial nomenclature* of plants and animals. He assigned to every known plant and animal a place in his system. This involved placing it first in a *class*, then in an *order*, then in a *genus*, then in a *species*. His great work are *Systema Naturae* (1735), *Genera plantarum* (1737), *Foundamenta Botanica* (1735) and *Bibliotheca Botanica* (1736).

Linnean system of classification contained the following classes—Mammals, Birds, Reptiles, Fishes, Insects and Vermes. The first four classes had already been grouped together by Aristotle. The remaining classes of Insects and Vermes contained all the groups of animals without vertebrae. Here he was behind Aristotle. Linnean arrangement of Animal Kingdom is given below.

Heart with 1 or 2 ventricles and 2 atria.

Blood red and warm:

1. Viviparous—Mammals.
2. Oviparous—Birds.

Heart with 1 ventricle and 1 or 2 atria.

Blood red and cold:

3. Breathing by lungs—Reptiles.
4. Breathing by gills—Fishes.

Heart with 1 ventricle and no atria.

Blood colourless and cold:

5. With antenna—Insects.
6. With tentacles—Vermes.

His influences. The work of Linnaeus marked a lasting influence on natural history. He brought into use the method of and means of *naming* animals and plants. The method is still in practice. The Systema Naturae is not merely a treatise on the structure of animals and plants, it is a catalogue of the productions of nature arranged most methodically.

The second important feature of his work is that they were done with much labour, extreme brevity and great clearness.

Despite his industry and force, Linnaeus gave natural history a one sided development and many disciplines of biosciences were given little attention. His students like their teachers became collectors and classifiers. "Thus in the zeal for naming and classifying, the higher goal of investigation, knowledge of the nature of animals and plants was lost sight of and the interest in anatomy, physiology and embryology lagged".

The general idea of Linnaeus concerning the nature of species was wrong. He held that species are constant and invariable. "There are as many species as they were created in the beginning" said Linnaeus. Such a concept contradicted many naturalists and consideration about the origin of species became burning question.

LYONET (1707–1789). Lyonet was born in the Hague in 1707. He had a good education and by profession he was a lawyer. His talents were great and varied. He was at the same time a painter, a sculpture, an engraver and a very gifted linguist with skill in eight languages.

Fig. 6.17. Lyonet.

His contributions. Lyonet (Fig. 6.17) had no training in anatomy but he had a flare for dissection. His interest in the subject was culminated in the publication of his treatise on minute anatomy of insects (Fig. 6.18). His monograph (1750) on the caterpillar of the goat moth displayed not only patience but great skill as dissector and sketch maker. He engraved his own figures on copper. He worked out in details the layers of muscles and could distinguish 4,041 separate muscles. As an artist he helped many naturalists by making drawings for them. He drew the figures for Lesser's *Theology of Insects* (1742) and for Trembley's famous treatise on *Hydra* (1744).

His influences. The extraordinary details shown in the drawings of Lyonet created a sensation. The existence of such complicated structures within the body of so small an insect was discredited. Some of

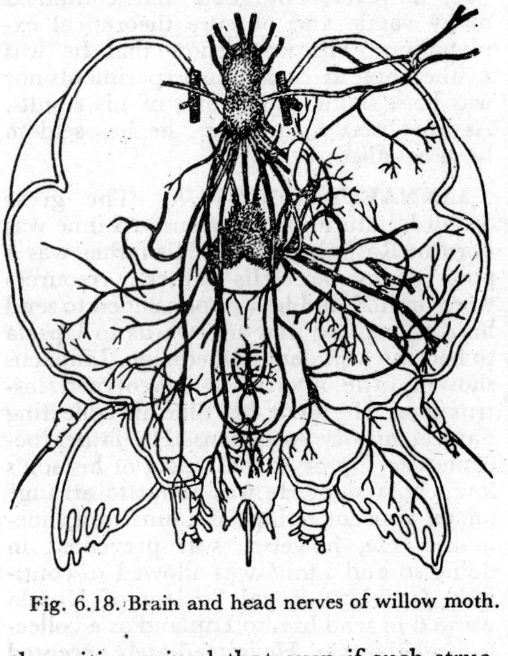

Fig. 6.18. Brain and head nerves of willow moth.

the critics opined that even if such structures exist they should be beyond the resolution power of human eye.

The work of Lyonet on minute anatomy might have induced other scientists following him to go for more minute things in the organisation of living system.

EHRENBERG (1795–1876). Ehrenberg (Fig. 6.19) was a German naturalist who held the chair of medicine in the University of Berlin. He was a great scientific traveller who attained great scientific height.

Fig. 6.19. Ehrenberg.

His contributions. He was one of the early observers of nerve fibres. His book "The Infusoria as Perfect Organisms" is a hall mark in the study of Protozoa. As Ehrenberg had a best microscope of his days at his disposal he could investigate upon minute animals. He made a faithful representation of the habit, feeding and discharging of undigested food of these protozoa. Ehrenberg was the first to demonstrate the existence of fossil forms of protozoa. His "Mikrogeologie" (1854) was for many years the standard reference work in this line. He also demonstrated that marine phosphorescence is due to the protozoa, *Noctiluca*.

His influences. Leeuwenhoek left little unnoticed of the microscopic world and it was he who made first records of unicellular animals. From his time onward these animals became favourite objects of microscopic studies. The first standard work in this line was done by Müller in 1786. Ehrenberg further extended the domain of these animalcules in the minds of biologists. Ehrenberg was, however, wrong in his interpretation of the nature of these organisms. He believed in the presence of many organs within the organism. His publication was almost simultaneous with the announcement of the *cell-theory* (1838–39). The wide acceptance of the theory overthrew his conception of protozoa.

The protozoa are the simplest expressions of life and afford a fertile field for experimentation. A separate division of biological study called "Protozoology" is now in operation. The germ of the subject was kindled by Ehrenberg.

CUVIER (1769–1832). Cuvier was the son of a Swiss Protestant officer in the French army. He was born near Belfort in 1769. He studied and in 1795 became an assistant at the Museé d' Histoire Naturelle at Paris.

He was selected by Napoleon to direct the reform of education in France.

His contributions. All work of Cuvier (Fig. 6.20) are guided by the "principle of correlation of parts". According to him organs do not function in nature as separate entities but they are certainly the parts of organic living wholes. In these living wholes there exists certain relation of parts which are fundamental to their modes of life. Thus feathers are always found in birds and not in other animals. Presence of feathers is related to certain formation of the forelimb so that it may be used as a wing. The forelimb is related to collar-bone, collar-bone to chest and so on. He

Fig. 6.20. Cuvier.

studied the fossils of elephants, mammals and many reptiles. His inferences on fossils were not correct. He believed in fixity of species. That is there is no evolution and there had been a succession of animal populations. He believed that the vast number of species, many no longer existing, had appeared on earth at different periods following catastrophes.

He developed considerable interest in the history of the development of sciences. He wrote a number of volumes on the rise of natural science. His historical writing can still be consulted with profit.

His influences. Cuvier has been called the "Dictator of Biology". He enjoyed a commanding position and was admitted as leader of science for years. He was a favourite of Napoleon Bonaparte and held the post of Director of the Higher Centres of Learning in France. But whatever prominence he attained in the Government he never lost his love for natural science.

Cuvier is best known by his book 'Le Regne Animal.......' which was published in 1817. It was the most comprehensive biological work since Linnaeus. It was the product of 25 years research. The work describes a species from every genus then recognised with illustrations. The ystem of classification divided animals into four great *embranchments*. Each embranchment was built on its own peculiar and definite plan. The major embranchments are:

1. Vertebrata—animals with back bone.
2. Mollusca—slugs, oysters, snails, etc.
3. Articulata—jointed animals as insects, lobsters, spiders.
4. Radiata—all remaining animals.

From 1801 to 1805 appeared his ".... Anatomie Comparee", a systematic treatise on the comparative anatomy of animals belonging to vertebrates and invertebrates.

He was interested on fossils and wrote a book in 1812 on the fossils about Paris, which laid the foundation stone of vertebrate palaeontology. He studied through his own industry to become an independent observer. He was certainly the founder of comparative anatomy.

His influence was stimulating to research. He had encyclopedic knowledge and profound energy. He formed vast scientific schemes and brought many of them to fruition. He had the gift of inspiring others and of causing them to work for him. Thus he formed a good team work.

A critical analysis of his work will show that he was often inaccurate in finer details of his subject. On the whole his intellectual leadership paved the way for further development of certain branches of Biology and his faults and adherence to a number of ideas retarded the progress of certain branches.

OWEN (1804–92). Owen was an English Biologist and was born in 1804 at Lancastar. He began his career as a surgeon's apprentice and then studied medicine at Edinburgh. In 1827 he became an assistant at the Hunterian Museum. Later he became director of the Natural History department of the British Museum. It was a place with immense wealth of material and Owen came up to the occasion in utilizing them.

His contributions. Owen (Fig. 6.21) was tremendously influenced by Cuvier. Most

Fig. 6.21. Owen.

of his work were published in the Scientific proceedings of the Zoological Society of London. His "Memoir on the pearly Nautilus" (1832) was his first significant contribution. Owen dissected a large number of animals. The dissection of rare animals like monotremata, marsupialia was recorded in "Catalogue of the physiological series of comparative anatomy contained in the Royal College" (in 5 vols.) as his monumental work. It is still of great value.

The credit of introducing the concept of 'Analogy' and 'Homology' in Anatomy goes to Owen.

Owen was one of the admitted masters of Palaeontology. He did immense investigations on the fossilised teeth of mammals. He wrote many monographs on extinct animals. Amongst these, the best known ones are on the extinct bird *Dinornis* of New-Zealand and on the giant extinct sloth *Mylodon* of South America (Fig. 6.22).

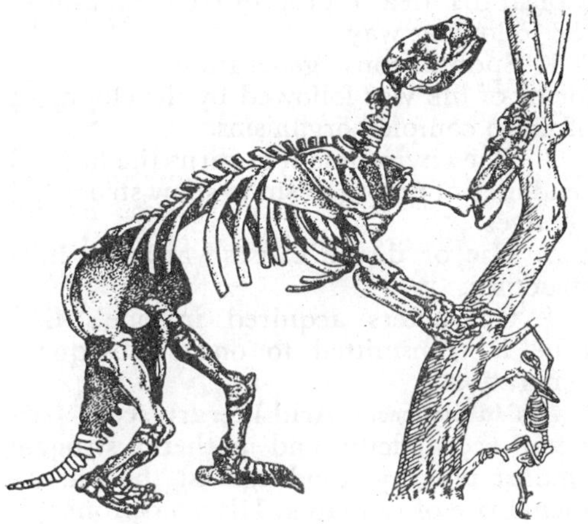

Fig. 6.22. Restoration of the skeleton of *Mylodon* by Owen.

He discovered the parasite *Trichinella spiralis* which causes the disease Trichinosis.

His influences. Owen was a prolific writer. He was singular in most of his thoughts and many of his generalizations could not stand the trial of time. His work on Palaeontology was creditable. He could lay stress on the fact that there lies an alliance of comparative studies with the doctrine of evolution and in seeking such an alliance attention should be focussed on structure and not on function. He, however, was an obstinate opponent of Darwinian concept of evolution.

BICHAT (1771–1801). Bichat (Fig. 6.23) was born in 1771 and died a premature death at the age of thirty-one. His father was a physician. Bichat showed brilliance from his early student days. He was distinguished in Latin and mathematics and showed inclination to natural history from his early life. He went to Lyons to study medicine.

The turbulent events of the French revolution forced him to go to Paris from Lyons. He got a good training in Paris. He became the Professor of Anatomy at the age of twenty-six.

Fig. 6.23. Bichat.

His contributions. He had a brief span of life. His industry was phenomenal. It is said that in a single winter he examined with care as many as six hundred bodies in pursuance of his research on pathological anatomy.

His treatises on the membranes, the phenomena of life and death and General Anatomy show the intensity and completeness of his investigations.

His influences. Bichat supplemented Cuvier and at the same time carried the organisation of animals to a deeper level. Bichat made intensive studies of tissues that make up organs. This particular branch is now called Histology. Thus he is called the father of Histology. He undoubtedly has drawn the attention of the world to the minute structure of the tissues within a productive period of seven years. He has published such work that has created an epoch and has laid an everlasting impression on the history of Biology.

LAMARCK (1744–1829). Jean Baptiste de Monet Lamarck (Fig. 6.24) was born in 1744 at Bazantin. As his father wished him to enter priesthood, he was sent to the Jesuit College at Amiens. As soon as his father died he left the college to join the French army then engaged against the Germans in the seven years war.

One of his comrades caused injury to the glands of his neck by lifting him by the head. This made him unfit for military life and he left the military service. Lamarck went to Paris to study medicine supporting himself with clerical work. He found special interest in Botany and abandoned the thoughts of a medical career. At the age of fifty he was forced to take a

Fig. 6.24 Lamarck.

chair in Invertebrate Zoology in the Museum d' Historire Naturelle. Though he had no training in Zoology at the time of his appointment, he took it as a challenge and mastered the subject. He devoted the rest of his life to Zoology.

Lamarck had a struggling life throughout. He had to fight against poverty and against his contemporaries who never could encourage or recognise his talents. Lamarck died in 1829 having been blind totally for ten years.

His contributions. Lamarck's first biological work was on the flora of France (Flora Francaise–1778). He was founder of the term 'Biology'. He believed that the animal and plant series, must at some remote point, were continuous with each other. He was of the opinion that living things should be studied as an entire entity—an attitude to which he was helped by his knowledge both in Botany and Zoology.

Lamarck's work on classification is probably his best. He separated spiders and crustaceans from insects. The class 'Vermes' created by Linnaeus was rather loose. Lamarck re-oriented the class by segregating the true worm like forms. He advanced to a considerable extent the classification of Echinoderms. The credit for classifying the animals into vertebrate and invertebrate goes to him.

Lamarck was the first man to give an idea of the causo-mechanical aspect of progressive evolution. He embodied these ideas in his book *Philosophic Zoologique* (1809). The conceptions outlined in this book is better known as Lamarck's 'Theory of use and disuse' or 'Inheritance of acquired characters'. Lamarck tried to establish his idea of progressive evolution in the following way:

1. Spontaneous generation of simple forms of life was followed by development of more complex organisms.
2. The environment governs the natural tendency of an organism to show structural changes.
3. Use or disuse causes variation in a structure.
4. Characters acquired in one's lifetime is transmitted to one's subsequent generations.

His influences. Arid literary style, personal eccentricity and pathetic struggle against poverty stood against the phenomenal rise of Lamarck. His over-proneness to speculation and poor attempts at Meteorology and at Chemistry discredited him to a great extent.

In his role as a systematist, he made important and lasting contributions.

CHARLES DARWIN (1809–1882). Darwin (Fig. 6.25) was the son of a doctor

Fig. 6.25. Charles Darwin.

and the grandson of Erasmus Darwin who himself was a doctor and wrote on evolution. From his very boyhood he was interested in nature and never liked conventional schooling. His father wanted to make him a doctor and sent him to Edinburgh in 1825 to study medicine. But he showed no taste for it. Gradually the influence of his father waned and his interest in medicine reached a vanishing point. He, in the next step, went to Cambridge to become a clergyman. He learned enough of Latin and Greek here. While he was wondering what to do next he got an offer to go as naturalist on the voyage of H.M.S. Beagle.

The Beagle 238 tons sailed in 1831 under Captain Fitzroy. The objective of the voyage was to survey South America to determine its longitude. Darwin was a naturalist without salary. Darwin worked in a narrow space with meagre equipments. The voyage ended in 1835. The voyage was a voyage of discovery for Darwin. He collected enormous amount of data and spent rest of his days with the idea of Natural Selection stirring in his mind.

His contributions. Some of Darwin's scientific papers appeared in the Journal of Researches (1839–40). These papers dealt with his observations on the very peculiar animal and plant inhabitants of various isolated oceanic islands which had never been connected with continental lands. These papers established him as a painstaking naturalist, remarkable for the general breadth of his outlook. He did commendable work on Barnacles and their allies, made investigations on mammalian fossil forms and wrote an admirable book on Coral reefs (1842).

In November 1859 Darwin's monumental work on Evolution appeared in the book "Origin of Species by means of Natural Selection." The book contains the conclusions arrived at by Darwin after twenty years of thinking and researches.

His influences. The origin of species created a revolution in biology and also in other departments of thought. It received a warm welcome soon after its publication and was accepted as a masterpiece by all thinking people. The book certainly suggested a simple and apparently universally acting biological relationship to explain the process of change of form. The late nineteenth century became saturated with the concept of natural selection. Scientists wanted to focus all their findings in the light of evolution and to quote Julian Huxley "Evolutionary studies became case books of adaptations."

Darwin's idea of Natural Selection has stood the trial of time though it has undergone tremendous metamorphosis.

SCHWANN (1810–1882). Theodor Schwann (Fig. 6.26) the German physiologist was born in Prussia in 1810. He was

Fig. 6.26. Schwann.

graduated in medicine from Berlin University in 1834. A few years after he joined the Berlin University as an assistant to Johannes Muller. Muller was a great teacher and greatest of all trainers of Anatomists and Physiologists. His stimulus was a great cause in the uplifting of Schwann. Schwann was given the chair of Professor in the University of Louvin after the publication of his cell theory. Later he was transferred to the University of Liege.

His contributions. In his doctor's dissertation he worked on the respiration of chick-embryos. He demonstrated the influence of organisms and lower fungi in fermentation and petrifaction. Thus it was an initial attempt to disprove spontaneous generation. He also demonstrated (1) organic nature of yeast and (2) the importance of pepsin in digestion. He further discovered the striped muscles on the upper part of oesophagus and the envelope of nerve fibres called Schwann cell.

His greatest work was his classic book "Microscopic investigations on the accordance in the structure and growth of plants and animals (1839)." In 1838, he together with his botanist friend **Schleiden** (1804–1881) advocated the *cell-doctrine.* The essence of the doctrine is that:

1. Life starts as a single cell.
2. All living bodies are made up of cells.
3. Cells are always derived from pre-existing cells.

The influence of cell-doctrine. The cell-theory in its original form was imperfect and contained many fundamental misconceptions. However, the great truth that all parts of animals and plants are built of similar units was well substantiated. It was certainly the "master-stroke in generalization". Moreover, it directed the attention of microscopists to the finer structures of the cell and paved the path of discovery of protoplasm, nucleus and chromosomes.

MENDEL (1822–1884). Gregor Johann Mendel (Fig. 6.27) was born in a peasant

Fig. 6.27. Mendel.

family in Silesia. He showed marked intelligence and ability in the village elementary school. He was sent to a good school at Leipnik. But poverty stood against him and he was forced to discontinue four years later. At the age of sixteen he tried to continue his education at Olmutz Philosophical Institute supporting himself by private tuition. With some initial difficulties he completed the two years course there.

He entered the monastery at Brunn in 1843 and was ordained a priest in 1847. He completed his theological studies and started working as a substitute teacher in Mathematics and Greek. The next year he took the examination for a teaching certificate and failed. In 1851 he was sent to the University of Vienna, where he studied science. After that he came back to his convent where he passed the rest of his life first as a teacher of physics and natural science and later as the Abbot.

His contributions. Mendel was fond of fruit-culture and gardening even from his boyhood. During his second term appointment as teacher, he performed a series of experiments with pea-plants in the garden of the monastery. His papers were published in 1866 and 1867 in the proceedings of the Natural History Society of Brunn.

His work contain experiments on the inheritance of individual (unit) characters in twenty-two varieties of garden peas. He selected certain constant and obvious characters of the pea plant such as length of stem, seed colour, seed form and made intensive cross-experiments to produce hybrids. And thereafter he examined the results of self-pollination amongst the hybrids. Results revealed that in cross-breeding the parental qualities are not blended but that they retain their individuality in the offsprings. Each inherited trait is determined by a pair of differentiating factors (later named Gene by Johansen). These factors must separate, segregate and independently assort in each generation.

His influences. Mendel is regarded today as a great scientist, the 'Father of Genetics'. He made epoch making investigations which added an entire realm to the world of Biology. His experiments have certainly laid the foundation for the science of Genetics.

The work of Mendel produced no immediate influence partly because of the fact that they were published in a journal with limited circulation and partly because of the fact that all attention was drawn at that time to the work of Darwin. His work remained unnoticed to the world and he never had the recognition due to him in his life-time.

In the year 1900 the great principles of heredity worked out by Mendel was re-discovered independently by three botanists, Correns, de-Vries and Tschermark. Their discovery was not a mere literary finding. It was a need-based and deliberate search of the literature for confirmatory evidence for explaining the sequences of events already found. Thus the unrecognised papers of Mendel were taken out from the grave and made known to the scientific world.

From the beginning of the twentieth century the extension of Mendelian principles has begun. Now it ranks as one of the greatest discoveries in the study of Genetics.

WEISMANN (1834–1914). August Weismann (Fig. 6.28) was born in Frankfort in 1834. He became a graduate in medicine in 1856 from Gottingen. He started practicing medicine for some time. He never liked the profession and started

Fig. 6.28. Weismann.

working with microscope on embryology and morphology. He was encouraged by Leuckart. He joined Freiburg in 1863 and became the Professor of Zoology there in 1867. Besides being a scientist he was an accomplished musician.

His contributions. Weismann developed with great theoretical skill his conception of the continuity of substance from parents to offsprings in his germ plasm theory and published it in his book "The Germplasm" (1893). Weismann thought that the offsprings resemble the parents because it is derived from the germ plasm of parents.

The body cells or somatoplasm act as vehicle for the conveyance of germ cells. The body cells nourish, protect and carry the germ cells to the offsprings and they do not convey any of their substances. The germ cells transmit an inconceivable complex inheritance. The germ substances pass through countless generations. In attempting to explain inheritance in detail he assumed the presence of distinct units embedded in the protoplasm of germinal elements. He called these units as *Biophores*.

The book "The evolution theory" (1904) is another important contribution of Weismann. The book contains according to Thomson the translator (*i*) Illumination of the process of evolution with a wealth of fresh illustrations. (*ii*) Germ plasm concept as valuable working hypothesis. (*iii*) Abandonment of the concept of inheritance of acquired characters. (*iv*) Further analysis of the origin of variation and (*v*) Extension of selection principle of Darwin.

He also wrote an autobiography "The Lamp" in 1903.

His influences. Weismann's germ plasm theory is primarily a theory of heredity. His views have been a field of controversy. Following on Weismann, scientists became divided into two schools. One school accepted the theory of Lamarck while the other school rejected it. The views of Weismann have passed through various stages of remodelling since its publication. It has become connected with other considerations to become the full-fledged theory of evolution known as Weismannism. He undoubtedly is the father of contemporary genetic theory.

DE VRIES (1848–1935) Hugo De Vries (Fig. 6.29) was born in 1848 at Haarlem,

Fig. 6.29. Hugo de Vries.

the Netherlands. He had his education at Leiden, Heidelberg and Wurgberg. He was appointed a lecturer at the University of Amsterdam and after few years held the chair of Professor of Plant Physiology. Following retirement in 1918 he kept himself engaged in research on mutation and evolution.

His contributions. He made extensive experiments with plants like evening primrose (*Oenothera Lamarkiana*) and was amazed to find the sudden appearance of different species of this plant. These sudden variants bred true to give rise to new forms. He termed the variations as *mutation*.

His book, "The mutation theory" (Die Mutation theorie) published in 1901 contains his notion that mutation is the universal source of origin of species. His mutation theory states that "species have not arisen through gradual selection accumulated for hundred or thousands of years but has appeared by sudden jumps and transformations."

Though he laid due importance on Natural Selection as its being the sole agency for perpetuation and improvement of favourable variations, his theory became somewhat antagonistic to that of Natural Selection. The reason behind this antagonism is that the conception of Natural Selection takes for granted that Natural Selection acts by accumulating slight, successive, favourable variation and not by great or sudden modifications.

In his other book "Intracellular Pangenesis" (1889) he developed a special theory of heredity related to that of Darwin. According to this theory hereditary characters in organisms may vary independent of one another or may combine with each other. He introduced *Pangenes* which were hypothetical material entities representing these individuals. His concept of Pangene is that the Pangenes remain within the nucleus, multiply and come out of the nucleus to determine the form and activity of the tissue.

He independently rediscovered Mendelian principles of heredity, verified them and could realise their real implications.

His influences. De Vries opened up a new period of investigation of the phenomenon of heredity. Towards the end of nineteenth century biologists had no idea about the cause of genetic variation. His mutation theory provided a first-hand information to this burning question.

His work was based on experiments, and gave a great stimulus to experimental studies.

He made important contributions to the study of origin of species and also pointed to the fact that in analysing the process of organic evolution many diverse factors are to be taken into consideration.

The theory of De Vries also provided an incentive towards exploration. Davenport, Tower and many others proved conclusively that species may arise by slow and steady accumulation of minute variations. They never denied the role of mutation in species formation but showed at the same time many evidences of evolution without mutation.

WOLFF (1738–94). Caspar Friedrich Wolff was born in Germany in 1738. Little is known about his life and early educations. Even his portrait is not available. He could not get a foothold in any of the German universities because of negligence and hostility to his work. In 1764, he went to the Academy of Sciences at St. Petersburg being invited by the Catherine[1] of Russia. He was there for thirty years.

His contributions. Wolff, at the age of twenty-one published his book, Theory of Generation (Theoria Generetionis–1759). The book consists of three parts—one devoted to the development of plants, one devoted to the development of animals and the third one to theoretical discussions. He showed that organs of plants like root, leave, etc., are developed by a process of differentiation from uniform tissue at the tip of growing root or shoot and thus indicating that the region of growth of these structures consists of an area of undifferentiated tissue.

The development of organs of animals also follows the same rule as that of plants, i.e., development from apparently homogeneous tissue which at the beginning is not stamped with its ultimate fate.

Wolff's best work is on the development of intestine of chick (De Formatione Intestinorum) published in 1768. The book shows his originality and strong power of observations and inferences. According to Von Baer "It is the greatest masterpiece of scientific observation which we possess".

1 **CATHERINE THE GREAT.** She continued the work of modernising Russia started by Peter the Great. She aimed at extending Russian territories and at making Russia European. She was known as one of the 'enlightened despots' of the age.

Wolff's name is still associated to a structure peculiar to the embryo. It is known as 'Wolffian body' or 'primitive kidney'. He could trace the development of this structure from undifferentiated tissue mass. The structure does not exist in adult stages of vertebrates excepting certain fish.

His influences. Wolff was the pioneer to advocate the process of *Epigenesis* in development. Epigenesis, as it stands, is an event in which something appears by development which was not before even in the rudimentary condition. Epigenesis concept was opposed to the then existing view of *Preformation*. The views of Wolff in his life time could gain no currency and as a result the progress of Embryology was retarded for another fifty years or so.

Wolff could very well establish the fact that development is a gradual process. But the explanation of his theory of development was unsatisfactory. He had no idea of germinal continuity, neither the idea had emerged in his time. He assumed a total lack of organisation at the beginning and ascribed development as a miraculous phenomenon through the action on the egg of a hyperphysical agent. Wolff thus made the same quandary as his predecessor when he tried to explain development.

Wolff was a man of great power and had much originality. His influence was retarded for a long time because of an uncongenial atmosphere. He should be recognized as a pioneer investigator who influenced future embryologists like Von Baer.

VON BAER (1792–1876). Karl Ernst Von Baer (Fig. 6.30) was born at Piep in Astonia. He intended to become a physician. He was, however, diverted from this aim by an interest in Anatomy and Comparative Embryology. After wandering through several universities, he stayed for some time in Wargburg. He then joined Konigsberg University in 1817 as Professor. In 1834 he accepted a chair at St. Petersberg and passed there the rest of his life. In the later part of his life he developed some interest in Anthropology.

Fig. 6.30. Von Baer.

His contributions. Von Baer's name is associated with the following advances in Embryology.

1. The discovery of mammalian ovum.
2. The proposition known as germ layer theory.
3. Von Baer's law concerning similarities in individual development and development of the race.
4. The discovery of notochord.

Of these the first appeared in his book "On the origin of the mammalian and human egg" (1827).

The other information were given in his book "Developmental History of Animals" (2 vols.—vol. 1–1828 and vol. 2–1837).

In the matter of germ layers he showed that during development in certain animals, the ovum divides to form layers of tissue. These layers give rise to various organs of the body. Further, in certain groups of animals specific organs develop from specific layers. He distinguished four such germ layers—an outer and an inner which develop first and the two intermediate layers that are later on budded off from these layers.

Van Baer's law consists of the following generalizations:

1. In development general characters appear ahead of special characters.
2. From the more general characters are developed the less general characters and finally the special ones.
3. During the course of its development the member of one species diverges continually from that of another species.
4. During development higher animals pass through stages resembling the stages of development of lower animals.

Von Baer discovered and described for the first time the development of notochord or chorda dorsalis.

His influences. Von Baer was instrumental in enriching Embryology in three counts.

First of all he set a higher standard for all work in embryology. Secondly, he advocated the germ layer theory. In the third plan, he brought the spirit of comparative embryology.

Von Baer's embryological doctrine is of tremendous historic and practical importance. At a later age when organic evolution received general acceptance, structural relationship came to be explained in the light of relationship with descent and as a corollary the process of development came to be the acid test of structural relationship.

Von Baer has been made dignified with the title of "Father of modern embryology"

HAECKEL (1834–1919). Ernst Haeckel (Fig. 6.31) was born in Jena in 1834. He studied with Johannes Muller, Virchow and Gegenbaur. Haeckel had a prodigious literary output. He held the Professorship

Fig. 6.31. Haeckel.

in Zoology at Jena till 1909. He kept himself busy with the great liberal intellectual movement[1] of his time and during the last part of his life he devoted himself to romantic natural philosophy.

His contributions. His first important contribution is on the Radiolaria of the Mediterranean (1860). He himself was a good artist and the illustrations of the memoir were done by him. It still serves the purpose of a standard reference book in this line.

He is best known for his "The Fundamental Biogenetic Law" or "The Theory of Recapitulation", which was embodied in his book "Generelle Morphologie", (1866). The theory may be briefly stated as 'Ontogeny is an epitome of Phylogeny".

Haeckel coined the term 'Ecology'. In 1872 he published a monograph on calcareous sponges. He tried to classify them on the basis of their canal system. The system is still followed by many. He also published another monograph on the group Monera. He tried to establish that this group lack nucleus. Nucleus, however, was found to be present in them by later workers.

Haeckel was the warmest exponent of Darwinism in Continental Europe.

His influences. It is almost impossible to estimate Haeckel's place in the history of biology. There is no denial that he had amazing influence on the biological thought in the late nineteenth and early twentieth centuries.

Few of his work have survived the test of science and many of them have been subjected to heavy criticism.

His work are of historical importance and now rest peacefully on the remote shelves of the Library. Yet they are rich with contributions which are still fundamental to the fabric of scientific thought.

BALFOUR (1851–1882). Francis M. Balfour (Fig. 6.32) was born in 1851. During his preparatory days in the university he was a good student. Being encouraged by his teachers he stepped into embryology. He entered into the subject with great intensity and by dint of his talent, devotion and hard labour soon became recognised. He became a Professor at Cambridge. He died untimely at the age of 31. He was a great stimulator of research and at the time of his death there were as many as twenty students working under him.

His contributions. His book "Comparative Embryology" (1880–81) in two volumes

[1] German Liberal Movement: Germany, at this time, was not one country, but a loose confederation of several states. Between 1815 and 1848 a liberal movement was born and grew in Germany both as a result of the French Revolution and as a reaction to Napoleonic conquests. The movement drew a large number of writers, philosophers, scientists and students among others. It alternated between the desire for a conservative, organic state and an emphasis on individual freedom and strengthened the movement for the ultimate unification of the country in 1871.

is a book of 'priceless value'. The volumes contain enormous information about the work of past embryologists.

His influences. At the period of Balfour, observations on the development of many animals accumulated in great number. But the information remained scattered in monograph, periodicals and transactions of learned societies. Balfour went through all these, assimilated them and took up the Herculean job to put them into an organised whole. He was successful in doing so and thereby hinted at the present tendencies in embryology.

Fig. 6.32. Balfour.

HERTWIG OSCAR (1849–1922) **and HERTWIG RICHARD** (1850–1872). These two brothers studied at Jena under Haeckel and secured degrees in medicine from Bonn. Oscar (Fig. 6.33) became the Professor of Anatomy first at Jena and then in Berlin. Richard held the chair of Professor of Zoology first at Konigsberg and then at Bonn. They cooperated closely on many investigations on anatomy and embryology.

Their contributions. The two together published many monographs. They studied the comparative anatomy of the coelom extensively and put forward the theory of coelom which contributed much to the understanding of comparative anatomy at the early stage of individual development.

Oscar studied fertilization and was the first to describe the phenomenon of fusion between sperm and ovum during fertilization.

Richard studied cytology and nucleus of protozoa and investigated the role of protozoan nucleus in vegetative reproduction.

Fig. 6.33. Hertwig Oscar.

ROUX (1850–1924). Wilhelm Roux was a student of Haeckel and studied medicine at Jena. For his doctorate degree he submitted a thesis on the haemodynamic factors that control the formation of blood vessels. The work of German Zoologist, August Weismann influenced him much and he devoted himself to the cause of experimental embryology.

His contributions. In a series of writings collected as "The developmental struggle of the parts within the organism" (1881) he presented his view of a mechanical basis for functional adaptation of the body parts to each other. Any investigation of function needs experimentations. So in order to seek a basis for his developmental mechanics (Entwicklungs mechanik) he switched over to experiments on embryo.

Making casual analysis as his aim he performed a simple type of experiment on frog's egg. When the frog's egg was at the two-celled stage of cleavage he injured one of the two first formed blastomeres. The remaining one after the experiment developed into a 'half-embryo'. Roux interpretated his results in the following way: There are two types of differentiation—self-differentiation (independent or mosaic development) and correlative dependent differentiation (interaction of cells or groups of cells). In his experiment, the development of one cell into a half embryo is suggestive of the fact that each cell develops independently of its neighbour and conversely total development represents a summation of partial mosaic developments. His paper "contribution to the developmental mechanics of the embryo. on the artificial production of half-embryos by destruction of one of the first two blastomeres and post generation of the missing half of the embryo" was published in 1888.

He established the journal "Archiv fur Entwicklungs mechanik der Organismen." The first volume of the journal was published in 1894–95. It is the first international journal for analytical embryology. The journal is in operation till date and now it has been named after him as "Roux Archiv . . . ."

His influences. The conclusions of Roux were later proved to be incorrect. But the experimental approach of Roux was significant on many counts. In the first place it brought about a new trend in method and attack on the process of embryonic development. In the second place his experiments shifted the emphasis from descriptive to experimental embryology. In the third place it brought the old problem of epigenesis in the lime light. The experimental embryology of today rests on the foundations laid by Roux.

SPEMANN (1860–1941). Hans Spemann was born in 1860 in Germany. He carried out extensive researches on developing newt's egg for over a period of more than twenty-five years and raised the standard of experimental embryology to a higher level.

His constriction experiments. During the year 1901–1903 he performed a series of experiments in which the blastomeres of the two-celled stage of newt's egg were separated by tightening a human hair along the first cleavage furrow. In some experiments each blastomere produced one miniature but whole embryo. In the majority of the cases, however, one of the blastomeres produced a small but whole embryo and the other one produced a ball of unorganised living cells. These differences in the results were due to the variable position of the first cleavage plane with reference to the median plane of the future embryo. Thus when the first cleavage plane appears in line with the median plane two dwarf embryos result. On the other hand, when the first cleavage furrow appears at right angles to the median plane, one blastomere gives a whole embryo and the other produces an abortive-cell mass. If constriction is made in early gastrula, a similar result of the second type is obtained. From these experiments Spemann concluded that as early as the two-celled stage there appears a qualitative change between the dorsal half and ventral half. The dorsal half is equipped with the quality to form an embryo whereas such a quality is absent in the ventral half.

In the year 1918 he made transplantation experiments with *Triton* embryos. When he transplanted a small piece of ectoderm of one early gastrula to another gastrula of the same age, the transplanted piece developed in accordance with its new position. But if the dorsal lip of the blastopore is grafted in the same manner, it would behave in a different way. The grafted dorsal lip will organise an embryo-like body with neural tube, somite and notochord in the new site. From these experiments Spemann inferred that the dorsal lip was already determined and represented a 'centre of differentiation'.

Spemann further wanted to know how much of self-differentiation ability is present in the grafted material (dorsal lip of blastopore) and how it was due to induction. The answer to this question was made available by grafting a dorsal lip of one species of Newt to another species having more darker pigments and which is supposed to behave as marker. Such experiments revealed that both the dorsal lip and the host embryo participated in the formation of the second embryo. The graft contributes the chorda-mesoderm and the host contributes the nervous system. He called the dorsal lip of the blastopore as the *organizer*.

The discovery of the organizer and the knowledge about its action opened up new avenues in experimental embryology. It demonstrated clearly that the steps in development are interrelated and one step is a necessary condition for the step above it. Thus the epigenetic concept of early embryologist proved true.

Spemann won Nobel prize in physiology and medicine in 1935.

MORGAN (1866–1945): Thomas Hunt Morgan was born in 1866 at Kentucky. After completing his undergraduate studies at the Kentucky College of Agriculture and Mechanical Arts, he entered John Hopkin's University in 1886 and studied Embryology. He obtained his Ph.D. degree in 1890. He then joined Brym Mowr as Associate Professor of Biology in 1891 and stayed there up to 1904. During his stay here he did his embryological work. He was made Professor of Experimental Biology at Columbia University. He held the chair here up to 1928 and then joined California Institute of Technology. He organised the Division of Biology and remained there till his death in 1945. He used to make summer trips to Woods Hole to utilise his vacation on the studies of marine animals.

Morgan has more than two hundred scientific papers and eight books to his credit. Morgan was interested in Embryology and the process of differentiation at the beginning of his research career. Later he turned to Genetics. Rats and mice were his first laboratory animals but recognising the advantages of the fruit fly Drosophila, he made the fly his experimental material and developed studies on heredity and variation.

The results of the research of Morgan and his associates like Sturtevant, Bridges and Muller led to a series of publications on genetical work. The classical book of this team. "The mechanisms of Mendelian heredity" (1915) is a wonderful attempt to explain Mendelian principle of heredity on chromosomal basis. The next edition of the book was published in 1923. It contained many revisions indicating the rapid pace at which accumulation of knowledge is taking place. The book "Sex linked inheritance in Drosophila" (1916) is stamped with the understanding and significance of linkage. "The theory of the gene" (1917) is another significant contribution of Morgan which has laid the foundation of basic research in genetics.

He also made contributions on regeneration and development of marine animals.

The work of Morgan and his associates are the most important ones in genetics after Mendel. Introduction of *Drosophila*, the Cinderella of genetics as a laboratory material gave a tremendous pace to genetical researches. Morgan formulated many principles which accelerated the investigations in hereditary mechanism.

Morgan was recipient of Darwin Medal (1924), Nobel Prize in Medicine (1933) and Copley Medal of Royal Society (1939).

MULLER (1890–1967). H. J. Muller was born in New York in 1890. He did his undergraduate and graduate work at Columbia University. In 1912 he joined Columbia University as a member of the faculty of Biology. In 1920, he became an Associate Professor of Zoology of the University of Texas. Later he held the chair of Professor of Zoology at the Indiana University and kept himself busy with his research on heredity.

During his days at Columbia he was associated with Morgan and was an active member of the "Fly-squad" of Morgan. Along with Morgan he did notable work on the linear arrangement of genes on the chromosomes and on the mechanism of heredity.

Muller was interested to know the underlying causes of gene change or mutation. He applied various means like heat, drug, poison, etc. on Drosophila with an idea to induce mutation. But mutations were exceedingly rare. In nature mutation occurs once in a million. So he wanted to find out a scientific device so that frequent mutations may be produced. Such induced mutations will provide an opportunity to make a probe into gene and chromosome.

But the genes lie in well protected condition. To reach the genes is a problem. Muller was struck by the idea that gene is a molecule and its properties depend upon its chemical composition and if that be so they can be shot or altered by electron bombardment from X-rays or rays of shorter length. Muller and his associates put hundreds of Fruit flies in gelatin capsules and subjected them to X-rays. The X-rayed flies were then mated with untreated mates and there were mutations, mutations and mutations.

The full significance of the discoveries of Muller became realised. For the first time man could know how new comes into the world and how the change or the new character is nursed by natural selection to fill most of the niches of the earth. Muller was awarded Nobel prize for his "Discovery of the production of mutations by means of X-ray radiation" in 1946.

WATSON (1928–). J. D. Watson was born in Chicago in 1928. He did his B.S. from Chicago University in 1947. After obtaining his Ph.D. degree in 1950 from Indiana University he accepted a Post-doctoral fellowship and came to Copenhagen in 1950. He joined the Cavendish Laboratories of Cambridge University in 1951 and stayed there up to 1953. Here in collaboration with Francis Crick he elucidated the double helical structure of the wonder molecule of DNA in 1953. Its discovery marked a new approach to scientist's attempt to solve the complexities of life. Now he is working as the Professor of Biology, Faculty of Harvard University. He is working out problems of protein synthesis and replication of virus.

He along with Crick and Wilkins was awarded Nobel Prize in Medicine and Physiology in 1962.

PAULING (1901–). Linus Pauling was born at Portland, Oregon. He graduated from Oregon State Agricultural College. He went to Munich to study under Arnold Sommerfeld and then to Copenhagen and Zurich. He had the honour of working with Neils Bohr and Erwin Schordinger. Under their stimulating influence he became interested on the forces that bind atoms into molecules. His famous book "The nature of the chemical bond" is an outcome of his initial interests in Physical Chemistry.

Later he worked on protein structure and could discover with his associates that the protein is another helical coil. He was awarded Nobel prize in 1954.

KHORANA (1922–). Hara Gobinda Khorana (Fig. 6.34) was born in Raipur, India in 1922. He became a graduate in 1943 from the Punjab University. He did his M.Sc. from the same university in 1945. For his Ph.D. degree he joined the Liverpool University and after obtaining the degree in 1948 he came back to India. He stayed in India for about eight years (1952–60) as Head of Organic Chemistry group of the B.C. Research Council. He was invited to Rockefeller Institute, New

Fig. 6.34. Khorana.

York as a visiting Professor in 1958. Later he settled in America. He worked in high positions in a number of American Universities.

Khorana's work in Khorana's language is "A Chemist's adventure in Biochemistry". In such an adventure he has already completed his work of chemical synthesis of naturally occurring co-enzyme-A. His book "Some recent developments of chemistry of phosphate esters of biological interest" is a new approach in this line. He has numerous research publications to his credit. His work on 'Genetic Code' is an epoch making discovery about the wonder molecule–DNA. On the basis of this work he has been awarded Nobel Prize in 1969. At present he is working on artificial synthesis of genes.

Glossary

Acquired characters: A character that originates during the life-time of an individual owing to environmental or any functional cause is designated as the acquired character. According to Lamarck, such a character acquired in the life-time of an individual is heritable. Inheritance of acquired characters is the essence of Lamarckian doctrine on the mechanism Organic Evolution.

Adaptation: This word is derived from two Latin words, *ad*=to+*apto*=fit. Adaptation signifies the permanent moulding in organisms to such an extent that they can live in a particular environment suitably. It involves the processes of becoming fitted in a particular environment by an entire organism or an organic structure.

Adaptative convergence: The word convergence (Latin, *cum*=together+*vergo* =incline) means the process of approaching towards a common point. In the realm of evolutionary dynamics many organisms belonging to quite distant and unrelated groups show close morphological and functional convergence (similarities) as a result of living in a similar environment. This phenomenon is known as adaptive convergence. All secondary aquatic vertebrates (Fishes, Whales, Seals, etc.) show adaptive convergence for living in watery medium.

Adaptive divergence: The word divergence (Latin, *di*=doubly + *vergo*=incline) signifies the process of going farther apart from a common point. In the biological process of adaptation many organisms after originating from the same stock exhibit adaptive divergence as a result of living in completely different environment.

Aerial: This word is derived from Greek word *aer* meaning air. Aerial means living in air. Most of the birds are aerial animals and they show perfect adaptation for living an aerial life.

Allantois: This word is derived from two Greek words, *allas*=stem and *eidos*= form. Allantois is an extra-embryonic structure which develops as an outpouching of the hindgut of the embryo to embrace the extra-embryonic coelom in reptiles, birds and mammals. Allantois is partly excretory in function and also serves as a reservoir for storing uric acid. It acts as a respiratory surface for the embryo. Allantois takes part in the formation of umbilical cord and fuses with the chorion in mammalian embryo to form the placenta.

Albinism: Albinism is derived from a Latin word *albus* meaning white. Absence of melanin pigment results in the condition known as Albinism. Albinism results from a failure on the part of the amino acid, *Tyrosine* to become metabolised to melanin, a black pigment.

Alcaptonuria: Alcaptonuria is a condition in which urine blackens upon exposure to air and the cartilage tissue becomes black and hardened. Alcaptonuria results from the absence of an enzyme to metabolise Homogentisic acid.

Allele (Allelomorph): This word is derived a Greek word, *allelon* meaning 'of one another' and signifies the alternative forms of a gene occupying the same locus in the homologous chromosomes. Alleles are also designated as Allelomorph

Amnion: Amnion is the innermost membrane enclosing the developing embryo of reptiles, birds and mammals. The amnion is a thin double (inner=ectoderm + outer=somatic mesoderm) membrane and remains filled with amniotic fluid. Amnion is protective in function and the amniotic fluid acts as a buffer and gives protection against any kind of shock. Amnion is discarded at the time of hatching or birth. Because of the presence of 'amnion', reptiles, birds and mammals are collectively called the Amniotes.

Amphidiploidy: Increase in the number of chromosomes due to crossing with other species is called the amphidiploidy.

Amphimixis: The word amphimixis signifies the process of mingling of germ-plasm of two individuals, *i.e.*, union of egg and sperm nuclei to form a zygote.

Analogy: The word analogy is derived from a Greek word *analogia* meaning ratio. Whenever the structures exhibting similarity in external features and/or functions but differ in structural construction and/or origin are known as analogy.

Anchallaxis: Deviation of the ontogenetic development from the racial pattern so that recapitulation becomes absent.

Aneuploid: When an individual instead of showing alternation in the complete set of chromosomes, sometimes possesses a single chromosome as extra or wanting.

Aniridia: Failure of development of a normal iris caused due to mutation of a gene which regulates the formation of normal iris in man.

Aquatic: This word is a derivative of a Latin word, *aqua* meaning *water*. Aquatic means living in or pertaining to water. Fish is an aquatic animal.

Arboreal: The term arboreal is a derivative of a Latin word, *arbor* meaning *tree*. This is related to living in tree or pertaining to trees. Tree-inhabiting animals are called the arboreal forms.

Autosome: Any chromosome besides the sex chromosomes is designated as autosome. Autosome is a derivative of two Greek words, *autos*=self + *soma*=body.

Autotroph: Organism capable of self-feeding.

Blastodisc: The term blastodisc is derived from two Greek words, *blastos*=germ + *diskcs*=platter. Blastodisc signifies the germinal area of a heavily yolked egg that gives rise to embryo.

Blastomere: This word is the derivative of two Greek words, *blastos*=germ + *meros*=part. Blastomeres mean the early cells produced by the process of cleavage of the ovum.

Blastocoel: The cavity inside the blastula is called the blastocoel. It is the derivative of two Greek words, *blastos*=germ + *koilos*=hollow.

Blastopore: The blastopore is derived from two Greek words, *blastos*=germ + *poros*=passage and means the mouth-like opening of gastrula.

Blastula: Blastula is the early stage of an embryo formed after the process of cleavage. It is a hollow cellular ball enclosing the blastocoel. It is derived from a Greek word, *blastos* meaning *germ*.

Cardiac: The derivative of a greek word, *kardia*=heart, cardiac means pertaining to heart.

Carnivorous: The word carnivorous means living or eating on flesh of other animals. It is derived from two Latin words, *caro*=flesh + *voro*=eat.

Cervical: The word cervical is a derivative of a Latin word *cervix*=neck and signifies anything pertaining to neck.

Chorion: The outermost extra-embryonic membrane surrounding the embryo of a mammal is called the chorion. It is a double membrane formed of ectoderm on the outerside and somatic mesoderm on the innerside. It unites with the allantois to participate in the formation of placenta. The term, chorion is usually reserved for the aforesaid membrane in mammalian embryo. The same membrane is also present in reptilian and avian embryos and is usually coined as 'serosa.'

Chromatin: It is a derivative of a Greek word, *chroma*=colour and means the deeply staining substance in the nucleus of a cell. Chromatin is conspicuous in the nuclear reticulum and in the chromosomes during cell-division.

Chromatophore: The term chromatophore is derived from two Greek words, *chroma*=colour + *phero*=bear and signifies the pigment cells containing the colouring granules or any colouring materials responsible for the colouration of most of the animals.

Chromomere: This word is derived from two Greek words, *chroma*=colour + *meros*=part. A chromomere means a chromatin granule in a chromosome.

Chromosome: Chromosome is derived from two Greek words, *chroma*=colour + *soma*=body. Thread-like deeply staining bodies found in the nucleus of a cell which become visible during cell-division. Chromosomes bears 'genes' (determiners of heredity) and are regarded as the physical and chemical basis of heredity.

Cleavage: The word cleavage signifies the division of an egg cell into many cells to form the blastula.

Colony: The word colony is derived from a Latin word, *colonus* meaning farmer. A group of organisms belonging to the same species live together for their survival and forms colony.

Commensalism: The word commensalism is derived from two Latin words, *cum*=together + *mensa*=table. Commensalism implies the association of two or more individuals belonging to different species. In this association the individuals are not harmed and one kind or more is benefited.

Community: The term community is derived from a Lalin word, *communis* meaning common. Community means a group of organisms belonging to one or more species living together to fulfil the environmental requirements.

Cornea: Cornea is a part of a vertebrate eye and is derived from a Latin word, *corneus* meaning horny. It is the rounded outer transparent coat of a simple eye.

Cross-fertilization: When an egg cell of an individual is fertilized by a spermatozoon of another individual is called cross-fertilization.

Deutoplasm: This word is derived from two Greek words, *deuteros*=second + *plasma*=form. Deutoplasm signifies the substances in an animal cell excepting the nucleus and cytoplasm. It usually implies the yolk material present in an egg cell.

Diploid: The term diploid is derived from two Greek words *diplous*=double + *blastos*=germ. It implies the somatic number of chromosomes (2n), *i.e.*, the normal number of chromosomes in all cells except the matured germ cells of an organism.

Dominant character: The term dominant is a derivative of a Latin word, *dominor* meaning rule. A character from one parent that expresses itself in an offspring to the exclusion of a recessive contrasted character from other parent is called the dominant character.

Dominant gene: A gene, responsible for a certain character manifests itself in the offspring to the exclusion of a contrasted recessive gene in the homologous chromosome pair is called the dominant gene.

Ecology: The term ecology is a derivative of two Greek words. *oikos*=home + *logos*=knowledge. Ecology is a branch of life-science which deals with the interrelationships between living organisms and their enviroments.

Ecosystem: Comprise of the abiotic physio-chemical environment and the biotic assemblage of plants, animals and microbes.

Ectoderm: The derivative of two Greek words, *ektos*=outside + *derma*=skin, the term ectoderm means the outer germ layer of an embryo.

Ectoparasite: A parasite that derives benefit from the exterior surface of its host.

Egg (**Ovum: The gamete** (**germ cell**) **produced by the matured functional female for the purpose of reproduction.**

Embryo: Embryo is an immature and newly forming young animal in the process of ontogenesis before birth or hatching.

Embryology: The branch of life-science which deals with the process of development of embryos.

Endoderm: The term endoderm is derived from two Greek words, *endon*= within + *derma*=skin. Endoderm implies the layer lining the archenteron or gastrocoel in a developing embryo.

Endoparasite: Parasite living within the body of its host.

Environment: The sum total of conditions (biological, physical, chemical etc.) surrounding an organism is called the environment.

Evolution: The term evolution is derived from a Latin word *evolvo* meaning unroll or unfold. Evolution is a process of changes undergone by the living organisms in course of time. It denotes that the present-day animals and plants are actually the modified descendants of somewhat different animals and plants which lived in times past. In essence, Evolution is process of descent with modifications.

Factor: According to Mendelism, factor implies an agent or cause of a hereditary character. Factor' is equivalent to 'gene' in modern genetics.

Fertilization: Fertilization is derived from two Latin words *fertil*=fruitful + *fero*=to bear. In higher forms it implies the process of union two dissimilar germ cells (egg and sperm) to form the zygote. Fertilization initiates the process of development of an individual.

Foetus (**Fetus**)**:** A stage of an embryo while remaining within the uterus or egg.

Fossil: The term fossil is derived from a Latin word, *fadio* meaning dig. Any relic of an animal or plant buried in the earth and preserved by natural causes in past geological time is called the fossil. Fossils are called the "nature's heiroglyphics" and provide the alphabets with which the language of the biological history is written in the strata of the earth.

Gamete: The term gamete means a mature germ cell (either sperm or ovum) and is derived from a Greek word, *gamos* meaning marriage.

Gametogenesis: Process of formation of gametes and is derived from two Greek words, *gamos*=marriage + *genesis*= generation. Gametogenesis includes two processes, one is spermatogenesis (formation of male gametes) and Oogenesis (formation of female gamete).

Gastrocoel: It is derived from two Greek words, *gaster*=stomach + *koilos*= hollow and indicates the primitive digestive cavity of a metazoan embryo formed by the process of gastrulation.

Gastrula: An early stage in embryonic development after blastulation. It is a crucial phage in embryonic development when the different tissues take up their future eventful position and the blue print of the future individual is laid down during this stage. It is derived from a Greek word, *gaster* meaning stomach.

Genes: The units of inheritance situated in the chromosomes are called the genes. They are the hereditary determiners and are transmitted from generation to generation through the gametes and control the emergence of characters in offsprings. The genes are composed of DNA and are equivalent to the hypothetical 'factors' of Mendel. They have the power of reproduction and transmutation.

Genotype: It is derived from two Greek words, *genos*=race and *typos*=form and implies the internal genetic constitution of an organism.

Germ Cell: A reproductive cell in a metazoan organism is called the germ cell.

Germ layer: One of the three fundamental cell-layers (ectoderm, endoderm and mesoderm) in an early metazoan embryo is designated as the germ layer. From the germ layers different tissues and organs of the adult individual are developed in course of development.

Germplasm: The gametes together with the tissues from which they develop are collectively known as germplasm.

Gonad: The reproductive organs (testis and ovary) from which gametes (sperms and ova) are produced. It is derived from a Greek word, *gonos* meaning seed.

Gregarious: A group of individuals that live habitually for protection and food-collection is called gregarious animals. It is derived from a Latin word *grex* meaning herd.

Gynandromorph: This term is derived from two Greek words, *gyne*=woman + *aner*=man. An individual of a dioecious (unisexual) species possessing one part of the body male and another part female in construction is called the gynandromorph. Gynandromorph often occurs in *Drosophila*.

Habitat: The natural dwelling place of an organism or a group of organisms is called the habitat. It is derived from a Latin word, *habitus* meaning condition.

Haemophilia: A sex-linked disease in which the clotting of blood, in case of injury, fails to occur.

Haploid: The halved number of chromosomes (n) as observed in matured germ cells is called as haploid. The word haploid is derived from two Greek words, *haplous*= single + *eidos*=form.

Herbivorous: Animals feeding exclusively on grasses, herb or any vegetable products are called herbivorous. It is derived from two Latin words, *herba*= grass + *voro*=devour.

Hereditary: Indicating transmission by inheritance from one generation to another.

Heredity: It is derived from a Latin word, *heres*=heir meaning transmission of characters from parents to offsprings so that the offsprings resemble their parents.

Heterotroph: Any organism whose nutritional requirements are met by feeding on other organism is called a heterotroph.

Heterozygous: An individual produced by union of two germ cells that contain a pair of unlike genes for a given character, one is dominant and the other is its recessive allele is called heterozygous. This term is derived from two Greek words, *heteros*=another + *zygon*=yoke.

Holoblastic: It is a derivative of two Greek words, *holos*=whole + *blastos*= germ and means the process of cleavage when an entire egg cells divides completely.

Homologous Chromosome: A pair of chromosomes in a cell, one coming from each parent, having similar structure and properties is designated as the homologous chromosomes.

Homology: The word homology is derived from two Greek words, *homos*= same + *lego*=speak and signifies fundamental similarity in structural construction of an organ or a part of it in an organism with the comparable portion in another. This has been resulted out of descent from a common ancestral form.

Homozygous: An individual produced by the union of two germ cells containing a pair of identical alleles for a given character is called homozygous. TT and tt in F_2 offspring of monohybrid cross of Mendel is the typical instance of homozygous individual in consideration of that particular character. It is a derivative of two Greek words, *homos*=like + *zygon*=yoke.

Host: The word host is a derivative of a Latin word, *hospes* meaning entertainer and implies an organism harbouring another organism as parasite.

Hybrid: The term hybrid is derived from a Latin word, *hybrida* meaning *mongrel*. The offsprings produced out of two parents differing in one or more heritable characters are called the hybrids. Hybrids are all heterozygous in nature. The tall pea plants containing (Tt) factors in Mendelian monohybrid cross are all hybrids.

Inheritance: Meaning the sum of all the characters which are transmitted from one generation to another by the germ cells. It is derived from two Latin words *in*=in + *heres*=heir.

Invagination: It is a morphogenetic event during gastrulation denoting folding in or inpocketing of the vegetal pole of the blastula to transform into a gastrula. In the gastrulation of Amphioxus and Frog, the wall of the blastoderm is pushed inside the blastocoel. This creates a new cavity called the archenteron which communicates with the exterior by the blastopore. It is derived from two Latin words, *in*= in + *vagina*=a sheath.

Larva: Denoting early active feeding stage in the life-history of an animal. The larva is unlike the adult in morphological and physiological aspects. It is derived from the Latin word meaning mask.

Linkage: Denoting inheritance of characters in groups because the genes for those characters are housed in the same chromosome. Linkage is an exception to Mendel's principle of independent assortment. Linked genes are housed in the some pair of chromosomes. Lesser the distance between the linked genes stronger is the linkage bond. Genes lying farther apart show less linkage. Linked genes tend to be transmitted together from one generation to the other. Separation of linked genes occurs very seldom by crossing over.

Marine: The term marine is derived from a Latin word *mare* meaning sea and denotes inhabiting or pertaining to sea and ocean.

Meroblastic: Denoting cleavage in an egg where the division is restricted only to the protoplasmic portion leaving the yolk portion uncleaved. Meroblastic cleavage is the characteristic of heavily yolked egg. This word is derived from two Greek words, *meros*=part + *blastos*=germ.

Mesenchyme: Denotes the loose gelatinous region between the inner and outer epithelia containing amoeboid cells. The term Mesenchyme is derived from two Greek words, *mesos*=middle + *chymos*= fluids.

Mesoderm: The primary germ layer between ectoderm and endoderm. It is derived from two Greek words, *mesos*= middle and *derma*=skin. From the mesoderm layer, muscles, bones, heart, kidney, gonads, connective tissues, blood etc. are developed.

Mimicry: Denoting a phenomenon which organisms bear to other animate or inanimate objects in form, colour and attitude.

Monohybrid: The term Monohybrid is a derivative of a Greek word, *monos*= single and a Latin word, *hybrida*=mongrel. Monohybrid implies the offsprings of parents that differ in one character.

Mutation: The term Mutation is derived from a Latin word, *mutatus*= changed and denotes the change in a gene or a group of genes resulting abrupt heritable modification/deviation of a character or characters. Mutation may occur naturally (Spontaneous mutation) or may be caused by artificial agencies (Induced mutation). While experimenting with *Oenothera Larmarckina* de Vries first established the fact of mutation and propounded the mutation theory of Evolution. Mutation is also coined as saltation.

Mutualism: Denoting the beneficial association between two or more individuals belonging to different species. It is derived from a Latin word, *mutuus*= exchanged.

Natural Selection: It is the main theme of Darwinian doctrine on Evolution. Natural Selection implies the elimination of unfit individuals in the struggle of life.

Nephron: The kidney is composed of numerous urineferous tubules. The tubules have similar structural organisation. A single urineferous tubule with all its structure is called the nephron. Nephron is regarded as the unit of excretory system. It is derived from a Greek word, *nephros*= kidney.

Nephrostome: The ciliated funnel-like opening of the nephron into the coelomic cavity is called the nephrostome. It is derived from two Greek words, *nephros*=kidney + *stoma*=mouth.

Neurone: A nerve cell with all its processes (*viz.* axon and dendrites) is called the neurone. Neurone is the unit of nervous system and the nervous impulses pass over the neurones.

Olfactory: Pertaining to the sense of smell. This word is derived fiom a Latin word, *olfacto*=smell.

Ontogeny: The term ontogeny is derived from two Greek words, *on*=being + *gen*=become and signifies the development of an individual.

Oogenesis: Involves the changes during the transformation of a primordial germ cell into a female gamete (ovum).

Optic: This word is a derivative of a Greek word, *optcs*=seen and implies the sense of sight or pertaining to eyes.

Otic: It is a derivative of a Greek word, *otikos* meaning everything pertaining to the ear.

Palaeontology: It is a branch of geology dealing with the study of fossils. It is derived from two Greek words, *palaios*=ancient + *logos*=discourse.

Parasite. The word parasite is a derivative of two Greek words, *para*=beside + *sitos*=food. A parasite is an organism that lives at the expense of another organism (host). Parasites are divided into two broad categories—Ectoparasites and Endoparasites.

Parthenogenesis: Parthenogenesis (Greek words—*parthenos*=virgin + *genesis* = origin) indicates the development of a new individual from an unfertilized egg. Parthenogenesis is opposite to fertilization.

Pelagic: The word pelagic is derived from a Greek word, *pelagos* meaning open sea and denotes everything pertaining to open sea.

Pericardium: (Greek words—*peri*= around + *kardia*=heart) indicates the membrane and also the cavity enclosing the heart.

Phenotype: The external appearance of an individual is called its phenotype (Greek—*phaino*=show + *typos*=type).

Phylogeny: The term phylogeny is a derivative of Greek words, *phylon*=race + *gen*=become and denotes the evolutionary history of a species.

Placenta: The word placenta means a flat cake in Latin and denotes the organ forming an organic bridge between the foetus and the uterine wall of mother for anchorage and physiological exchange of materials necessary for the metabolism of the developing foetus specially in eutherian mammals.

Plantigrade: Mode of walking on the surface of the earth on the whole of sole in bipedal animals especially in human beings. It is derived from two Latin words, *planta*=sole + *gradior*=walk.

Polyploid: When the basic number of chromosomes get multiplied, *viz.* Trilploid (3n), Tetraploid (4n), this condition is called the polyploid.

Predator: Predator (Latin *praeda* = booty) is an animal which preys upon other animals for food.

Recessive Characters: In heredity when a character from one of the parents remains undeveloped in offsprings in presence of corresponding dominant character of the other parent. The word recessive comes from a Latin word, *recessus*=a going back.

Retina: The innermost photosensitive layer of the eye-ball of a vertebrate eye. This layer contains the photosensitive cells, rods and cones.

Scansorial: Pertaining to climbing and is a derivative of a Latin word, *scando*= climb.

Sex Chromosomes: The special chromosomes (X and Y in human and *Drosophila*) which are concerned in the determination of sex in organisms.

Sex-linked Character: A gene or a group of genes for a character may be located in the sex-chromosomes. Inheritance of such characters (Colour-blindness, Night-blindness, Haemophilia etc.) is associated with the process of sex determination.

Spermateleosis: The changes involved in the differentiation of a spermatid into a functional spermatozoan.

Spermatogenesis: Involves the differentiation of the primordial germ cells into the functional male gametes.

Spermatozoa: The functional matured male gametes are called the spermatozoa.

Symbiosis: The word symbiosis (Greek *syn*=together + *bios*=life) indicates interrelationship between two organisms belonging to different species.

Synapsis: The term synapsis (Greek—*syn*=together + *hapto*=unite) means the temporary union between the homologous chromosomes during first meiotic division for the purpose of segmental interchange.

Terrestrial: The term terrestrial Latin—*terra*=earth) pertaining to living on the ground or earth's surface.

Umbilical Cord: The cord-like structure containing blood vessels and connective tissues uniting the foetus with the uterine wall of the mother of eutherian mammals. It is the derivative of a Latin word, *umbilicus* meaning navel.

Unit Characters: A character which behaves as a unit in heredity and may be inherited independently of other characters.

Vestigial: The term vestigial is a derivative of a Latin word, *vestigium* meaning a trace. An incompletely developed small or degenerated form representing a structure that was fully formed and functional in earlier forms. These rudimentary structures have no use to their possessors but resemble very closely and correspond to the useful organs of other forms. As many as 100 vestigial organs have been recorded in man: vermiform appendix, wisdom teeth (third molars), caudal vertebrae, ear muscles are some of the vestigial organs in man.

Villus (plural=**Villi**): The word *villus* is a Latin word meaning shaggy hair and denotes the minute finger-like projections arising from the trophoblast of the foetus into the trophospongia of the uterine wall for anchorage and physiological exchange of materials. Such finger-like projections are also present in the inner wall of small intestine.

Yolk: The nutrients stored in egg is called yolk. Yolk is composed of fats and other materials and provides nourishment for the developing embryo. Yolk exerts tremendous influence on the development of embryo. The size of egg (female gamete) depends exclusively on the quantity of yolk present within the egg.

Y-Chromosome: Chromosome associated with sex in many male animals. A human male has two sex-chromosomes in every cell, one is X-chromosome and the other is Y-chromosome.

Zygote: This word is derived from a Greek word *zygon* (yoke). A fertilized egg resulted from the union of two opposite gametes (ovum and sperm) is designated as zygote. Every bisexual animal starts its life from the zygote.

Bibliography

The following books and journals are listed for further reading in this part of Zoology. We acknowledge that the facts and figures used in compiling this book are largely based on these work. We feel that by consulting the undermentioned books and journals our readers will be benefited.

Balinisky, B. I., 1970 *An Introduction to Embryology*, W. B. Saunders Company, Philadelphia.

Beadle, G. W. and Tatum, E. L., 1945, *Neurospora* II. *Methods of producing and detection of Mutation concerned with nutritional requirements*, *Am. Jour. Botany*, 32.

Berrill, N. J., 1955, *The Origin of Vertebrates*, Heinemann Educational Book Ltd., London, U.K.

Bodemer, C. W., 1968, *Modern Embryology*, Holt, Rinehart and Winston Inc., New York.

Brachet, J., 1950, *Chemical Embryology*, Interscience Publishers, New York.

Colbert, E. H., 1955, *Evolution of the Vertebrates*, John Wiley & Sons, Inc., New York, U.S.A.

Child, C. M., 1941, *Patterns and Problems of Development*, University of Chicago Press, Chicago.

Crick, F.H.C., 1962, *The Genetic Code*, Scientific American, **207**: 66-74.

Dehaan, R.L. and Ursprung, H., 1965, *Organogenesis*, Holt, Rinehart and Winston, New York.

Dobzhansky, T., 1972, *Evolution, Genetics and Man*, Wiley Eastern Private Ltd., New Delhi.

Dobzhansky, T., Ayala, F.J., Stebbins, G.L. and Velentine, J.W., 1977, *Evolution*, W. H. Freeman and Company, San Francisco.

Dodson, E.O., 1960, *Evolution: Process and Product*, Affiliated East-West Press Pvt. Ltd., New Delhi, India.

Dowdeswell, W.H., 1959, *Animal Ecology*, Harper & Brothers, New York.

Ebert, J.D. and I.M. Sussex, 1970, *Interacting System in Development*, Holt, Rinehart and Winston, Inc., New York.

Elton, C., 1966, *Animal Ecology*, Sidgwick & Jackson, London.

Ford, E.B., 1971, *Ecological Genetics*, Chapman & Hall, London.

Garrod, A.E., 1963, *Inborn Errors of Metabolism. Reprinted with a supplement by Harry Harris.* Oxford University Press, New York.

Gray, P., 1961, *The Encyclopedia of the Biological Sciences*, Reinhold Publishing Corporation, New York.

Gregory, W.K., 1957, *Evolution Emerging*, The Macmillan Company, New York, U.S.A.

Haldane, J.B.S., 1932, *The Causes of Evolution*, Longmans, London.

Holtfreter, J., 1944, *Experimental Studies on the Development of the Pronephros.* Review on Cancer Biology, Vol. 3, pp. 220-250.

Hamilton, W.J. and H.W. Mossman, 1952, *Human Embryology*, Cambridge University Press, U.S.A.

Horstadius, S., 1950, *The Neural Crest*, Oxford University Press, New York.

Huettner, A.F., 1953, *Fundamentals of Comparative Embryology of the Vertebrates*, The Macmillan Company, New York.

Huxley, J.S., 1942, *Evolution—The Modern Synthesis*, Allen & Unwin, London.

Ingram, V.M., 1963, *Haemoglobin in Genetics and Evolution*, Columbia University Press.

Jacobson, A. G., 1961, *Heart Determination in the Newt*, Journal of Experimental Zoology, U.S.A., Vol. 146, pp. 139–151.

Kendeigh, C.S., 1961, *Animal Ecology*, Prentice Hall Inc., Englewood, Cliffs, New Jersey.

Knight, C.B., 1965, *Basic Concepts of Ecology*, The Macmillan Company, New York.

Kurz, S.M., 1958, *The Electron · Microscopy of the Developing Human Renal Glomerulus*, Experimental Cell Research, Vol. 14, pp. 355–367.

Lillie, F.R., 1919, *The Development of Chick*, Henry Holt, New York.

Lillie, F.R., 1919, *Problems of Fertilization*, University of Chicago Press, Chicago, U.S.A.

Locy, W.A., 1949, *Biology and its makers*, Henry Holt & Company, New York.

Lull, R.S., 1958, *Organic Evolution*, The Macmillan Company, New York, U.S.A.

Moody, P.A., 1964, *Introduction to Evolution*, Harper & Row, New York.

Moore, R., 1965, *The Coil of Life*, Dell Publishing Co. Inc., New York.

Needham, J., 1942, *Biochemistry and Morphogenesis*, University Press, Cambridge.

Nirenberg, N.W., 1963, *The genetic code II.* Scientific American, March (1963).

O'Connor, R.J., 1938, *Experiments on the Development of the Pronephric Duct*, Journal of Anatomy, Vol. 73, pp. 145–154.

Pearse, A.S., 1939, *Animal Ecology*, McGraw-Hill Book Company Inc., New York.

Romanoff, A.L., 1960, *The Avian Embryo, Structural and Functional Development*, The Macmillan Company, New York.

Romer, A.S., 1945, *Vertebrate Palaeontology*, University of Chicago Press, Chicago, U.S.A.

Singer, C., 1962, *A History of Biology*, Abelard-Schuman, London.

Stebbins, G.L., 1970, *Processes of Organic Evolution*, Prentice-Hall of India Private Limited, New Delhi.

Storer, T.I. and R.L. Usinger, 1965, *General Zoology*, McGraw-Hill Company, New York, U.S.A.

Waddington, C.H., 1952, *The Epigenetics of Birds*, University Press, Cambridge.

Waddington, C.H., 1957, *The Strategy of the Genes*, George Allen & Unwin Ltd., London.

Waddington, C.H., 1966, *Principles of Development and Differentiation*, The Macmillan Co., New York.

Watson, J.D., 1968, *The Double Hilex*, Alheneum Press, New York.

Weiss, P., 1939, *Principles of Development*, Holt, Rinehart and Winston, New York.

Willier, B.H. and J.M. Oppenheimer, 1968, *Fundamentals of Experimental Embryology*, Prentice-Hall of India Private Limited, New Delhi.

Wolpert, L. and E.H. Mercer, 1961, *An Electron Microscope study of fertilization of sea urchin egg (Psammechinus miliaris)*. *Expt. Cell Res.*, **22**: 44-55.

Index

U

V

W

X

Y

Z